STRUCTURAL ENGINEERING
AND
STRUCTURAL MECHANICS

STRUCTURAL ENGINEERING AND STRUCTURAL MECHANICS

A Volume Honoring
EGOR P. POPOV

KARL S. PISTER, *Editor*

PRENTICE-HALL, INC. *Englewood Cliffs, N.J.* *07632*

Library of Congress Cataloging in Publication Data

Main entry under title:

Structural engineering and structural mechanics.

Papers presented at a symposium honoring
Professor Popov held at the University of
California, Berkeley, Aug. 11–12, 1977.
"E. P. Popov: bibliography": p.
1. Structural engineering—Congresses.
2. Popov, Egor Paul.   3. Popov, Egor Paul—
Bibliography.   I.   Popov, Egor Paul.
II.   Pister, Karl S.
TA630.S82   624′.1   79-11281
ISBN   0-13-853671-6

C150037198
624·1

Editorial/production supervision and interior design
    by Barbara A. Cassel
Cover design by Edsal Enterprises
Manufacturing buyer: Gordon Osbourne

Printed in the United States of America

10   9   8   7   6   5   4   3   2   1

PRENTICE-HALL INTERNATIONAL, INC., *London*
PRENTICE-HALL OF AUSTRALIA PTY. LIMITED, *Sydney*
PRENTICE-HALL OF CANADA, LTD., *Toronto*
PRENTICE-HALL OF INDIA PRIVATE LIMITED, *New Delhi*
PRENTICE-HALL OF JAPAN, INC., *Tokyo*
PRENTICE-HALL OF SOUTHEAST ASIA PTE. LTD., *Singapore*
WHITEHALL BOOKS LIMITED, *Wellington, New Zealand*

*The Symposium Committee acknowledges with gratitude support
of the Symposium provided by the National Science Foundation
under Grant No. ENV77-06518 A01 to the University of California,
Berkeley, California.*

CONTENTS

PREFACE

On August 11 and 12, 1977, a Symposium on Structural Engineering and Structural Mechanics honoring Professor Egor Paul Popov was held at the University of California, Berkeley, California. The technical program encompassed eighteen papers covering a broad range of topics: engineering education, applied mechanics, structural theory, and structural engineering practice. It is the hope of the Symposium Committee that these papers, presented by some of Professor Popov's former students and colleagues, will be viewed not only as significant contributions in their own right but also as being reflective of the wide range of interests and versatility of the man whom the Symposium honored.

The volume begins with a biographical sketch and bibliography of Professor Popov, along with a list of his doctoral students up to the present.

Speaking for former students and colleagues who have had the privilege and pleasure of working with Professor Popov over the years, we express our thanks for inspiring teaching and scholarship, as well as leadership in enhancing structural engineering practice. It is our hope that this volume will serve as a similar inspiration to a wider audience of readers.

The Symposium Committee

M. S. AGBABIAN
V. V. BERTERO
H. KRAWINKLER
S. J. MEDWADOWSKI
K. S. PISTER

STRUCTURAL ENGINEERING
AND
STRUCTURAL MECHANICS

Egor Paul Popov

EGOR P. POPOV: BIOGRAPHY

Egor Paul Popov was born in Kiev, Russia, in 1913. After spending his early years in his native country, his secondary education began in Manchuria, China, and was completed in San Francisco, California. This was followed by graduation with a B.S. degree from the University of California, Berkeley, and the award of an M.S. degree from the Massachusetts Institute of Technology, Cambridge, Massachusetts. Advanced graduate studies were initiated at the California Institute of Technology, followed by completion of the Ph.D. degree at Stanford University, Palo Alto, California, under the direction of Professor Stephen Timoshenko.

For the past thirty-two years he has been a member of the faculty of the Department of Civil Engineering of his alma mater, University of California, Berkeley. Prior to his initial appointment in 1946, Dr. Popov was engaged in professional engineering practice for approximately nine years. This work included extensive activity in the fields of structural engineering and jet propulsion, the latter occurring during World War II, when he was employed as a design engineer with the Aerojet Corporation. This wide experience qualified him for professional licenses in civil, mechanical, and structural engineering in the State of California, and provided an invaluable preparation for subsequent teaching and research.

1

At the University, Dr. Popov's principal concern has been with teaching, which he considers to be his primary responsibility. In 1977 his long and effective career as a teacher was recognized by the Academic Senate's Distinguished Teaching Award. A concomitant of his teaching efforts has been the publication of two highly successful textbooks: *Mechanics of Materials*, which appeared in 1952, and *Introduction to Mechanics of Solids*, published in 1968. These books continue to be widely used in the United States and abroad, and have been translated into foreign languages.

A strong believer in the necessity of interrelating graduate teaching and research, Dr. Popov became involved in engineering research early in his career. Twenty-five doctoral and numerous master's theses have been completed under his supervision. The range of topics covered by the works is significantly broad, including the inelastic behavior of materials and structures, especially shells, and earthquake engineering. The latter interest has been an outgrowth of his early experience in the late 1930s with aseismic design and reinforcement of school buildings in the Los Angeles area following the Long Beach earthquake. Over the years Dr. Popov's research has been generously supported by such agencies as the National Science Foundation, National Aeronautics and Space Administration, American Iron and Steel Institute, and American Institute of Steel Construction.

More than 150 technical papers and reports have been authored or coauthored by Dr. Popov. The quality of this research has been recognized by numerous awards over the years. In 1967–1968 he was appointed Miller Research Professor in the prestigious Miller Institute for Basic Research in Science on the Berkeley Campus of the University of California. In that same year he was corecipient with one of his graduate students, M. S. Lin, of the Hetenyi Award of the Society for Experimental Stress Analysis for the paper "Buckling of Spherical Shells." In 1971 he was the recipient of the first Theodore R. Higgins Lectureship Award of the American Institute of Steel Construction. This award was given for work resulting in the paper "Cyclic Yield Reversal in Steel Building Connections," coauthored with a former graduate student, R. B. Pinkney.

In 1976 Dr. Popov was elected to membership in the National Academy of Engineering for his "contributions in mechanics of solids and the inelastic cyclic behavior of structural systems." In the same year the American Society of Civil Engineers conferred on him the Ernest E. Howard Medal, with the citation "For his unique activities over a period of four decades, which demonstrated that theoretical abstraction does relate to the real world and for the results of his extensive research that are used widely in the design of modern structures."

He holds membership and has actively participated in many honorary and professional societies and is a Fellow of the American Academy of Mechanics and of the American Society of Civil Engineers. On the Berkeley campus of the University of California, in addition to teaching and research activities, he has served in a number of administrative positions in the Department of Civil Engineering. These include appointment as the first Chairman of the Division of Structural Engineering and Structural Mechanics and as Director of the Structural Engineering Laboratory.

LIST OF DOCTORAL STUDENTS OF EGOR P. POPOV
(1950–present)

M. S. Agbabian	Per Larsen
Bayard S. Wilson	S. Nagarajan
John Brotchie	O. K. Kiciman
Zung-An Lu	Y. F. Dafalias
James Goodman	S. Y. M. Ma*
Mahmond Khojasteh-Bakht	T. Y. Wang*
John Abel	C. W. Roeder
M. K. S. Rajan	J. Vallenas*
J. Vasquez	S. Viwathanatepa*
M-S Lin	D. Soleimani*
S. Yaghmai	V. Zayas
P. Sharifi	D. Manheim
H. Krawinkler*	

E. P. POPOV: BIBLIOGRAPHY

Technical papers

[1] POPOV, E. P., "Stresses in Turbine Disks at High Temperatures," *Journal of the Franklin Institute*, May, 1947, pp. 365–389.

[2] POPOV, E. P., "Correlation of Tension Creep Tests with Relaxation Tests," *Journal of Applied Mechanics*, Vol. 14, 1947, pp. A135–A142 and (Author's Closure) A355; also in *Transactions of the ASME*, Vol. 69, same pages.

[3] POPOV, E. P., "Limit Design of Prismatic Beams," *Engineering News-Record*, Vol. 140, No. 20, May 13, 1948, pp. 89–91.

[4] POPOV, E. P., "Bending of Beams with Creep," *Journal of Applied Physics*, Vol. 20, March, 1949, pp. 251–256.

[5] POPOV, E. P., "Successive Approximations for Beams on an Elastic Foundation," *Proceedings, ASCE*, Vol. 76, Separate No. 18, 1950; *Transactions of the ASCE*, Vol. 116, 1951, pp. 1083–1095 and (discussion) 1096–1108.

*Denotes Co-supervisor.

[6] Popov, E. P., "A Modified Sequence in Teaching Mechanics of Materials," *Journal of Engineering Education*, Vol. 41, No. 8, April, 1951, pp. 474–477.

[7] Agbabian, M. S., and Popov, E. P., "Unsymmetrical Bending of Rectangular Beams Beyond the Elastic Limit," *Proceedings*, First U.S. National Congress of Applied Mechanics, 1951, pp. 579–584.

[8] Mohammed, I. A., and Popov, E. P., "Effective Stiffness of a Plate Subjected to a Local Edge Moment," *Proceedings*, Second U.S. National Congress of Applied Mechanics, 1954, pp. 423–426.

[9] Hirsch, E. G., and Popov, E. P., "Analysis of Arches by Finite Difference," *Proceedings, ASCE*, Vol. 81, No. 829, Nov., 1955, pp. 829–1 to 18; Closure in Vol. 82, No. ST6, November, 1956.

[10] Popov, E. P., "Earthquake Stresses in Spherical Domes and in Cones," *Journal Structural Division, ASCE*, Vol. 82, No. ST3, May, 1956.

[11] Mohammed, I. A., and Popov, E. P., "Beam Restraints Provided by Walls with Openings," *Transactions of the ASCE*, Vol. 121, No. 2830, 1956, pp. 1034–1053.

[12] Popov, E. P., "Plastic Theory," *Proceedings*, Structural Engineering Conference on Plastic Design in Structural Steel, University of Utah, October 21–22, 1957.

[13] Popov, E. P., and Willis, J. A., "Plastic Design of Cover Plated Continuous Beams," *Journal of the Engineering Mechanics Division, ASCE*, Paper 1495, January, 1958.

[14] Popov, E. P., and McCarthy, R., "Deflection Stability of Frames Under Repeated Loads," *Journal of the Engineering Mechanics Division, ASCE*, Vol. 86, No. EM1, January, 1960, pp. 61–78.

[15] Popov, E. P., and Medwadowski, S. J., "Membrane Stresses in Hyperbolic Paraboloid Shells Circular in Plan," *Publications*, International Association for Bridge and Structural Engineers, Zurich, Switzerland, October, 1960.

[16] Apeland, K., and Popov, E. P., "Analysis of Bending Stresses in Translational Shells," *Proceedings*, International Colloquium on Simplified Calculation Methods, Brussels, Belgium, September 1–4, 1961, pp. 9–43 (International Association for Shell Structures).

[17] Tocher, J. L., and Popov, E. P., "Plastic Analysis of Rigid Frames by Modified Linear Programming," *Symposium on the Use of Computers in Civil Engineering*, Lisbon, Portugal, October, 1962, Paper 9.

[18] Tocher, J. L., and Popov, E. P., "Incremental Collapse Analysis of Rigid Frames," *Proceedings*, Fourth U.S. National Congress of Applied Mechanics, ASME, New York, 1962, pp. 727–731.

[19] Popov, E. P., Penzien, J., and Lu, Z. A., "Finite Element Solution for Axisymmetrical Shells," *Journal of the Engineering Mechanics Division, ASCE*, No. EM5, October, 1964, pp. 119–145.

[20] Popov, E. P., and Becker, E., "Shakedown Analysis of Some Typical One-Story Bents," *Journal of the Structural Division, ASCE*, October, 1964, pp. 1–9.

[21] BERTERO, V. V., AND POPOV, E. P., "Effect of Large Alternating Strains of Steel Beams," *Journal of the Structural Division, ASCE*, No. ST1, February, 1965, pp. 1–12.

[22] POPOV, E. P., AND LU, Z. A., "Computer Analysis of Axisymmetrically Loaded Shells of Revolution," *Proceedings*, IASS Symposium, Budapest, Hungary, September, 1965, Vol. 2, pp. 531–539.

[23] LACHANCE, L., AND POPOV, E. P., "Limitations of Shallow Shell Equations for Spherical Shells," *Proceedings*, IASS Symposium, September, 1965, Vol. 2, pp. 389–403.

[24] POPOV, E. P., PENZIEN, J., AND RAJAN, M. K. S., "Stress Concentrations in Thin Spherical Shells," *ASME Transactions*, Vol. 87(D), 1965, p. 231.

[25] POPOV, E. P., AND FRANKLIN, H. A., "Steel Beam-to-Column Connections Subjected to Cyclically Reversed Loading," *Proceedings*, 34th Annual Convention of Structural Engineers Association of California, 1965, pp. 79–86; republished by AISI in February, 1966.

[26] POPOV, E. P., "Behavior of Steel Beam-to-Column Connections Under Repeated and Reversed Loading," *Guest Lectures*, 1965 Summer Conference on Plastic Design of Multi-story Frames, Lehigh University, 1966, pp. 241–260.

[27] POPOV, E. P., AND CHOW, H. Y., "Addendum" to "The Linear Elastic Dynamic Analysis of Shells of Revolution by the Matrix Displacement Method," *Proceedings*, Conference on Matrix Methods in Structural Mechanics, 1965, Wright-Patterson Air Force Base, Dayton, Ohio, AFFDL-TR-66-80, pp. 317–328.

[28] LU, A. A., AND POPOV, E. P., "Stresses in Cylindrical Shell Segments Fixed Along Four Edges," *IASS Symposium on Large-Span Shells*, September 6–9, 1966, Leningrad, USSR, Vol. 1, pp. 454–466.

[29] POPOV, E. P., "Analysis of Shells of Revolution by Finite Elements," *Congreso Internacional Sobre la Aplicación de Estructuras Laminares en Arquitectura*, Mexico, September, 1967.

[30] KHOJASTEH-BAKHT, M., AND POPOV, E. P., "Analysis of Elastic–Plastic Shells of Revolution," EMD Specialty Conference, November 8–10, 1967, N.C. State University at Raleigh, pp. 87–90; full paper, *Journal of the Engineering Mechanics Division, ASCE*, Vol. 96, No. EM3, June, 1970, pp. 327–340.

[31] POPOV, E. P., KHOJASTEH-BAKHT, M., AND YAGHMAI, S., "Analysis of Elastic–Plastic Circular Plates," *Journal of the Engineering Mechanics Division ASCE*, Vol. 93, No. EM6, December, 1967, pp. 49–65. Digest: *Transactions of the ASCE*, Vol. 134, 1969, p. 315.

[32] POPOV, E. P., KHOJASTEH-BAKHT, M., AND YAGHMAI, S., "Bending of Circular Plates of Hardening Material," *International Journal of Solids and Structures*, Vol. 3, 1967, pp. 975–988.

[33] POPOV, E. P., "Performance of Steel Beams and Their Connections to Columns During Severe Cyclic Loading" *Final Report, IABSE Eighth Congress*, New York, September 9–14, 1968, pp. 657–665.

[34] ABEL, J. F., AND POPOV, E. P., "Static and Dynamic Finite Element Analysis of Sandwich Structures," *Proceedings*, Second Conference on Matrix Methods in Structural Mechanics, October, 1968, Wright-Patterson Air Force Base, Dayton, Ohio, AFFDL-TR-68-150, pp. 213–245.

[35] GOODMAN, J. R., AND POPOV, E. P., "Layered Beam Systems with Interlayer Slip," *Journal of the Structural Division, ASCE*, Vol. 94, No. St11, November, 1968, pp. 2535–2547; digest: *Transactions of the ASCE*, Vol. 134, 1969, p. 947.

[36] POPOV, E. P., AND RAJAN, M. K. S., "Nonlinear Membrane Shapes for Shell Roofs," *IASS Madrid Colloquium*, October 1969, in volume of U.S. and Canadian contributions.

[37] LIN, M. S., AND POPOV, E. P., "Buckling and Spherical Sandwich Shells," *Experimental Mechanics*, Vol. 9, No. 10, October, 1969, pp. 433–440.

[38] POPOV, E. P., AND PINKNEY, R. B., "Reliability of Steel Beam-to-Column Connections Under Cyclic Loading," *Proceedings*, Fourth World Conference on Earthquake Engineering, Santiago, Chile, January 13–18, 1969, pp. B3-15 to B3-30.

[39] POPOV, E. P., AND PINKNEY, R. B., "Cyclic Yield Reversal in Steel Building Connections," *Journal of the Structural Division, ASCE*, Vol. 95, No. ST3, March, 1969, pp. 327–353.

[40] POPOV, E. P., BERTERO, V. V., AND WATABE, M., "A Test Frame for Simulating Gravity and Ground Motion on Structural Assemblages," *Proceedings*, International Symposium on Methodology and Technique of Testing Structures, September 9–11, 1969, Bucharest, Rumania, pp. 53–80.

[41] POPOV, E. P., AND YAGHMAI, S., "Linear and Nonlinear Static Analysis of Axisymmetrically Loaded Thin Shells of Revolution," *Proceedings*, First International Conference on Pressure Vessel Technology, Delft, The Netherlands, September, 1969, pp. 237–244; discussion: Part III, pp. 79–80.

[42] KHOJASTEH-BAKHT, M., SHARIFI, P., AND POPOV, E. P., "Elastic–Plastic Analysis of Some Pressure Vessel Heads," *Journal of Engineering for Industry*, Vol. 92, Series B, No. 2, May, 1970, pp. 309–316.

[43] POPOV, E. P., AND BERTERO, V. V., "Research on Seismic Behavior of High-Rise Moment-Resisting Steel Frames," *Proceedings*, Structural Engineers Association of California, 39th Convention, Lake Tahoe, October 15–17, 1970, pp. 76–87.

[44] BERTERO, V. V., POPOV, E. P., AND KRAWINKLER, H., "Half-Scale Tests of Sub-assemblages of Multi-story Unbraced Steel Frames Under Cyclic Loading," *Proceedings*, Third Japan Earthquake Engineering Symposium, Tokyo, Japan, November 17–20, 1970, pp. 654–666.

[45] POPOV, E. P., "Kelvin's Solution of Torsion Problem," *Journal of the Engineering Mechanics Division, ASCE*, Vol. 96, No. EM6, December, 1970, pp. 1005–1012.

[46] SHARIFI, P., AND POPOV, E. P., "Nonlinear Buckling Analysis of Sandwich Arches," *CANCAM*, Jan. 7, 1971.

[47] BERTERO, V. V., AND POPOV, E. P., "Testing Facility for Frame Subassemblages," *Experimental Mechanics*, Vol. 11, No. 8, August, 1971, pp. 337–345.

[48] YAGHMAI, S., AND POPOV, E. P., "Large Deflection Elastic–Plastic Analysis of Shells of Revolution," *First International Conference on Structural Mechanics in Reactor Technology*, Berlin, Germany, September, 1971, Bundesanstalt für Material Prüfung (BAM), Berlin Commission of European Communities Brussels, in Vol. J2/2; preprints: Vol. 4, Part J, Paper J2/2, pp. 1–28.

[49] BERTERO, V. V., POPOV, E. P., AND KRAWINKLER, H., "Effects of Local Instabilities in Beam-Column Subassemblage Behavior," *Proceedings*, International Symposium on Experimental Analysis of Instability Problems on Reduced and Full-Scale Models, RILEM, Buenos Aires, Argentina, September 13–15, 1971.

[50] SHARIFI, P., AND POPOV, E. P., "Nonlinear Buckling Analysis of Sandwich Arches," *Journal of the Engineering Mechanics Division, ASCE*, Vol. 97, No. EM5, October, 1971, pp. 1397–1412 (abstracts also published in CANCAM, 1971).

[51] YAGHMAI, S., AND POPOV, E. P., "Incremental Analysis of Large Deflections of Shells of Revolution," *International Journal of Solids and Structures*, Vol. 7, No. 10, October, 1971, pp. 1375–1393.

[52] LARSEN, P. K., AND POPOV, E. P., "Elastic–Plastic Analysis of Thick Walled Pressure Vessels with Sharp Discontinuities," *Journal of Engineering for Industry*, Vol. 93, Series B, No. 4, November, 1971, pp. 1016–1020.

[53] POPOV, E. P., AND SHARIFI, P., "A Refined Curved Element for Thin Shells of Revolution," *International Journal for Numerical Methods in Engineering*, Vol. 3, No. 4, October–December, 1971, pp. 495–508.

[54] SHARIFI, P., AND POPOV, E. P., "Nonlinear Finite Element Analysis of Sandwich Shells of Revolution," AIAA/ASME/SAE 13th Annual Structural Conference April, 1972, San Antonio, Texas, AIAA Paper 72-356; expanded version in *AIAA Journal*, Vol. II, No. 5, May, 1973, pp. 715–722.

[55] BERTERO, V. V., POPOV, E. P., AND KRAWINKLER, H., "Beam-Column Subassemblages Under Repeated Loading," *Journal of the Structural Division, ASCE*, Vol. 98, No. ST5, May, 1972, pp. 1137–1159.

[56] SHARIFI, P., AND POPOV, E. P., "Nonlinear Buckling Analysis of Sandwich Shells," *Proceedings*, IASS Symposium on Shell Structures and Climatic Influences, University of Calgary, July 3–6, 1972, pp. 119–128.

[57] POPOV, E. P., "Low Cycle Fatigue of Connections and Details," State-of-the-Art Report 3, Committee 18, *ASCE-AIBSE International Conference on Tall Buildings*, August, 1972; reprints: reports Vol. II-18, pp. 21–35.

[58] POPOV, E. P., "Recent Cyclic Tests of Structural Steel and Reinforced Concrete Members," *Proceedings*, Structural Engineer's Association of California, 41st Annual Convention, Monterey, California, October 5–7, 1972, pp. 45–58.

[59] POPOV, E. P., AND BERTERO, V. V., "Cyclic Loading of Steel Beams and Connections," *Journal of the Structural Division, ASCE*, Vol. 99, No. ST6, June, 1973, pp. 1189–1204.

[60] KRAWINKLER, H., AND POPOV, E. P., "Hysteretic Behavior of Reinforced Concrete Rectangular and T-Beams," *Fifth World Conference on Earthquake Engineering*, Paper 28, Rome, Italy, June, 1973.

[61] POPOV, E. P., BERTERO, V. V., AND KRAWINKLER, H., "Moment-resisting Steel Subassemblages Under Seismic Loadings," *Fifth World Conference on Earthquake Engineering*, Paper 184, Rome, Italy, June, 1973.

[62] VASQUEZ, J., POPOV, E. P., AND BERTERO, V. V., "Earthquake Analysis of Steel Frames with Non-rigid Joints," *Fifth World Conference on Earthquake Engineering*, Paper 219, Rome, Italy, June, 1973.

[63] POPOV, E. P., "Experiments with Steel Members and Their Connections Under Repeated Loads," *IABSE Symposium on Resistance and Ultimate Deformability of Structures Acted on by Well-Defined Repeated Loads*, Lisbon, September, 1973, pp. 125–135.

[64] NAGARAJAN, S., AND POPOV, E. P., "Nonlinear Elastic–Plastic Analysis Using the Finite Element Method," *Proceedings*, International Conference on Computational Methods in Nonlinear Mechanics, The University of Texas at Austin, September 23–25, 1974, pp. 979–988.

[65] NAGARAJAN, S., AND POPOV, E. P., "Elastic–Plastic Dynamic Analysis of Axisymmetric Solids," *Computers and Structures*, Vol. 4, No. 6, December, 1974, pp. 1117–1134.

[66] LARSEN, P. K., AND POPOV, E. P., "A Note on Incremental Equilibrium Equations and Approximate Constitutive Relations in Large Inelastic Deformations," *Acta Mechanica*, Vol. 19, 1974, pp. 1–14.

[67] LARSEN, P. K., AND POPOV, E. P., "Large Displacement Analysis of Viscoelastic Shells of Revolution," *Computer Methods in Applied Mechanics and Engineering*, Vol. 3, 1974, pp. 237–253.

[68] POPOV, E. P., SHARIFI, P., AND NAGARAJAN, S., "Inelastic Buckling Analysis of Pipes Subjected to Internal Pressure, Flexure, and Axial Loading," *Pressure Vessels and Piping: Analysis and Computers*, ASME, 1974, pp. 11–23.

[69] NAGARAJAN, S., AND POPOV, E. P., "Non-linear Finite Element Dynamic Analysis of Axisymmetric Solids," *International Journal of Earthquake Engineering*, Vol. 3, No. 4, April–June, 1975, pp. 385–399.

[70] DAFALIAS, Y. K., AND POPOV, E. P., "A Plasticity Theory with Nonlinear Kinematic Hardening for Stress Reversals," *Proceedings*, Fifth Canadian Congress of Applied Mechanics, University of New Brunswick, Frederickton, N.B., Canada, May, 1975, pp. 91–92.

[71] POPOV, E. P., BERTERO, V. V., AND CHANDRAMOULI, S., "Hysteretic Behavior of Steel Columns," *Proceedings*, U.S. National Conference on Earthquake Engineering, University of Michigan, Ann Arbor, Michigan, June, 1975, pp. 245–254.

[72] BERTERO, V. V., AND POPOV, E. P., "Hysteretic Behavior of R/C Flexural Members with Special Web Reinforcement," *Proceedings*, U.S. National Conference on Earthquake Engineering, University of Michigan, Ann Arbor, Michigan, June, 1975, pp. 316–326.

[73] BERTERO, V. V., POPOV, E. P., AND WANG, T. Y., "Seismic Design Implications of Hysteretic Behavior of Reinforced Concrete Elements Under High Shear," *U.S.–Japan Conference of Earthquake Engineering*, Hawaii, August 18–19, 1975.

[74] DAFALIAS, Y. F., AND POPOV, E. P., "A Simple Constitutive Law for Artificial Graphite-like Materials," *Transactions*, Third International Conference on Structural Mechanics in Reactor Technology, Vol. 1, C1/3, London, September, 1975.

[75] CLOUGH, R. W., BERTERO, V. V., BOUWKAMP, J. G., AND POPOV, E. P., "Experimental Study of Structural Response to Earthquakes," *Transactions*, Third International Conference on Structural Mechanics in Reactor Technology, Vol. 4, K4/2, London, September, 1975.

[76] POPOV, E. P., BERTERO, V. V., AND VIWATHANATEPA, S., "Analytic and Experimental Hysteretic Loops for R/C Subassemblages," *Proceedings*, Fifth European Conference on Earthquake Engineering, Istanbul, Turkey, September, 1975, Paper 89.

[77] KRAWINKLER, H., BERTERO, V. V., AND POPOV, E. P., "The Beam-Column Joint—An Important Element in the Seismic Response of Moment Resisting Steel Frames," *Proceedings*, Fifth European Conference on Earthquake Engineering, Istanbul, Turkey, September, 1975, Paper 93.

[78] POPOV, E. P., AND BERTERO, V. V., "On Seismic Behavior of Two R/C Structural Systems for Tall Buildings," in *Structural and Geotechnical Mechanics* (honoring N. M. Newmark), W. J. Hall, ed., Prentice-Hall, Englewood Cliffs, N.J., 1977, pp. 117–140.

[79] POPOV, E. P., AND BERTERO, V. V., "Repaired R/C Members Under Cyclic Loading," *Earthquake Engineering and Structural Dynamics*, Vol. 4, No. 2, October–December, 1975, pp. 129–144.

[80] KRAWINKLER, H., BERTERO, V. V., AND POPOV, E. P., "Shear Behavior of Steel Frame Joints," *Journal of the Structural Division, ASCE*, Vol. 101, No. ST11, November, 1975, pp. 2317–2336.

[81] NAGARAJAN, S., AND POPOV, E. P., "Plastic and Viscoplastic Analysis of Axisymmetric Shells," *International Journal of Solids and Structures*, Vol. 11, 1975, pp. 1–19.

[82] DAFALIAS, Y. K., AND POPOV, E. P., "A Model of Nonlinearly Hardening Materials for Complex Loadings," *Acta Mechanica*, Vol. 21, 1975, pp. 173–192.

[83] NAGARAJAN, S., AND POPOV, E. P., "Nonlinear Dynamic Analysis of Axisymmetric Shells," *International Journal for Numerical Methods in Engineering*, Vol. 9, 1975, pp. 535–550.

[84] DAFALIAS, Y. F., AND POPOV, E. P., "Rate-Independent Cyclic Plasticity in a Plastic Internal Variables Formalism," *Mechanics Research Communications*, Vol. 3, 1976, pp. 33–38.

[85] MA, S. M., POPOV, E. P., AND BERTERO, V. V., "Cyclic Shear Behavior of R/C Plastic Hinges," *ASCE-EMD Specialty Conference*, UCLA, March, 1976, pp. 352–362.

[86] BERTERO, V. V., POPOV, E. P., ENDO, T., AND WANG, T. Y., "Pseudodynamic Testing of Wall Structural Systems," *ASCE-EMD Specialty Conference*, UCLA, March, 1976, pp. 286–297.

[87] PETERSSON, H., POPOV, E. P., AND BERTERO, V. V., "Practical Design of R/C Structural Walls Using Finite Elements," *Proceedings*, IASS World Congress on Space Enclosures, Montreal, Canada, July, 1976, pp. 771–780.

[88] ADAMS, C. J., AND POPOV, E. P., "Thru Thickness Fatigue Properties of Structural Steel," *Welding Research Council Bulletin*, No. 217, July, 1976, pp. 1–13.

[89] DAFALIAS, Y. F., AND POPOV, E. P., "Plastic Internal Variables Formalism of Cyclic Plasticity," *Journal of Applied Mechanics*, Vol. 43, December, 1976, pp. 645–651.

[90] BERTERO, V. V., POPOV, E. P., WANG, T. Y., AND VALLENAS, J., "Seismic Design Implications of Hysteretic Behavior of Reinforced Concrete Structural Walls," *Proceedings*, Sixth World Conference on Earthquake Engineering, New Delhi, India, January, 1977, Vol. 5, pp. 5-159 to 5-165.

[91] POPOV, E. P., BERTERO, V. V., GALUNIC, B., AND LANTAFF, G., "On Seismic Design of R/C Interior Joints of Frames," *Proceedings*, Sixth World Conference on Earthquake Engineering, New Delhi, India, January, 1977, Vol. 5, pp. 5-191 to 5-196.

[92] POPOV, E. P., AND STEPHEN R. M., "Capacity of Columns with Splice Imperfections," *Engineering Journal*, Vol. 14, No. 1, First Quarter, 1977.

[93] PETERSSON, H., AND POPOV, E. P., "Substructuring and Equation System Solutions in Finite Element Analysis," *Computers and Structures*, Vol. 7, No. 2, April, 1977, pp. 197–206.

[94] DAFALIAS, Y. F., AND POPOV, E. P., "Cyclic Loading for Materials with Vanishing Elastic Region," *Nuclear Engineering and Design*, Vol. 41, No. 2, April, 1977, pp. 293–302.

[95] BERTERO, V. V., AND POPOV, E. P., "Structural Research in Reducing Earthquake Damage," *Forefront*, Research in the College of Engineering, University of California, Berkeley, July 1, 1975– June 30, 1976, pp. 111–114.

[96] PETERSSON, H., AND POPOV, E. P., "Generalized Loading Constitutive Relations," *Proceedings*, Annual Engineering Mechanics Division Specialty Conference, ASCE, North Carolina State University, Raleigh, N.C., May 23–25, 1977, pp. 354–356.

[97] PETERSSON, H., AND POPOV, E. P., "Constitutive Relations for Generalized Loadings," *Journal of the Engineering Mechanics Division, ASCE*, Vol. 103, No. EM4, August, 1977, pp. 611–627.

[98] POPOV, E. P., BERTERO, V. V., AND MA, S. M., "Model of Cyclic Inelastic Flexural Behavior of Reinforced Concrete Members," *Transactions*, Fourth International Conference on Structural Mechanics in Reactor Technology, San Francisco, August, 1977, Vol. K(a), pp. K3/14-1 to 3/14-12.

[99] BERTERO, V. V., AND POPOV, E. P., "Seismic Behavior of Ductile Moment-Resisting Reinforced Concrete Frames," *Reinforced Concrete Structures in Seismic Zones*, Publication SP-53, American Concrete Institutes, Detroit, 1977.

[100] ROEDER, C. W., AND POPOV, E. P., "Elevated Tank Supports with Hysterstic Damping," Preprints, *International Conference on Finite Elements in Nonlinear*

Solid and Structural Mechanics, Geilo, Norway, August, 1977, Vol. 2, pp. F05.1–F05.22.

[101] POPOV, E. P., AND STEPHEN, R. M., "Tensile Capacity of Partial Penetration Welds," *Journal of the Structural Division, ASCE*, Vol. 103, No. ST9, September, 1977, pp. 1721–1729.

[102] POPOV, E. P., AND ROEDER, C. W., "A Structural Support System for Long Span Roofs in Seismic Regions," *Proceedings*, International Conference on Light-Weight Shells and Spatial Structures for Normal and Seismic Zones, Alma-Ata, Kazakhstan, USSR, September, 1977, Vol. 2, pp. 274–281, Mir Publishers, 1977.

[103] ROEDER, C. W., AND POPOV, E. P., "Eccentrically Braced Steel Frames for Earthquakes," ASCE Fall Convention, Preprint 2924, San Francisco, October, 1977; to be republished in *Journal of the Structural Division, ASCE*.

[104] POPOV, E. P., AND PETERSSON, H. B., "Constitutive Relations for Inviscid Plasticity," ASCE Fall Convention, Preprint 2932, San Francisco, October, 1977, pp. 1–17.

[105] KICIMAN, O. K., AND POPOV, E. P., "A General Finite Element Model for Shells of Arbitrary Geometry," *Computer Methods in Applied Mechanics and Engineering*, in press.

[106] KICIMAN, O. K., AND POPOV, E. P., "Post-Buckling Analysis of Cylindrical Shells," *Journal of the Engineering Mechanics Division, ASCE*, in press.

[107] POPOV, E. P., "Mechanical Characteristics and Bond of Reinforcing Steel Under Seismic Conditions," Workshop on Earthquake-Resistant R/C Building Construction, University of California, Berkeley, July 11–15, 1977, in press.

[108] ROEDER, C. W., AND POPOV, E. P., "Cyclic Shear Yielding of Wide Flange Beams," *Journal of the Engineering Mechanics Division, ASCE* (in press).

[109] POPOV, E. P., LE, D. Q., AND PETERSSON, H., "A Finite Element Program for Structural Walls," *ACI Journal* (in review).

[110] POPOV, E. P., AND PETERSSON, H., "Constitutive Relations for Cyclic Plasticity," *Journal of the Engineering Mechanical Division, ASCE* (in review).

[111] BERTERO, V. V., POPOV, E. P., AND VIWATHANATEPA, S., "Bond of Reinforcing Steel: Experiments and a Mechanical Model," IASS Symposium, Darmstadt 1978, (in press).

[112] COWELL, A. D., BERTERO, V. V., AND POPOV, E. P., "Repair of Bond in R/C Structures by Epoxy Injection," Sixth European Conference on Earthquake Engineering, Dubrovnik, Yugoslavia September, 1978, (in press).

Books

[1] *Mechanics of Materials*, Prentice-Hall, 1952, 500 pp.

[2] Problems and Solutions for *Mechanics of Materials*, translation into Hebrew by E. Steg, 1969.

[3] *Introduction to Mechanics of Solids*, Prentice Hall, 1968, 571 pp.

[4] *Introduction to Mechanics of Solids* in MDS units (Prentice-Hall of India, Eastern Economy Edition), 1973.

[5] *Introduction to Mechanics of Solids* (Prentice-Hall), published in Japanese, 1975, by Baifudan Company, Tokyo, 2 vols. (345 pages and 320 pages).

[6] *Introductión a la Mecánica de Sólidos* (Limusa, 1976) 652 pages, translation into Spanish of *Introduction to Mechanics of Solids* (Prentice-Hall).

[7] *Mechanics of Materials*, 2nd edition, Prentice-Hall, 1976.

[8] *Mechanics of Materials*, 2nd edition, SI version, Prentice-Hall, 1978.

Technical reports

[1] "Analysis of Thin Spherical Shells Under Steady State Accelerations," with J. F. Brotchie and J. Penzien, University of California, *Institute of Engineering Research Report 142*, September, 1959.

[2] "Analysis of Plates of Linear Strain Hardening Material," with J. F. Brotchie and J. Penzien, University of California, *Institute of Engineering Research Report, Series 100*, Issue 7, November, 1960.

[3] "Analysis of Stress Concentrations—Thin Rotational-Shells of Linear Stress-Hardening Material," with J. F. Brotchie and J. Penzien, University of California, *Institute of Engineering Research Report, Series 100*, Issue 11, September, 1961.

[4] "Behavior of Reinforced Concrete Frames Subjected to Repeated Reversible Loads," with V. Bertero and F. McClure, University of California, *Institute of Engineering Research Report, Series 100*, Issue 18, January, 1962.

[5] "Analysis of Stress Concentrations in Thin Spherical Shells," with M. K. S. Rajan and J. Penzien, University of California, *Institute of Engineering Research Report, Series 100*, Issue 22, December, 1962.

[6] "Design Tables for Translational Shells," with K. Apeland, *Struc. Engineering Laboratory Report 63-1*, University of California, April, 1963.

[7] "Finite Element Solution for Thin Shells of Revolution," with Z. A. Lu and J. Penzien, University of California, Institute of Engineering Research, Structures and Materials Research, *SESM 63-3*, September, 1963.

[8] "Finite Element Solution for Thin Shells of Revolution," with Z. A. Lu and J. Penzien, University of California, *NASA CR-37*, Washington, D.C., July, 1964.

[9] "Edge Disturbances in Axisymmetrical Shells of Revolution," with Z. A. Lu, University of California, Institute of Engineering Research, Structures and Materials Research, *SESM 64-3*, September, 1964.

[10] "Behavior of Steel Building Connections Subjected to Repeated Inelastic Strain Reversal—Experimental Data," with R. B. Pinkney, University of California, *SESM 67-31*, December, 1967; republished as *AISI Bulletin 13*, November, 1968.

[11] "Behavior of Steel Building Connections Subjected to Repeated Inelastic Strain Reversal," with R. B. Pinkney, University of California, *SESM 67-30*, December, 1967; republished as *AISI Bulletin 14*, November, 1968.

[12] "Computer Program for Elastic–Plastic Analysis of Axisymmetrically Loaded Shells of Revolution," University of California, *SESM 68-3*, April, 1968.

[13] "Static and Dynamic Analysis of Sandwich Shells with Viscoelastic Damping," by J. Abel, University of California, *SESM 68-9*, August, 1968.

[14] "Refined Finite Element Analysis of Elastic–Plastic Thin Shells of Revolution," with P. Sharifi, University of California, *SESM 69-28*, December, 1969.

[15] "Cyclic Loading of Full-Size Steel Connections," with R. M. Stephen, Earthquake Engineering Research Center, University of California *EERC Report 70-3*, July, 1970; republished as *AISI Bulletin 21*, February, 1972.

[16] "Elastic–Plastic Analysis of Axisymmetric Solids Using Isoparametric Finite Elements," with P. K. Larsen, Structural Engineering Laboratory, University of California, *SESM 71-2*, January, 1971.

[17] "Inelastic Behavior of Steel Beam-to-Column Subassemblages," with H. Krawinkler, V. Bertero, Earthquake Engineering Research Center, University of California, *EERC Report 71-7*, October, 1971.

[18] "Cyclic Behavior of Three R. C. Flexural Members with High Shear," with V. V. Bertero and H. Krawinkler, Earthquake Engineering Research Center, University of California, *EERC Report 72-5*, October, 1972.

[19] "Elastic–Plastic Dynamic Analysis of Axisymmetric Solids," with S. Nagarajan, Structural Engineering Laboratory, University of California, *UCSESM 73-9*, July, 1973.

[20] "Further Studies on Seismic Behavior of Steel Beam-Column Subassemblages," with V. Bertero and H. Krawinkler, University of California, *EERC Report 73-27*, December, 1973.

[21] "Nonlinear Finite Element Dynamic Analysis of Axisymmetric Solids," with S. Nagarajan, Structural Engineering Laboratory, University of California, *UCSESM 74-9*, July, 1974.

[22] "Hysteretic Behavior of Reinforced Concrete Flexural Members with Special Web Reinforcements," with V. V. Bertero and T. Y. Wang, Earthquake Engineering Research Center, University of California, *EERC Report 74-9*, August, 1974.

[23] "Hysteretic Behavior of Ductile Moment-Resisting Reinforced Concrete Frame Components," with V. V. Bertero, University of California, *EERC Report 75-16*, April, 1975.

[24] "Hysteretic Behavior of Steel Columns," with V. V. Bertero and S. Chandramouli, University of California, *EERC Report 75-11*, September, 1975.

[25] "Experimental and Analytical Studies on the Hysteretic Behavior of Reinforced Concrete Rectangular and T-Beams," with S. M. Ma and V. V. Bertero, Earthquake Engineering Research Center, University of California, *EERC Report 76-2*, May, 1976.

[26] "Structural Steel Bracing Systems: Behavior Under Cyclic Loading," with K. Takanashi and C. W. Roeder, Earthquake Engineering Research Center, University of California, *EERC Report 76-17*, June 1976.

[27] "Capacity of Columns with Splice Imperfections," with R. M. Stephen and

R. Philbrick, Earthquake Engineering Research Center, University of California, *EERC Report 76-21*, September, 1976.

[28] "Tensile Capacity of Partial Penetration Welds," with R. M. Stephen, Earthquake Engineering Research Center, University of California, *EERC Report 76-28*, October, 1976.

[29] "Hysteretic Behavior of Reinforced Concrete Framed Walls," with T. Y. Wang and V. V. Bertero, Earthquake Engineering Research Center, University of California, *EERC Report 75-23*, December, 1975.

[30] "Subwall—A Special Purpose Finite Element Computer Program for Practical Elastic Analysis and Design of Structural Walls with Substructure Option," with D. Q. Le and H. Petersson, Earthquake Engineering Research Center, University of California, *UCB/EERC Report 77-09*, March, 1977.

[31] "Inelastic Behavior of Eccentrically Braced Steel Frames Under Cyclic Loadings," with C. W. Roeder, Earthquake Engineering Research Center, University of California, *UCB/EERC Report 77-18*, August, 1977.

[32] "Concrete Confined by Rectangular Hoops and Subjected to Axial Loads," with J. Vallenas and V. V. Bertero, Earthquake Engineering Research Center, University of California, *UCB/EERC Report 77/13*, August, 1977.

Miscellaneous

[1] Discussion of, "A Direct Method of Moment Distribution," by T. Y. Lin, *Transactions of the ASCE*, Vol. 102, 1937, pp. 578–579.

[2] Discussion of, "Tension and Torsion Creep Properties of Cloth Laminates," by J. Marin, *ASTM Bulletin 143*, December, 1946, p. 47.

[3] "Stress Analysis and Creep," Abstracts of Dissertations, *Stanford University Bulletin*, December 1, 1946, 00. 39–41.

[4] Discussion of "Critical Stresses in a Circular Ring," by E. A. Rippenberger and N. Davids, *Proceedings, ASCE*, Vol. 72, 1946, pp. 1261–1294; *Transactions of the ASCE*, Vol. 112, 1947, pp. 631–634.

[5] Discussion of "Theory of Inelastic Bending with Reference to Limit Design," by A. Hrennikoff, *Proceedings, ASCE*, Vol. 74, 1948, pp. 227–280; *Transaction of the ASCE*, Vol. 113, pp. 258–261.

[6] Discussion of "Elastic Foundations Analyzed by the Method of Redundant Reactions," by Z. Levinton, *Proceedings, ASCE*, 1948, pp. 1174–1176; *Transaction of the ASCE*, Vol. 114, 1949, pp. 54–56.

[7] Discussion of, "Applied Column Theory," by F. R. Shanley, *Proceedings, ASCE*, Vol. 75, 1949, pp. 1397–1399; *Transactions of ASCE*, Vol. 115, 1950, pp. 736–738.

[8] "Strength of Materials," with K. S. Pister, University Extension, Correspondence Course Civil Engineering XB 108A, 1950, University of California, Berkeley.

[9] Book Review: E. E. Sechler, *Elasticity in Engineering* (Wiley, 1952), *Zeitschrift für Angewandte Matematik und Physik*.

[10] Book Review· V. V. Novozhilov, *Theory of Thin Shells, Applied Mechanics Reviews*, Vol. 9, No. 3, March, 1956, pp. 101–102.

[11] Synopsis of a paper, "Incremental Collapse Analysis of Rigid Frames," with J. L. Tocher, *Proceedings*, Fourth U.S. National Congress of Applied Mechanics, *J. Applied Mechanics*, Vol. 29, Series, E. No. 2, June, 1962, p. 238.

[12] "Stephen P. Timoshenko—1878–1972," *The Russian Review*, Vol. 32, No. 1, January, 1973, pp. 110–112.

[13] Discussion of paper "Collapse Load Analysis Using Linear Programming," with J. L. Tocher, *Structural Division Journal, ASCE*, Vol. 90, No. ST11, November, 1972, p. 2626.

[14] Discussion of "Inelastic Behavior Under Cyclic Loading," *Planning and Design of Tall Buildings*, Vol. 2, ASCE–IABSE, New York, 1972, pp. 383–385.

[15] Discussion of "Additional Items Affecting Frame Instability," *Planning and Design of Tall Buildings*, Vol. 2, ASCE–IABSE, New York, 1972, 595–597.

SYMPOSIUM WELCOME

*John B. Scalzi**

I welcome you to this symposium and I am extremely happy that the National Science Foundation is able to cosponsor this conference, which combines a technical program with a tribute of appreciation to an individual who has contributed greatly to each of the themes to be presented today.

I am a strong believer that the engineering profession, as a whole, should sponsor award programs of this type each year to honor those engineers who have made significant contributions to the advancement of their field of activity for the benefit of society.

I propose that the engineering awards be sponsored by the National Academy of Engineering, with endorsement by the Professional Societies. Assuming that this meeting is the first of such programs, I imagine the presentation could take the following format.

Today, it is my great honor and privilege to present the most cherished engineering award for the year, the "da Vinci Award," named for the great master of multidisciplines, Leonardo da Vinci. The qualifications for this award are the most stringent of all awards and the award is not granted every

*Manager, Earthquake Engineering Program, National Science Foundation.

Berkeley Symposium, August 11, 1977.

year, but only when someone rises high above the individual discipline awards.

The specific qualifications are as follows:

1. The individual must be recognized throughout the world for his contributions in several fields of endeavor, in the manner of Leonardo da Vinci.

2. He must possess those unique qualities of theoretical and experimental pursuits exemplified by a Galileo.

3. He must be endowed with the art and skill of communications in both the written and spoken word. The written word to be in the manner of a Henry Wadsworth Longfellow or a Jack London. The spoken word should have the oratorical eloquence of a Daniel Webster, the impact of a Verdi opera, *Rigoletto*, and be expressed in the soothing tones of a Sigmund Romberg operetta *Desert Song*.

4. He must be a great teacher and sincere educator in the manner of a Horace Mann.

5. He must be a master craftsman in the eyes of his apprentices as in the tradition of a Michaelangelo.

6. This individual must be a living personification of his professional pursuits, by practicing what he preaches, by serving as a consultant's consultant to solve difficult problems encountered in the practice of engineering in the fashion of a John Roebling.

7. This individual must possess the sensitivities and artistry of a silversmith artisan reminiscent of a Paul Revere or an artist in the style of a Raphael.

8. And finally these qualities would be insufficient without those personal characteristics which make all else possible: honesty, integrity, and loyalty, topped by the ever-present personal appeal of an Arnold Palmer.

There is no need for a nominee list, because the members of the profession were unanimous in their nomination and acclaim. With the greatest of pride and pleasure, I congratulate the winner:

Dr. Egor P. Popov

Dr. Popov, I can think of no greater honor than the unanimous acclaim and recognition by one's peers. Today's program, which delves into the subjects of education, theoretical and experimental pursuits, and professional practice, is a fitting tribute and recognition of your many talents and achievements. This honor is bestowed upon you today, in the early evening of your career, because we have the feeling that your activities will not simply cease, but, instead, may increase in the coming years. We view this symposium, which is a tribute to your past accomplishments, as simply a prologue of the many achievements yet to come.

I am pleased to know you and to be associated with this symposium and wish you the very best of health and energy for your future activities.

DESIGN OF MATERIAL
AND STRUCTURAL MODELS

———— *Melvin L. Baron* and Ivan S. Sandler*** ————

INTRODUCTION

The title "Design of Material and Structural Models" covers the entire process of solving certain physical problems by the engineer. Several steps are required in this process, as shown in Fig. 1. Engineers, using judgment and experience, must first transform a physical problem into a related mathematical problem. During this process, a mathematical model of the material and the structure must be evolved. The evolution of this model is a key requirement in the process and, in fact, is often the most difficult step in the entire solution procedure. The degree of complexity of the mathematical model must be such that it gives, over its range of application, answers that conform to the physics of the situation. On the other hand, many models are made far more complex than that which would be required for the adequate solution of the specific engineering problem at hand, thus greatly increasing the time and cost required for the solution of the problem. Engi-

*Partner and Director of Research, Weidlinger Associates, Consulting Engineers, New York, and Adjunct Professor of Civil Engineering, Columbia University.
**Associate Partner, Weidlinger Associates, Consulting Engineers, New York, and Adjunct Associate Professor of Civil Engineering, Columbia University.

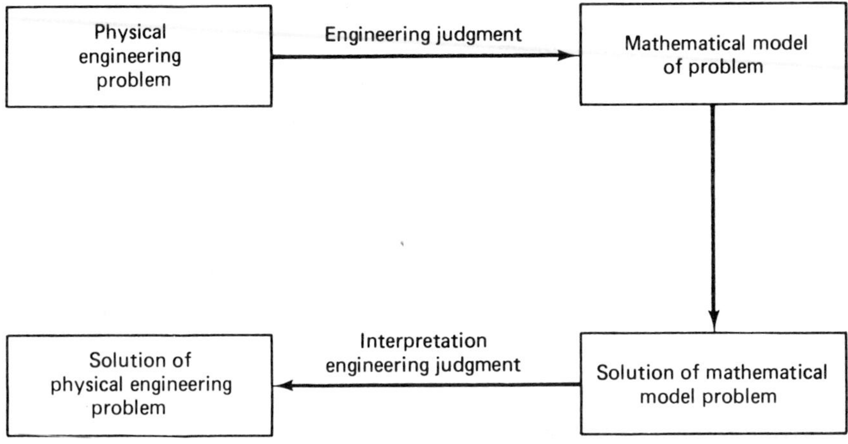

Figure 1

neers must carefully distinguish between two situations insofar as the complexity of the mathematical model is concerned: (1) the development of a complex model that is not required for the solution of the problem or problems at hand (e.g. a three-dimensional elasticity solution for the simple bending of a beam); and (2) the development of a complex model that is applied to the relatively simple problem of immediate interest, but which has been developed for later use in more complex problems (e.g., an elastic–nonideally plastic soil model that will be used in a range in which only the elastic portion of the model may be exercised).

While each of the three steps shown in Fig. 1 requires careful engineering judgment, the first step, in which the physical problem is transformed into a mathematical problem, is of particular concern here. Although much of what will be said applies to many different areas of engineering, the main applications to be reviewed involve the modeling of materials and structures. Such modeling can be thought of as an iterative procedure in which the engineer must weigh the requirements of the model in the light of the problem to be solved, the accuracy required, and the experimental data that are available for fitting the model. The constant iteration between the mathematical description and the experimental data for fitting is the cornerstone of the modeling procedure. The most sophisticated model will be of little or no use if sufficiently accurate data are not available for fitting the parameters of the model. Consequently, as the mathematical development of advanced models progresses, it is imperative that detailed experimental studies also take place. It should be realized, however, that in many instances, cases will be considered for which the experiments required to fit a model adequately for a problem with complex phenomenology are just too difficult to perform. For example, there are many cases in the modeling of material behavior in which the stress paths relevant to the problem of interest cannot be reproduced in the laboratory or in small in-situ experiments. In such cases, peripheral experiments along other stress paths (related if possible) may be made and the mathematical model is fit to these stress paths, in the hope that it will then conform, in at least an approximate manner, to the very complicated stress paths of direct interest. To illustrate this, consider the mathematical modeling of soil and rock materials for the analysis of the ground response to high-intensity explosions on a layered soil/rock medium. The models for such an analysis are generally fit to a series of laboratory tests, namely uniaxial strain and triaxial compression, as well as to whatever in-situ tests are available. One such test, the cylindrical in-situ test CIST, itself involves the detonation of a high explosive in the medium. It must be well understood by the engineer/analyst that the stress paths to which the material models are fit are considerably simpler than the very complex paths that will occur during the postulated event of interest in the soil/rock layered medium. Such a complete stress path, however, can be realistically simulated only by the event itself.

MODEL SELECTION

Some of the alternatives that must be considered in constructing mathematical models for different engineering problems are shown in Fig. 2. In many problems, the choice of the proper modeling procedures will be evident. However, questions involving the choice of material models and computer codes are becoming increasingly complicated as new and more complicated engineering problems are encountered, and their solutions are attempted over wider and more difficult ranges of application. A major cause of trouble often occurs in the use of computer codes. Unfortunately, in many cases there appears to be a distinct temptation to use large, general-purpose codes, such as NASTRAN, as "black boxes" to solve all problems. The use of any code by persons unfamiliar with the workings of that code is extremely dangerous. The authors have seen many cases in which completely erroneous results were obtained by such procedures. In addition, in some cases in which the codes were run correctly, the results were obtained at a cost in computer time of as much as 5 to 10 times that which would have been incurred had smaller, specialized codes, specifically oriented to the solution of the problem, been used. The point that must be made is threefold: (1) there are a great number of problems in which large, general-purpose codes are required and are quite efficient; (2) there are a great number of problems for which smaller specialized codes can be more efficiently used; and (3) no code should be used as a "black box," particulary by those unfamiliar with the code and its internal components. The third point seems so obvious that it is almost superfluous to mention it, however, it is often disregarded.

PHYSICAL PROBLEM
TIME SCALE
(a) Static
(b) Quasi-static
(c) Dynamic
(1) Strain rate dependent
(2) Strain rate independent
NATURE OF PHENOMENA
(a) Linear
(b) Nonlinear
(1) Material nonlinear
Geometrically linear
(2) Material linear
Geometrically nonlinear
(3) Both nonlinear
MATERIAL MODEL
(a) Linear
(b) Hysteretic
(1) Pressure-volume
(2) Shear
(elastic-plastic)
(c) Stability and
uniqueness
requirements
COMPUTER CODE
(a) Finite difference
(b) Finite element
(c) Hybrid
General purpose
Specialized stability
OUTPUT REQUIRED
(a) Detailed time histories
(1) Frequency content
(b) Peak values
(c) Sensitivity information
Interpretation
Solution of
physical
problem

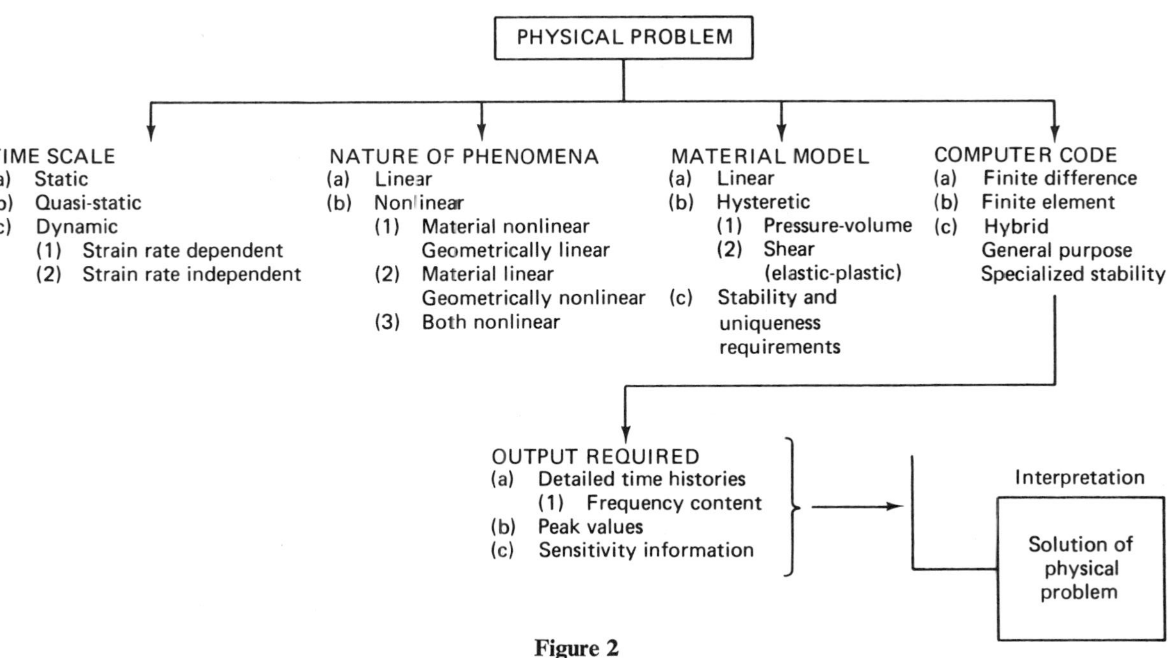

Figure 2

All that has been said to this point is basic to the modeling and solution procedure. Several examples will now be given to illustrate some specific points in the modeling of materials and structures.

MODELS FOR GEOLOGICAL MATERIALS

To illustrate the processes involved in the development of material models, and, in particular, the interrelationship between theory and experiment in such a development, let us consider the case of mathematical models for geological materials (soil and rock) for use in problems involving transient loadings of such media. Specifically, let us consider ground shock problems in which an explosion produces a moving surface overpressure wave that decays as it expands over the space surrounding the explosion point. The reader is referred to [1] for a detailed description of the evolution of such models.

Linear elastic models were originally utilized in the analysis of air-induced ground shock effects at low surface overpressure levels (say, 20 to 50 psi). It was erroneously postulated at that time that the behavior of such materials at these low overpressures was essentially elastic. As experimental laboratory results from uniaxial strain and triaxial compression tests became available and indicated the presence of hysteresis in loading and unloading cycles for most soils and rocks, and as technological changes necessitated the design of hardened structures at higher overpressures, inelastic effects were introduced into the analysis. The first approaches to inelastic models were elastic–ideally plastic formulations, in which a yield condition of the form suggested by Drucker and Prager,

$$\sqrt{J_2'} = k + \alpha J_1 \tag{1}$$

was introduced into the model. Two special cases were also used in the early inelastic model: (1) the von Mises yield condition, $\sqrt{J_2'} = k$, and (2) the Coulomb yield condition, $\sqrt{J_2'} = \alpha J_1$. These early elastic–plastic models were generally assumed to be linearly elastic below the yield surface and ideally plastic on the surface. Figure 3 shows the stress–strain diagram in uniaxial strain for these early inelastic models. Of particular interest is case (c), in which a Drucker–Prager material is cyclically loaded considerably beyond the elastic range, leading to a permanent set that is extensile. This set can be considered to be a one-dimensional analog of the three-dimensional phenomenon of dilatancy (i.e., inelastic volume increase under shear loading). Dilatancy, which can be quite significant in the behavior of soil and rock models, will be discussed later as one of the reasons for the development of the currently used series of advanced elastic–nonideally plastic cap models.

At this point a word is in order regarding the question of uniqueness in

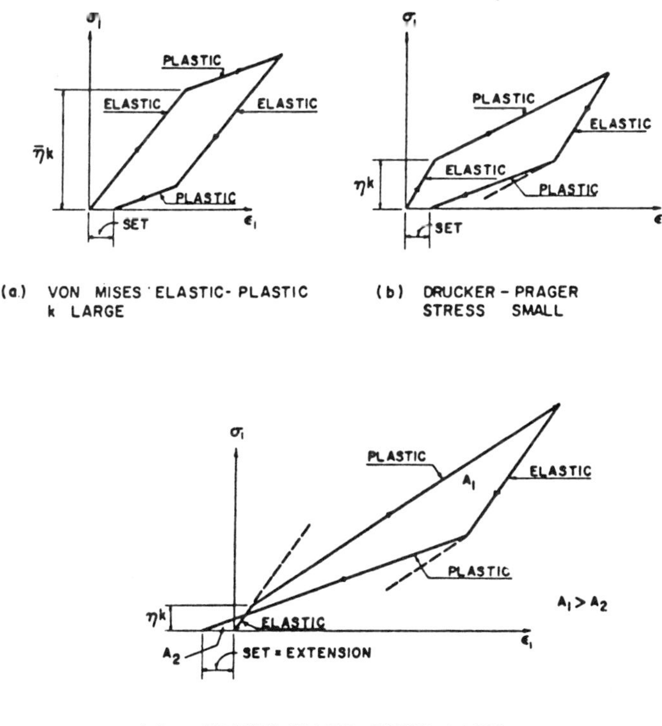

Figure 3 Uniaxial strain, von Mises and Drucker–Prager models.

elastic–plastic problems. In mathematical plasticity, the plastic strain rate is derived from a plastic potential function ϕ, by means of a flow rule

$$\dot{\epsilon}_{ij}^{p} = \lambda \frac{\partial \phi}{\partial \sigma_{ij}} \tag{2}$$

When the plastic potential ϕ is identical to the yield condition, eq. (2) is called an "associated" flow rule; otherwise, it is considered nonassociated. When an associated flow rule is used and the yield surface is convex, unique solutions are assured in dynamic continuum problems in which the governing partial differential equations are hyperbolic.

Problems quickly arose with respect to the suitability of the early inelastic models. At that time few of the typical experimental soil data (Fig. 4), and rock data (Fig. 5) were available. As more experimental data were obtained in response to the need for harder structures located at higher overpressure levels, it became evident that major modifications were required to match these new data, particularly in the areas of (1) a generalization of the

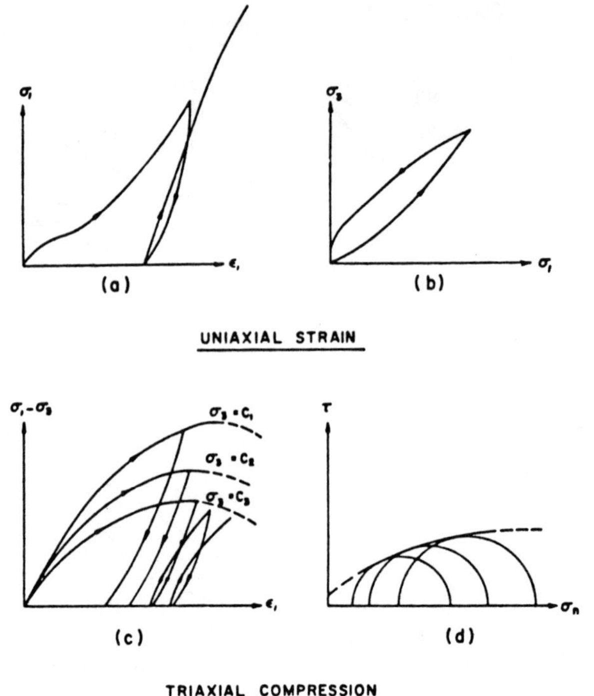

Figure 4 Typical experimental results, soil.

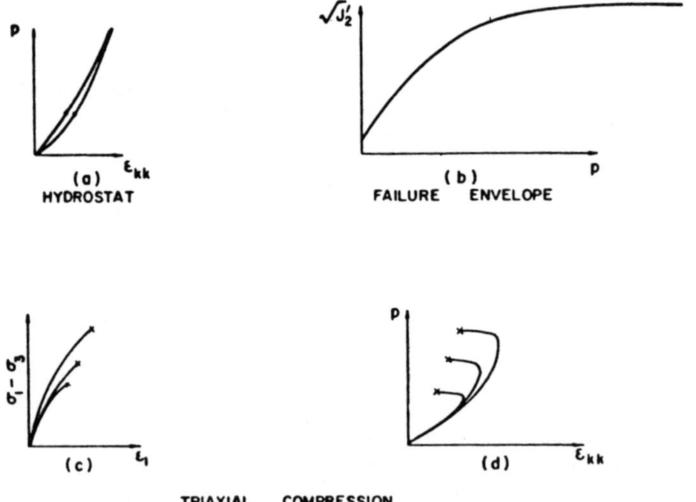

Figure 5 Typical experimental results, rock.

yield condition to more closely match the failure envelope from triaxial comparison tests, (2) the use of a nonlinear pressure–volume relation, and (3) the use of different pressure–volume relations in loading and unloading for most soils. The last item was of major importance insofar as it produced pressure–volume hysteresis in the soil models. Hence, the two main types of rate-independent hystereses, pressure–volume and shear (plasticity), were incorporated into the new soil and rock models. This was done by fitting both stress–strain and stress path information from uniaxial strain and triaxial comparison tests, as well as Mohr circle failure envelope data from the latter. Again, the iterative nature of the model development is noted as a continuous interplay between the theoretical development and the increasing need for complete experimental information, at least for the several stress paths obtainable in the laboratory.

The next series of major modifications in the models consisted of the introduction of a generalized yield condition of the form

$$J_2' = f(J_1) \tag{3}$$

which approximates the Prager–Drucker material at low overpressure and the von Mises material at higher overpressures (Fig. 6). The transition region

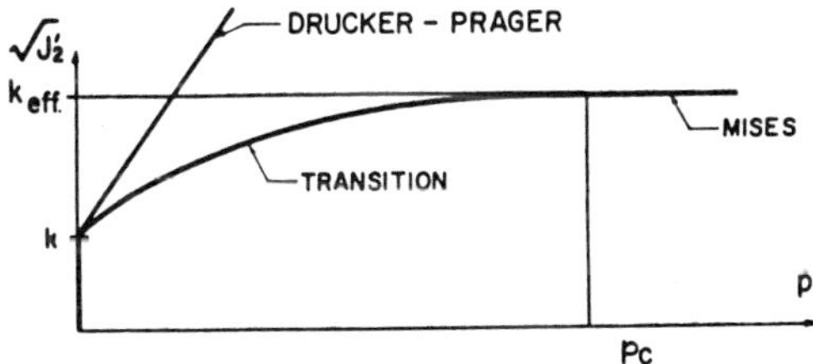

Figure 6 Yield condition, alluvium–playa model.

was fit from failure data obtained from triaxial comparison tests. At the same time, the introduction of a nonlinear pressure–volume relation or, alternatively, a nonlinear bulk modulus, which could differ for loading and unloading, enabled the engineer to fit both portions of the stress–strain curves from uniaxial strain tests, and thus to control, independent of the yield condition the amount of pressure–volume hysteresis in a loading–unloading cycle. This gave rise to a series of advanced elastic–ideally plastic models, in which both pressure–volume and shear (plastic) hysteretic mechanisms were utilized.

The problems of modeling remained, for many cases of interest, unsolved by this advanced series of models. The use of a yield condition containing the first stress invariant J_1 resulted in dilatancy when the associated flow rule was used. Although dilatancy is exhibited by rocks [Fig. 5(d)] and very dense soils, it is not present to any noticeable extent in most soils. Moreover, its presence in a mathematical model can produce large horizontal displacements in ground shock problems. To eliminate the effects of dilatancy in the soil models, a nonassociated flow rule was often utilized. This procedure raised doubts about the uniqueness of the solution. Even for cases in which dilatancy was physically present, the advanced, elastic–ideally plastic models did not allow the engineer to fit the effect on the basis of test results, because it was rather arbitrarily introduced into the analysis through the use of an associated flow rule, and no free parameter was included in the model for fitting the amount of dilatancy. In addition, the use of elastic–ideally plastic models did not allow the fitting of stress–strain results in triaxial compression, because after failure, the material as described by such models merely flowed, as shown in Fig. 7. To overcome these difficulties, and at the same time control dilatancy for soils and rocks, the elastic–nonideally plastic cap-type

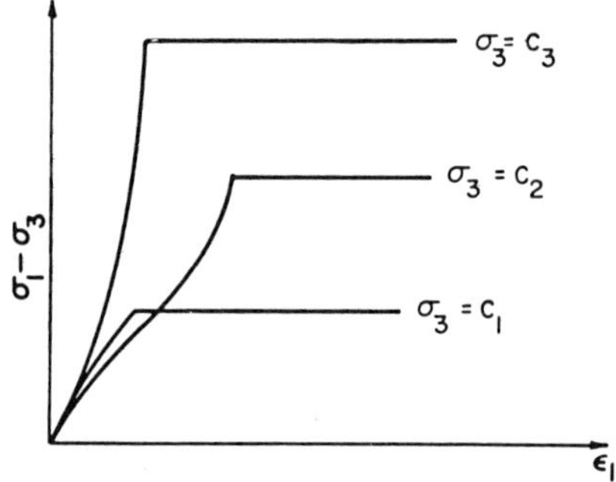

Figure 7   Triaxial compression test, elastic–ideally plastic model.

models were developed and are now widely used in ground shock applications [2,3]. The yield surface consists of a generalized condition of the form of eq. (3), together with a strain–hardening "cap" (Fig. 8), which "expands" or "contracts" as the plastic volumetric strain increases or decreases. The model is capable of predicting observed laboratory data for all available tests, while satisfying the most restrictive theoretical continuity and uniqueness requirements. The cap surface is determined by a relation of the form

$$f_2(J_1, \sqrt{J'_2}, \epsilon^p) = 0 \tag{4}$$

Because the movement of the cap is controlled by the increase or dec ease of plastic volumetric strain, strain hardening can be reversed in the cap soil model and the mechanism can effectively control dilatancy (which can be kept very small, effectively zero, as required for a soil model). A somewhat

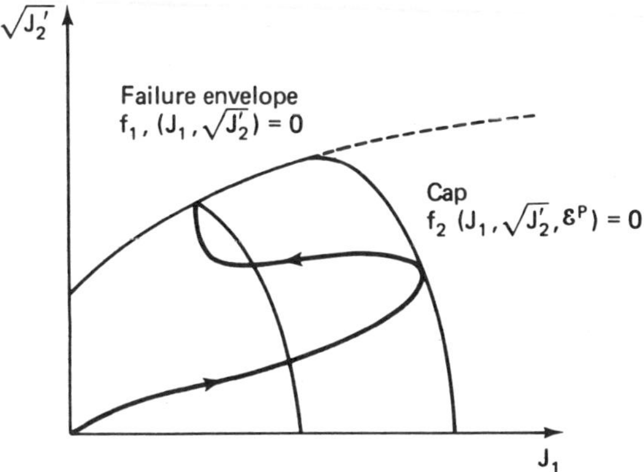

Figure 8   Yield condition for soil cap model.

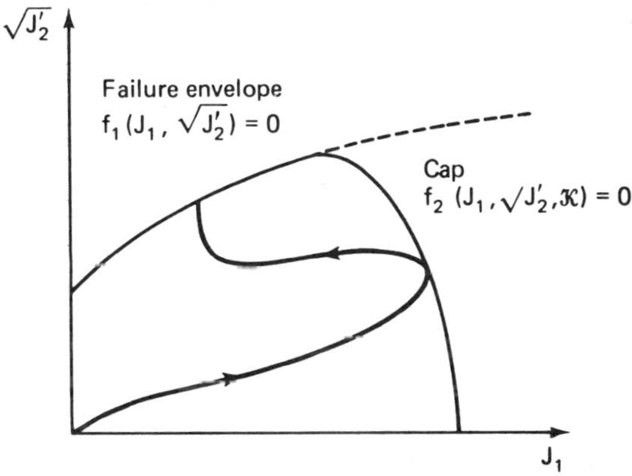

Figure 9   Yield condition for rock cap model.

similar plastic-hardening material model with a capped yield surface (Fig. 9) has also been developed for rock. The capped yield surface in this case is irreversible and is controlled by a hardening parameter κ, which is a given function of the stress–strain history of the rock

$$f_2(J_1, \sqrt{J_2'}, \kappa) = 0 \tag{5}$$

The model utilizes an associated flow rule and allows—and controls—dilatancy. The pressure–volume curves from a triaxial comparison test simulation, together with those from this model, show the proper behavior [Fig. 5(d)].

Concurrent with the development of advanced models in response to fitting phenomena observed in laboratory tests, it became obvious that laboratory tests on small specimens were insufficient to reproduce the in-situ behavior of many materials. Consequently, standard procedures now consist of developing preliminary models from laboratory data and refining the constants in the models on the basis of in-situ test data.

For certain applications, anisotropic as well as strain-rate-dependent cap models have also been developed for use in special situations for which they are appropriate.

An interesting example of the necessity for coupling both theory and experiment in developing mathematical models lies in the problems encountered by the "variable modulus" soil model that was developed some years ago on a primarily experimental data-fitting basis. In such models, both the bulk and shear moduli are functions of the stress and/or strain invariants and there is no explicit yield condition. Such models do a good job of fitting the results from available experiments. They do, however, suffer from a severe theoretical deficiency for cases of neutral loading (i.e., $\sqrt{\dot{J_2'}} = 0$) insofar as continuity conditions are concerned. The problem is illustrated in Fig. 10, where point A lies on the surface $\sqrt{\dot{J_2'}} = 0$ (i.e., $\sqrt{J_2'} = $ constant).

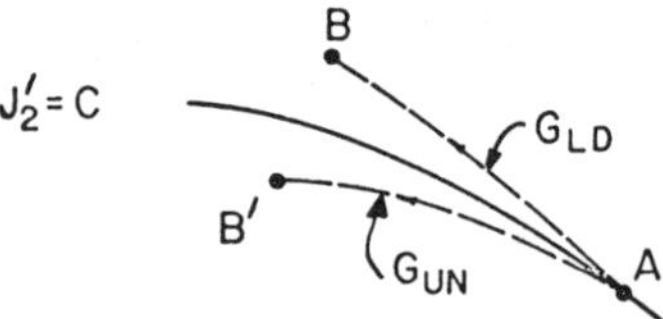

Figure 10   Continuity condition.

Let us consider two paths, AB and AB', arbitrarily close but on opposite sides of such a surface. In this model, path AB is considered loading (in shear) while path AB' is unloading. On the former path, the shear modulus $G = G_{LD}$ applies while on the latter path G_{UN} would apply. Therefore, there will be a finite difference in strain even when B and B' are infinitesimally close, thus violating the theoretical requirement of continuity (in a manner similar to the deformation theories of plasticity). While this violation of continuity perhaps does not sound particularly serious, it has been shown to be quite important in general ground shock problems, and variable moduli models are no longer utilized for problems in which neutral loading conditions are likely to be encountered [4]. Here, a basic dependence on fitting experiments without a corresponding effort at satisfying theoretical requirements led to an unsatisfactory model.

A somewhat similar theoretical problem is encountered in the use of certain recently proposed endochronic material models [5]. In these models the deformation history of the material is represented by means of a monotonically increasing quantity called intrinsic time. This time is then used in the equations of general viscoelasticity in place of the real time. The resulting equations can be used to represent many aspects of inelastic material behavior without the use of a yield or loading surface.

The difficulty with the use of endochronic models in dynamic problems is that any definition of intrinsic time that does not use a yield condition or inelastic barrier type of concept must necessarily lead to undue sensitivity of the solution to infinitesimal high-frequency disturbances. Such disturbances are almost always present in numerical calculations of dynamic problems (in the form of numerical errors) as well as in the real world. Therefore, endochronic models are too sensitive (unstable) to be used for general material modeling in dynamic problems.

A structural modeling problem

An illustrative problem involving the modeling of a structure will now be considered. Let us consider the problem of the elastic response to a traveling shock wave of a shell with internal structure submerged in an infinite acoustic medium (Fig. 11). Of particular interest in this problem is the evaluation of the velocity versus time histories at points of the shell and, more important, at points on the selected internal equipment in order to ascertain the shock loading on this equipment. The reader is referred to [6] for a detailed description of this problem.

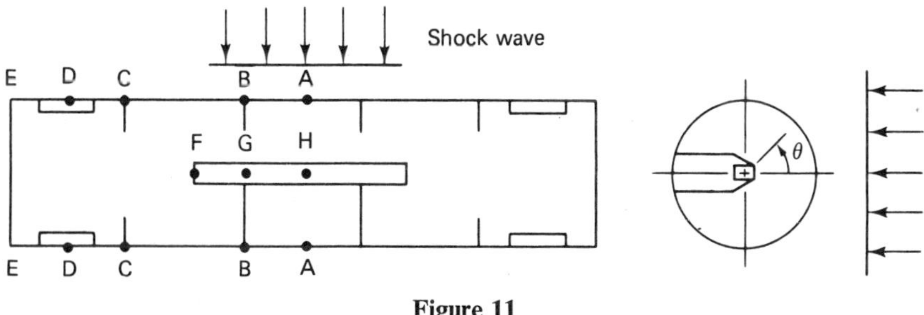

Figure 11

In deciding on the mathematical model for this structural analysis problem, several points were of paramount importance. First, a decision had to be made regarding the treatment of the fluid–structure interaction. This, as will be seen later, would be accounted for by means of an approximate relation between the radiated fluid pressure and velocity. Second, a structural analysis scheme had to be determined. Several such approaches were possible. A complete finite element direct integration approach, using a general-purpose code such as NASTRAN to model the shell/internal structure configuration, was one possibility. On the other hand, a special-purpose cylindrical shell analysis code, BOSOR-4, and a finite element description of the internal structure, SAP-IV, was another possibility in which an *in-vacuo* modal analysis could be quite efficient. Inherent in this modal structural analysis was the use of subpartitioning between the shell and the internal structure; that is, separate *in-vacuo* modes were used for the shell and for the internal structure, and the response of these structures was obtained through the use of time-dependent boundary conditions at the connection points. Because the purpose of the analysis was to obtain the shock response of internal equipment (the amount of which could be considerable in a given problem), it became evident at an early stage in the development that the use of the special-purpose shell code, coupled with a finite-element code description of the interior equipment, would be far more economical than and preferable to the use of a single general-purpose code for the combined shell/ internal structure analysis. The computer storage space saved by not requiring a full finite element analysis of the shell provided extra capacity for modeling additional internal structures. Moreover, the use of a modal approach, including subpartitioning between the shell and the internal structure, was an advantageous technique, used often in aircraft and missile analyses, and appeared to lend itself well to the present problem.

The outline of the structural problem can be written in terms of a series of *in-vacuo* shell modes $\phi_j(s)$ and internal structure modes $\phi_{\sigma j}(s)$ as follows. Let the *in-vacuo* shell modes $\phi_j(s)$ have associated generalized masses μ_j, generalized coordinates $q_j(t)$, and natural frequencies ω_j. The shell displacements can be written

$$\mathbf{d}(s, t) = \sum_j q_j(t)\boldsymbol{\phi}_j(s) \tag{6}$$

Let the fixed-base modes of the internal structure, $\phi_{\sigma j}(s)$, have associated generalized masses $\mu_{\sigma j}$, generalized coordinates $q_{\sigma j}(t)$, and natural frequencies $\omega_{\sigma j}$. The internal equipment displacements can be written

$$\mathbf{d}_\sigma(s, t) = \sum_j q_{\sigma j}(t)\boldsymbol{\phi}_{\sigma j}(s) + \mathbf{g}(s, t) \tag{7}$$

where the function $\mathbf{g}(s, t)$ is chosen to satisfy the time-dependent boundary conditions at connection points S_k: for example,

$$\mathbf{d}_\sigma(S_k, t) = \mathbf{d}(S_k, t) \tag{8}$$

This separate use of shell and internal structural modes, and their coupling at connection points, is the technique known as substructuring.

With Lagrange's equations, the equations of motion of the generalized coordinate of the shell q_j and the internal structure $q_{\sigma j}$ are written as

$$\mu_j\ddot{q}_j + \mu_j\omega_j^2 q_j = Q_{Tj} + Q_{\sigma j} \tag{9}$$

$$\mu_{\sigma j}\ddot{q}_{\sigma j} + \mu_{\sigma j}\omega_{\sigma j}^2 q_{\sigma j} = Q_{Sj} \tag{10}$$

where Q_{Tj} are the generalized forces due to the fluid pressure on the shell, $Q_{\sigma j}$ the generalized forces due to the reactions exerted by σ on S, and Q_{Sj} the generalized forces due to the reactions exerted by S on σ. Initial rest conditions on the generalized coordinates are specified:

$$\dot{q}_j(0) = q_j(0) = \dot{q}_{\sigma j}(0) = q_{\sigma j}(0) = 0 \tag{11}$$

The evaluation of Q_{Tj}, the generalized force due to the fluid–structure interaction, is the key to the solution of the problem. The total fluid pressure on the shell and the total normal fluid velocity are written in terms of an incident (I) and a radiated component,

$$p_T(s, t) = p_I(s, t) + p(s, t) \tag{12}$$

$$U_T(s, t) = U_I(s, t) - U(s, t) \tag{13}$$

so that the generalized force Q_{Tj} becomes

$$Q_{Tj}(t) = \int_A p_T(s, t)\phi_j^n(s) \, dA = Q_{Ij}(t) + Q_j(t) \tag{14}$$

where $\phi_j^n(s)$ is the normal component of the mode shape and

$$Q_{Ij}(t) = \int_A p_I(s, t)\phi_j^n(s) \, dA \qquad \text{(incident)} \tag{15}$$

$$Q_j(t) = \int_A p(s, t)\phi_j^n(s) \, dA \qquad \text{(radiated)} \tag{16}$$

Continuity of normal velocities at the fluid–shell interface lead to the relation

$$U_T(s, t) = \dot{d}_n(s, t) \tag{17}$$

or

$$U(s, t) = U_I(s, t) - \dot{d}_n(s, t) \tag{18}$$

As mentioned earlier, it was decided to use an approximate treatment for the radiated fluid effects in order to eliminate the need to perform calculations in the fluid itself (e.g., finite elements in the fluid). Several such approximations had been developed and used in various investigations. They essentially decouple the interaction problem by allowing the approximation of the fluid interaction as an additional force-like loading on the *in-vacuo* structure.

Three such approximate treatments were available. The first, known as the plane-wave approximation, is valid at early times (high frequencies) and was introduced by Mindlin and Bleich [7]. In this approximation, it is assumed that at each point the structure radiates a plane wave; that is,

$$p = \rho c U = \rho c (U_I - \dot{d}_n) \tag{19}$$

so that

$$Q_j(t) = \int_A p(s, t)\phi_j^n(s)\, dA = \sum_k \alpha_{jk}(U_{Ik} - \dot{q}_k) \tag{20}$$

where

$$\alpha_{jk} = \rho c \int_A \phi_k^n \phi_j^n \, dA = \alpha_{kj} \tag{21}$$

$$U_{Ik}(t) = \frac{1}{\mu_k} \int_A m(s) U_I(s, t)\phi_k^n(s)\, dA \tag{22}$$

The equations of motion for the shell thus become

$$\boldsymbol{\mu}\ddot{\mathbf{q}} + \boldsymbol{\alpha}\dot{\mathbf{q}} + \boldsymbol{\mu}\omega^2\mathbf{q} = \mathbf{Q}_I + \boldsymbol{\alpha}\mathbf{U}_I + \mathbf{Q}_\sigma \tag{23}$$

It is seen that this approximation, which is valid for high frequencies and short times, adds a term $\boldsymbol{\alpha}\dot{\mathbf{q}}$ to the differential equation on the generalized coordinates $\mathbf{q}$ and thus leads to a radiation damping effect.

The second treatment, the virtual mass approximation, was developed by Chertock and is valid at late times (i.e., low frequencies) [8]. The radiated pressure is assumed to be in phase with the shell displacement, and the pressure and the acceleration in the radiated wave are related by the virtual mass matrix, thus yielding the relation

$$Q_j = \sum_k \gamma_{jk}(\dot{U}_{Ik} - \ddot{q}_k) \tag{24}$$

where γ_{jk} are the elements in the virtual mass matrix $\boldsymbol{\gamma}$. The equations of motion for this approximation become

$$\boldsymbol{\mu}\ddot{\mathbf{q}} + \boldsymbol{\gamma}\ddot{\mathbf{q}} + \boldsymbol{\mu}\omega^2\mathbf{q} = \mathbf{Q}_I + \boldsymbol{\gamma}\dot{\mathbf{U}}_I + \mathbf{Q}_\sigma \tag{25}$$

The effect of this approximation is the term $\boldsymbol{\gamma}\ddot{\mathbf{q}}$ on the left-hand side of the differential equation, thus representing an added or "virtual" mass in the problem.

The third, and most versatile approximation, developed by Geers in the United States [9] and Mnev and Pertsev in the USSR [10], is called the doubly asymptotic approximation. It is given by the relation

$$\boldsymbol{\alpha}^{-1}\dot{\mathbf{Q}} + \boldsymbol{\gamma}^{-1}\mathbf{Q} = \dot{\mathbf{U}}_I - \ddot{\mathbf{q}} \tag{26}$$

where $\boldsymbol{\alpha}$ and $\boldsymbol{\gamma}$ are the damping and virtual mass matrices, respectively. This approximation was designed in an attempt to extend the range of applicability of the fluid–structure interaction procedure. At early times it reduces to the plane-wave approximation, while at late times it reduces to the virtual mass approximation. A smooth transition between the early and late approximations is thus effected. When the incident wave has a zero rise time, an integrated form of eq. (26) is used,

$$\boldsymbol{\alpha}^{-1}\mathbf{Q} + \boldsymbol{\gamma}^{-1}\int_0^t \mathbf{Q}\, dt = \mathbf{U}_I - \dot{\mathbf{q}} \tag{27}$$

where

$$\int_0^t \mathbf{Q}\, dt \bigg]_{t=0} = 0 \tag{28}$$

The choice of the fluid–structure interaction approximation to be used in the analysis is again a modeling decision to be made by the engineer. In the present problem, the requirement that the velocity–time histories at points on the shell and the internal structure be analyzed dictated that the doubly asymptotic approximation be used, since a description of the motion over the full time regime was required.

At this point, the modeling decisions for this structure–medium interaction problem were complete. This did not, however, end the complications of the problem because additional decisions on the numerical techniques for solving the boundary value problem and an analysis of the accuracy of the solution based on these techniques were also required.

The efficient solution of the problem involved the use of an improved numerical technique, using surface expansion functions that are orthogonal over the wetted shell area. Preliminary studies indicated that the matrices $\boldsymbol{\alpha}$ and $\boldsymbol{\gamma}$, when formulated in terms of dry modes, are poorly conditioned for inversion and lead to numerical problems. To circumvent these difficulties, surface expansion functions ψ_j were introduced. These functions are orthogonal over the wetted shell area,

$$\int_A \psi_j(s)\psi_k(s)\, dA = \hat{\mu}_j \delta_{jk} \tag{29}$$

In the fluid portion of the analysis, these functions were utilized in place of the shell modes, thus resulting in a diagonal damping matrix $\boldsymbol{\alpha}$ and a virtual mass matrix $\boldsymbol{\gamma}$, which, although not diagonal in itself, is dominated by diagonal elements. In this manner, the poorly conditioned equations were eliminated and faster and more efficient computations resulted.

Finally, a decision was required on the number of modes that were to be used for each value of n, as well as a decision on the number of terms n in the series itself. This decision was dependent on the frequency range for which the solution was desired and is discussed in [6].

The major steps required for the solution of the problem are now complete. Specification of the input parameters, that is, the details (geometric and time history) of the incoming shock wave, the dimensions of the shell and the internal structure, the time integration steps for each mode, and so on, is required for the specific problem to be solved. Specification of output parameters (i.e., velocity versus time histories at selected points on the shell and on the internal structure), together with their associated shock spectra, is also required. These details, together with results for a specific configuration, are discussed and presented in [6].

CONCLUSIONS

Some problems that arise in the mathematical modeling of materials and structures have been discussed and illustrated in the preceding pages. A rational approach to the solution of such problems largely depends on the individual investigator. If the investigator is willing to take the time to outline the approach rationally and to consider the various possibilities relative to such items as generality of solution required and computer code complexity, then many of these problems will easily be mastered. It is worth repeating that the modeling of materials must be considered an iterative procedure in which experimental data and theoretical considerations (uniqueness, stability) play major roles.

If the investigator will indeed consider the study of such problems of planning as a necessary first step in the overall engineering studies, appropriate procedures for solving the tasks should become readily evident.

REFERENCES

[1] NELSON, I., BARON, M. L., AND SANDLER, I. S., "Mathematical Models for Geological Materials for Wave Propagation Studies," *Shock Waves and the Mechanical Properties of Solids*, Syracuse University Press, Syracuse, N.Y., 1971.

[2] DiMAGGIO, F. L., AND SANDLER, I. S., "Material Model for Granular Soils," *Journal of the Engineering Mechanics Division, ASCE*, Vol. 97, No. EM3, Proc. Paper 8312, June 1971, pp. 935–950.

[3] SANDLER, I. S., AND DiMAGGIO, F. L., "Material Models for Rocks," *Report DASA 2595*, Defense Atomic Support Agency, Weidlinger Associates, New York, October 1970.

[4] NELSON, I., AND BALADI, G. Y., "Outrunning Ground Shock Computed with Different Models," *Journal of the Engineering Mechanics Division, ASCE*, Vol. 103, No. EM3, Proc. Paper 12975, June 1977, pp. 377–393.

[5] SANDLER, I. S., "On the Uniqueness and Stability of Endochronic Theories of Material Behavior," *Jour. Appl. Mech.,* Vol. 45, No. 2, June 1978, p. 263.

[6] RANLET, D., DiMAGGIO, F. L., BLEICH, H. H., AND BARON, M. L., "Elastic Response of Submerged Shells with Internally Attached Structures to Shock Loading," *International Journal of Computers and Structures,* Vol. 7, pp. 355–364, Pergamon Press, 1977.

[7] MINDLIN, R. D., AND BLEICH, H. H., "Response of an Elastic Cylindrical Shell to a Transverse, Step Shock Wave," *Journal of Applied Mechanics,* Vol. 20, 1953, p. 189.

[8] CHERTOCK, G., "Transient Flexural Vibrations of Ship-Like Structures Exposed to Underwater Explosions," *Journal of the Acoustical Society of America,* Vol. 48, No. 1, Part 2, 170, 1970.

[9] GEERS, T. L., "Residual Potential and Approximate Methods for Three-Dimensional Fluid-Structure Interaction Problems," *Journal of the Acoustical Society of America,* Vol. 49, No. 5, Part 2, 1971, p. 1505.

[10] MNEV, Y. N., AND PERTSEV, A. K., "Hydroelasticity of Shells," *Translation FTD-MT-24-119-71,* Translation Division, Foreign Technical Division, Wright Patterson Air Force Base, Dayton, Ohio.

CENTRAL PROBLEMS IN STRUCTURAL LOADING BY THE NATURAL WIND

*Robert H. Scanlan**

INTRODUCTION

The reasonably accurate prediction of structural loadings by wind proceeds through a series of steps from the general to the particular. For a given site there is, first, the need to assess the direction and velocity of the highest winds and the probability of their occurrence. This necessitates an excursion into climatology and meteorology. At this point micrometeorology and planetary boundary-layer theory and data are required for a clear definition of the nature of local winds. Local orographic effects may be taken into account at this time. A definition of the mean high wind, its velocity profile with height, and the full character (spectra and scales) of its turbulence (gustiness) are obtained.

If the structure under consideration is important it will very likely be the subject of both analytical and experimental (wind tunnel) studies from this point on. In principle, no particular wind loading studies at all can be made in the absence of experimental information, although some useful general information of this type is now available.

---

*Department of Civil Engineering, Princeton University.

In order for wind tunnels to supply the further necessary information, laws of similitude must be observed. These laws apply first to the duplication of the atmospheric boundary layer itself and second to the relation that the modeled structures bear to the wind. In brief, structural model geometric scales and natural frequencies must be compatible with those characteristic of the simulated natural wind.

A number of wind tunnel simulation problems are intrinsic to the fact that, under most test conditions, both Reynolds and Rossby numbers are necessarily violated. Both "long" and "short" wind tunnels are in use, the long being associated with a naturally developed thick, turbulent boundary layer, while the short are associated with more artificially "stirred" boundary layers. The latter are established by means of fixed mechanical elements such as stationary spires and roughness elements or by jets entering either laterally or longitudinally. Knowledge of expected wind spectra are needed as target values, at least for short-tunnel simulations. To date it is not certain that any of the wind tunnel simulations fully duplicates atmospheric winds in all desirable details.

Theoretical wind response calculations, based in part on knowledge both of the atmospheric characteristics and those of flows previously observed around structures, refer back to the theory of elastic structures excited by forces that are random in space and time. While theory is in good condition here, hypotheses about inputs have to date been somewhat strained. It is usually experimental data that must be called upon to clarify the character of the randomness. Thus, calculations are still strongly dependent upon some form of experimental input; they nontheless become an important link in the structural design chain.

In what follows, the areas of interest or concern that have been alluded to above will be briefly reviewed and state-of-the-art information on them referenced.

CLIMATOLOGICAL DATA ON WIND

Maximum wind velocities for sites within the United States have been specified for given "return periods" by Thom [1]. Such results, which should be subjected to further upgrading as newer data and statistical methodology are brought to bear, are of strong primary value for design. Additionally, on-site or near-site data must be employed if directional probabilities of the wind are to be assessed.

Winds high aloft, above the earth's boundary layer, follow a bivariate Gaussian distribution (Brooks and coworkers [2]), and when magnitude only, independent of direction, is considered, the speeds of such winds follow a Rayleigh distribution. Under these conditions the probability of wind speed exceeding a given velocity U takes the form $e^{-(U/C)^2}$, where C is a constant. Closer to the ground probability P of the wind speed exceeding a value U has been somewhat more accurately fitted with a Weibull distribution:

$$P(>U) = Ae^{-(U/C)^K} \tag{1}$$

where A, C, and K are parameters to be fitted to the data, K usually being in the neighborhood of 2, however.

The Weibull distribution, with A, C, and K considered functions of the azimuth angle θ, can be conveniently employed to incorporate local wind data into prognoses for a given site. Data from local sources (airports, government buildings, etc.) are first gathered and normalized to standard height, typically 10 m. (Details on wind velocity as a function of height appear later in this article. The various forms that raw wind data may take will not be discussed here.) For a fixed azimuth sector (22.5° is representative) cumulative probabilities are calculated from the available wind speed data for that sector. Results are then typically plotted on "probability paper," consisting of provision, in anticipation of using eq. (1) in double logarithmic form, for a plot of $\ell n \, [-\ell n \, P(>U)]$ versus $\ell n \, U$. Figure 1 suggests the aspect of such a plot, which usually results in an array of points to which a straight line can be fitted, either by eye or by use of a more sophisticated regression technique, and from which the constants A, C, and K can be estimated.

A plot of this type, for each 22.5° sector about a given site, results in the definition of the probability $P(>U)$ for any assigned value of velocity U.

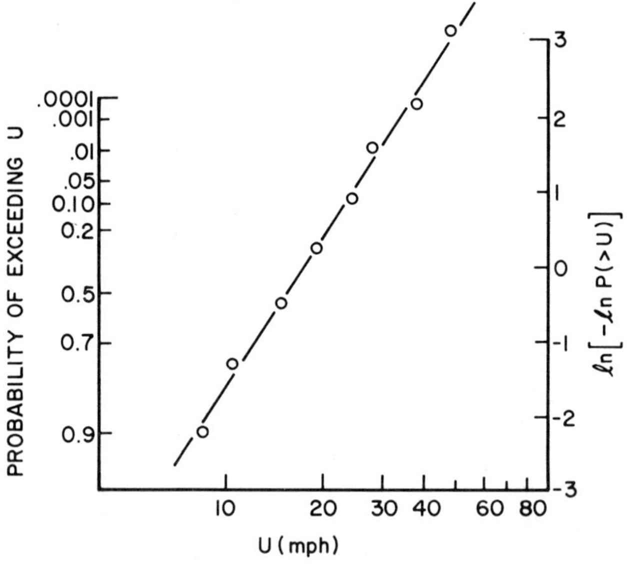

Figure 1   Appearance of plotted wind data.

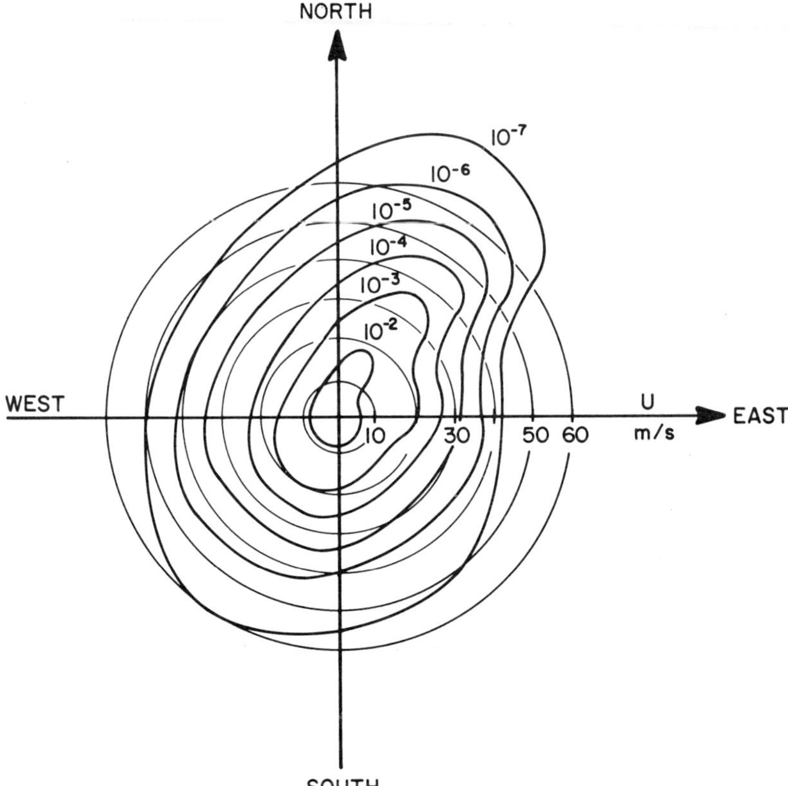

Figure 2   Directional distribution of wind speed for various probabilities of exceedance in 22.5° sectors for hypothetical site.

From this the directional distribution of wind about the site in question may be presented as in Fig. 2, where contours of equal probability of exceedance are suggested on a plot of wind speed for different azimuth angles at the site.

This type of plot alone may suffice for design use, but a prediction of highest winds is also presently made on the basis of the statistical theory of extremes [3]. The nature of the variate (wind velocities following a parent Weibull-type distribution) suggests that the cumulative probability distribution $P(U_{\text{extreme}} < U)$ of the *largest* values, taken for example to be the largest in given time intervals such as 1-year periods, might reasonably be represented by

$$P(U_{\text{extr}} < U) = \exp\{-[U - U_L))/U_s]^{-\gamma}\} \tag{2}$$

(sometimes called a "Type II distribution") where U_L and U_S are referred to as *location* and *scale* parameters, respectively, and $\gamma > 0$. Studies [4] have indicated, however, that the limiting form of eq. (2) as $\gamma \to \infty$ is more appropriate to wind data, at least for situations in the northern hemisphere in which tropical storms are not involved.

This leads to the alternative and more appropriate distribution

$$P(U_{\text{extr}} < U) = \exp\{-\exp[U - U_L)/U_S]\} \tag{3}$$

commonly known as the "Type I distribution." It then follows that the probability of an extreme wind velocity U_{extr} exceeding a given velocity U in any one year is expressed by

$$P(U_{\text{extr}} > U) = 1 - \exp\{-\exp[-(U - U_L)/U_S]\} \tag{4}$$

The *return period* (mean recurrence interval) R is defined as

$$R = \frac{1}{P(U_{\text{extr}} > U)} \tag{5}$$

and it follows from eq. (4) that the wind speed corresponding to a return period of R years is given by

$$U = U_L - U_S \ell\text{n}\left[-\ell\text{n}\left(1 - \frac{1}{R}\right)\right] \tag{6}$$

The parameters U_L and U_S are shown in the associated statistical theory to bear the following relations to the mean $\bar{U}$ and standard deviation σ_U:

$$\bar{U} = U_L + 0.5772U_S \tag{7}$$

$$\sigma_U = \frac{\pi}{\sqrt{6}}U_S \tag{8}$$

Parts of the discussion of this section have been borrowed from [4] and [5]. Good results have been obtained to date by the approaches outlined, but the subject is still an area where varied methods of attack are being explored.

CHARACTERISTICS OF PLANETARY BOUNDARY LAYER

The wind profile

Boundary-layer theory, and certain simplifications thereof which are beyond the scope of the present paper, result in a definition of the mean horizontal wind velocity U as a function of altitude z according to the formula

$$U(z) = \frac{1}{\kappa}u_* \ell\text{n}\frac{z - z_d}{z_0} \tag{9}$$

where $\kappa \cong 0.4$ is Kármán's constant, z_d is the average height of surrounding buildings (as defined subsequently), z_0 is a roughness length characteristic of the terrain fetch over which the wind arrives at the point in question, and u_* is a "friction" velocity characteristic of the flow. These quantities are defined below; z_0 must be selected from information such as that abridged in Table 1.

Table 1

Type of Surface	z_0 *(cm)*
Sea	0.0003 (calm)–0.5 (windy)
Grass	0.1 (mowed)–1.0 (high)
Brush	10–30
Pine forest (15 m high)	90–100
Suburbs	20–40
Centers of towns	35–45
Centers of large cities	60–80

Zero-plane displacement is

$$z_d = \bar{H} - \frac{z_0}{0.4} \tag{10}$$

where $\bar{H}$ is average height of surrounding buildings (if any). Friction velocity is obtained from reference information,

$$u_* = \frac{0.4 U(z_{\text{ref}})}{\ell \text{n}\,[(z_{\text{ref}} - z_d)/z_0]} \tag{11}$$

where $U(z_{\text{ref}})$ is wind velocity specified at some reference height z_{ref}. (Typically $z_{\text{ref}} = 10\,\text{m}$ for weather data.) Outdoor mean wind velocity measurements generally conform well to eq. (9), except for cases involving strong local effects. Figure 3 suggests the form of the profile of mean wind velocity with altitude.

Autospectra of the turbulence

The total horizontal wind velocity may be expressed as $U + u$, where U is the mean wind velocity defined by the boundary-layer profile given above, and $u = u(t)$ is the time-dependent or gust velocity, that is, the horizontal component of the turbulence. The lateral and vertical velocity components are designated as $v(t)$ and $w(t)$, respectively.

The principal cause of that turbulence which interests the civil engineer in the atmospheric boundary layer is the mechanical stirring of the wind as it passes over and around objects on the earth's surface. Formulas developed from both theory and outdoor measurements for the distribution of wind turbulent energy have been developed.

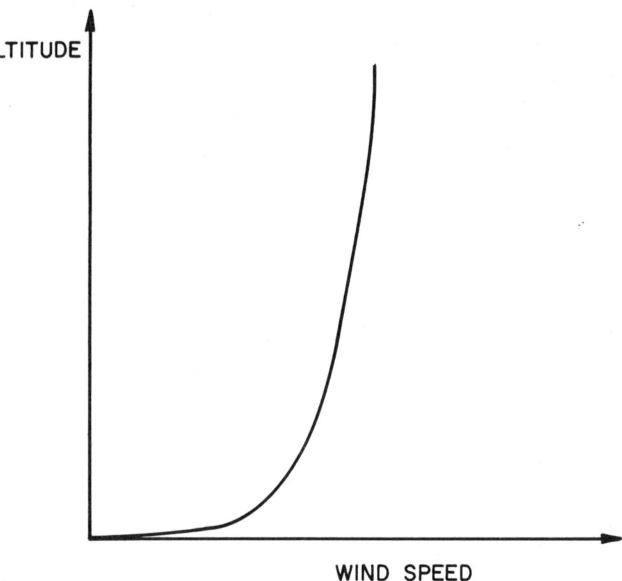

Figure 3

The power spectral density S_u (autospectrum) of horizontal gust velocities in strong winds has been given [6, 7] in the form

$$\frac{nS_u(z, n)}{u_*^2} = \frac{200f}{(1 + 50f)^{5/3}} \tag{12}$$

where n is frequency in hertz, and

$$f = \frac{nz}{U(z)} \tag{12}$$

The spectrum of vertical velocity fluctuations up to about 50 m may be estimated by the analogous formula [8]

$$\frac{nS_w(z, n)}{u_*^2} = \frac{3.36f}{1 + 10f^{5/3}} \tag{14}$$

For lateral velocity fluctuations a corresponding formula [6] is

$$\frac{nS_v(z, n)}{u_*^2} = \frac{15f}{(1 + 9.6f)^{5/3}} \tag{15}$$

(Note that the dimension of S_u, S_v, or S_w is m²/s.)

Cross-spectra of the turbulence

Horizontal velocity will be considered first. The cross-spectrum of two wind records, $u_1(t)$ and $u_2(t)$, taken at two points, 1 and 2, a distance r apart is designated

$$S_{u_1 u_2}^{Cr}(r, n) = S_{u_1 u_2}^{C}(r, n) + i S_{u_1 u_2}^{Q}(r, n) \tag{16}$$

where $i = \sqrt{-1}$ and S^C and S^Q are the co— and quad spectra, respectively. The coherence function $\mathrm{Coh}(r, n)$ is defined by

$$[\mathrm{Coh}(r, n)]^2 = \{[S_{u_1 u_2}^{C}(r, n)]^2 + [S_{u_1 u_2}^{Q}(r, n)]^2\}/S_{u_1} S_{u_2} \tag{17}$$

where S_{u_1} and S_{u_2} are the autospectra at points 1 and 2, respectively. In homogeneous turbulence, and approximately in atmospheric turbulence, S^Q vanishes.

Let (y_i, z_i) be the respective horizontal and vertical coordinates of a point i in a vertical plane; the following empirical expression for coherence of horizontal wind velocities has been offered [9, 10]:

$$\mathrm{Coh}(r, n) = e^{-\hat{f}} \tag{18}$$

where

$$\hat{f} = \frac{n[C_{1z}^2(z_1 - z_2)^2 + C_{1y}^2(y_1 - y_2)^2]^{1/2}}{\tfrac{1}{2}[U(z_1) + U(z_2)]} \tag{19}$$

and $C_z \simeq 10$, $C_y \simeq 16$ are approximations representing present knowledge. Actually, these coefficients are probably dependent upon fetch conditions as well as height above ground. An alternative form of the result (18) for engineering calculations is

$$S_{u_1 u_2}^{C}(r, n) = S_{u_1}^{1/2}(z_1, n) S_{u_2}^{1/2}(z_2, n) e^{-\hat{f}} \tag{20}$$

The cross-spectrum between vertical velocity fluctuations at two points may be expressed [11,12] by

$$S_{w_1 w_2}(\Delta y, n) = S_w(z_1, n) e^{-8n\Delta y/U(z)} \tag{21}$$

where Δy is the horizontal separation of the points 1 and 2.

Cross-spectra of lateral velocity fluctuations may tentatively be assumed [13] to be given by an expression like (20) with $C_y \simeq 7$, $C_z \simeq 10$.

The present summary of velocity profile and spectral information is taken from [4].

Turbulence intensities and scales

Perceived of as a time-varying function, the wind velocity may be considered to be made up of a steady part, of value U, plus Fourier components of sinusoidal nature associated with various frequencies n ranging from a low of 2 or 3 cycles per hour to high frequencies of 10 or 15 cycles per second. If the concept of "wavelength" λ is introduced in the form $\lambda = U/n$, some measure of "gust size" is obtained thereby. A representative, or average, alongwind wavelength of atmospheric gusts (normally calculated from the autocorrelation function) is termed the *scale* and varies in the atmosphere from about 65 m to 200 m, depending upon altitude, the longer wavelengths occurring at higher altitude.

Turbulence intensity I is defined as the ratio of the square root of the mean-square value of the gust component in question to the mean velocity; for example, the longitudinal intensity is

$$I_u = \frac{\sqrt{\overline{u^2}}}{U} \tag{22}$$

where $\overline{u^2}$ is the mean of $u^2(t)$ over some suitable averaging period. Longitudinal turbulence intensity is dependent upon fetch (upstream roughness conditions) and may vary from 3 or 4% to as high as 20%. Typical values may be 8 to 12% in high winds.

Although much has been accomplished up to the present relative to the definition of atmospheric wind effects as they bear upon the problems of the structural engineer, it is fair to state that many of the effects cited remain subject to continued study and/or differences of opinion. For example, the longitudinal wind spectrum of the atmosphere has not been universally accepted in the form given above.

The exact location of the peak of this spectrum, the formula for the location, and the character of its low-frequency side have recurrently been discussed. The definition of the vertical wind profile in logarithmic form has not been universally accepted for engineering applications. The coefficients affecting lateral and longitudinal wind correlations are subject to further assessment and refinement. However, what has been cited in this section has been consonant with recently published trends in meteorology. Considerable work remains to be done relative to the character of winds in particular storm conditions, notably hurricanes, tornadoes, and thunderstorms.

WIND TUNNEL SIMULATIONS

Although a few effects of turbulent wind upon structures can be estimated by direct calculation (as will be discussed later), the flow over bluff civil engineering structures is in general so complex as to require simulation by modeling. This requires, first, that an acceptable simulation of the natural wind be provided, usually by means of wind tunnels.

Similitude considerations

Most wind tunnel testing is carried out in air and at wind velocities which are of necessity much reduced with respect to full scale. Also, limited test space inevitably implies that the geometric scale of models is greatly reduced compared to prototypes. Under these conditions Reynolds and Rossby numbers[1] are habitually not properly simulated. Reynolds number affects the development of wake flows, drag on bodies, and turbulence. Rossby number affects the depth of the boundary layer and large-scale circulations. Space does not permit discussion of the full implications of failure to duplicate these two nondimensional numbers, but it may be stated, in brief, that present state-of-the-art testing remains provisionally acceptable with these shortcomings. Some criticisms are expressed at a later point in the paper.

Fixed geometric scale of models is conventionally respected in all three mutually perpendicular directions. Since this scale is, in turn, dependent upon the length scales of turbulence which can be achieved, the basic geometric scale usually employed ranges from $\frac{1}{500}$ to $\frac{1}{100}$. Larger scales have, to date, rarely been achieved.

An important consideration which must be respected, both for frequencies of turbulent eddies and for structural frequencies, is the reduced frequency relation requiring that

$$\left(\frac{U}{NB}\right)_{\text{model}} = \left(\frac{U}{NB}\right)_{\text{prototype}} \tag{23}$$

where U is velocity, N is any frequency, and B any representative structural dimension.

Roughness elements, simulating the earth's surface, tripping the turbulent boundary layer, and establishing the spectra of turbulence, are normally duplicated to geometric scale, except for certain special devices to be discussed in relation to short tunnels, below.

Long boundary layer tunnels [14,15]

A relatively stable, well-established turbulent boundary layer with duplicated mean velocity profile can be achieved with a 20- to 30-m-long tunnel having a floor covered with appropriately spaced "roughness elements" often con-

---

[1] Reynolds number $Re = \rho UB/\mu$, where ρ and μ are air density and viscosity, respectively; U is mean wind velocity, and B is a typical structural dimension. Rossby number $Ro = U(z)/zf_c$, where $U(z)$ is the mean wind velocity at height z; f_c is the Coriolis parameter $f_c = 2\omega \sin \phi$, where ω is the angular rate of rotation of the earth and ϕ is the angle of latitude.

sisting of cubic blocks. At the end of a long fetch of such blocks, a test model (often a building immersed in geometrically duplicated building models of the surrounding city) is installed. Pressure taps in a fixed model or accelerometers in an aeroelastic model pick up desired information.

Short-tunnel devices [16–19]

Various schemes have been introduced to create the desired turbulence and/or boundary-layer conditions within short distances inside either an aeronautical-type tunnel or a special, short tunnel created for the purpose. The main devices used to set up the desired turbulent boundary layer are (1) air jets parallel to the general flow, (2) air jets counter to the general flow, and (3) spires or other "trip" devices. Each of the devices has its particular attributes.

The longitudinal jets permit, in principle, fine control of the velocity profile, although with considerable adjustment required because of fairly unpredictable jet interaction. The spires set up a boundary layer about equal in depth to the spire height. They are followed by a surface covered with blocks of sizes and locations dictated by test explorations of the boundary layer. Finally, the counter-jet technique is a device that in essence replaces the spires as a turbulent boundary layer trip device, through forcing air jets from the floor of the tunnel diagonally upward with a slight upstream component. All short-tunnel devices may be conceived of as setting up a turbulent boundary layer by trial and error to achieve assumed "target" profiles and turbulence spectra. In the absence of known target criteria, the short-tunnel methods could not be effective. All turbulence scales are relatively short, the largest (achieved in a 30-foot-diameter aeronautical tunnel) being of the order of $\frac{1}{100}$ of full scale.

A final device that should be mentioned is the simple grid of bars placed across the wind tunnel [20,21]. Grid-induced turbulence is usually characterized by small-scale spectra which reasonably approximate homogeneous turbulence but without a boundary-layer velocity profile.

New turbulence-induction devices

In some instances actively driven flap systems have been employed to create large-wavelength turbulence. The development of such devices is only in its infancy, but it offers a new avenue of research promising to extend the scope of wind tunnel testing. For example, Princeton University and Colorado State University are at present studying a device to create long-wavelength, two-dimensional variations in longitudinal vertical wind gusts for purposes of the two-dimensional testing of bridge deck section models. Active flaps have been used in Japan [22] and in England [23] for special gust effects. They are also used to produce gusts for aircraft models in the NASA Langley transsonic tunnel at Hampton, Virginia.

Criticisms of wind tunnel testing

How well does a long boundary-layer wind tunnel simulate the natural wind? The usual device in such tunnels has been the passage of air over 20 to 30 m of roughness elements simulating the desired terrain at a small scale, Reynolds and Rossby numbers being violated. Under these conditions reasonable duplications of wind velocity profiles and of turbulence spectra have been accomplished, although few tunnels appear to have duplicated to date the available velocity spectra corresponding to the latest meteorological data. The spectra generated appear to exhibit the appropriate Kolmogoroff energy cascade "roll-off" at the high-frequency end (with wind energy decreasing as $n^{-2/3}$). The modeled dependence of such spectra on height (as demonstrated in meteorology) has not to date always been clearly emphasized in tests. The shift of spectral peaks to the right or left (and consequent rise or fall in the spectral tail of 100 or 200% or more in important regions of structural dynamic response) are points not always clearly delineated. The significance of the aeroelastic response of an elastic structural model relative to expected prototype action remains, in many tests, to be fully interpreted.

Because vortex shedding from bluff objects is Reynolds-number-dependent, it is not yet absolutely certain that the response of aeroelastic models of circular-section towers in the wind tunnel fully duplicates prototype action. Furthermore, the modeled wakes from such objects may not fully resemble prototype wakes; in the cases where these wakes affect downstream bodies their effects may require interpretation. Since mechanical turbulence is tripped off by wakes, the interpretation of its effects on structural models is thus still open to some question, particularly if these models are aeroelastic in nature.

It is shown in meteorological boundary layer theory [24,25] that the depth δ of the boundary layer is a function of the Rossby number:

$$\delta \cong 0.25 \frac{u_*}{f_c} \tag{24}$$

which implies that in nature δ is of the order of several kilometers. For example, in open terrain ($z_0 \cong 5$ cm), if U at 10 m is 25 m/s and $f_c \cong 10^{-4}$/s (at latitude 45° north), then $\delta \cong 5$ km. Within this depth boundary-layer theory shows that the profile of the natural wind velocity versus height is logarithmic for the first 10% (i.e., of the order of 500 m).

Turbulent boundary-layer theory further demonstrates an intimate relationship between the velocity profile and the associated spectra of turbulence. This relationship is a function of the momentum transfer and dissipation mechanisms of the flow. It is not possible to reproduce these mechanisms exactly at reduced model scales. Consider, for example, a prototype building 100 m high which is to be modeled at $\frac{1}{400}$ scale in a wind tunnel wherein the achievable boundary-layer depth is 1 m. In full scale the logarithmic wind profile, with its associated spectral characteristics of turbulence, will be 500 m deep and will encompass the building. In model scale the logarithmic part of the velocity profile will be only 0.1 m deep and will fail to encompass the model, which will be 0.25 m high. Another question, not fully explored to date, is the extent to which boundary layers built up on the sides of wind tunnels (rather than the floors) affect the flows over models. Although good simulations are nonetheless often achieved, considerations of the type mentioned leave questions open for further interpretation and investigation.

Reference [26], considered a rather good comparison between model and full-scale results, demonstrates pressure distributions obtained in both cases for a large building in Toronto. Figures 4 and 5 are reproduced from this source. While the steady pressure distributions at a point on the test building are quite reasonably reproduced, important discrepancies persist in the fluctuating part of the results.

A question of recurring importance is the treatment of the surfaces of models in order to bring pressure coefficients developed over them more nearly into agreement with full-scale results. In general, it has been found

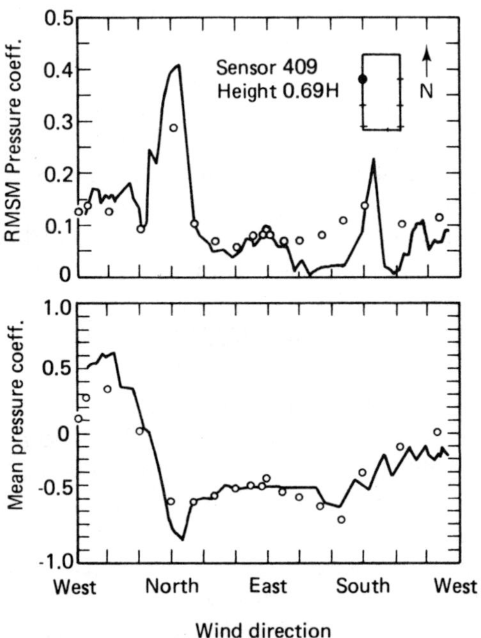

Figure 4   Results from full-scale (solid line) and model (circles) for pressure coefficients at a point (•) on a tall building (after [26]). (RMSM, root mean square about the mean.)

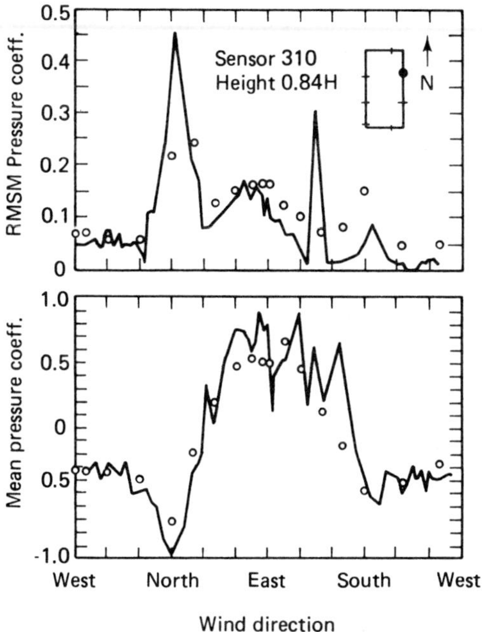

Figure 5 Results from full-scale (solid line) and model (circles) for pressure coefficients at a point (·) on a tall building (after [26]). (RMSM, root mean square about the mean.)

that surface roughnesses of models require exaggerated scales to bring pressure distributions into alignment with high Reynolds number results. No uniform methodology appears yet to have been developed in this area.

The response of aeroelastic models, particularly small-scale ones, to simulated turbulence appears to demand great caution as to interpretation, as the detailed natures of both the incident and the signature turbulence are open to some question, and both affect the structural response.

CALCULATIONS OF WIND EFFECTS

Although, as stated earlier, most effects of wind on structures depend strongly on wind tunnel testing for their confirmation, a few theoretical, or quasi-theoretical, approaches to the definition of wind effects have been made. Notable among these have been methodologies for the prediction of the alongwind response of tall buildings and towers and the prediction of the performance of long, suspended-span bridges and other line-like structures under wind.

Broadly speaking, the approaches to these problems are all based on the theory of random space- and time-dependent loadings of elastic structures.

Both problems assume the wind to have stationary random properties about its mean steady velocity, and both make use of empirical observations, either about the wind alone, or the particular manner in which the wind affects the structure in question.

Along-wind response of tall buildings

The problem of the alongwind response of tall buildings of simple prismatic form has been the subject of quasi-theoretical calculation [10,27–30]. The along-wind response, calculated as though the wind were directed normally to a face of the building, has experimentally been found generally to be the greatest of the possible responses to wind, and so it has certain design usefulness. Space does not permit development of the entire theory here, but allusion will be made to the conceptual form of the problem and to the variables entering into the final results obtained.

The wind, of velocity $U + u(t)$, is assumed to impinge normally on the building face; the mean wind U is a function of height z: $U = U(z)$. The pressure p against the face is assumed to vary as $u(t)$, since wind dynamic pressure is proportional to $\frac{1}{2}\rho U^2 + \rho U u(t)$, when $u^2(t)$ is legitimately neglected. Pressure coefficients C_w on the windward face and suction coefficients C_l on the leeward face of the building are assumed known, or estimated. Building dimensions are taken as width B and height H, with y and z being corresponding local coordinates. The variable wind forces being proportional to $u(t)$, the direct and co-spectra of u between points (y_1, z_1) and (y_2, z_2) play key roles in the response definition. The direct spectrum of u is designated by $S_u(z, n)$ for any height z. Lateral coherence effects of wind are accounted for by the introduction of a coherence function $\text{Coh}(y, z, n)$. These parameters account for the main effects of the wind. Most depend to a greater or lesser extent upon empirical observations. Along-wind coherence is accounted for by a function $N(n)$. Full details will not be given.

In addition, building response to wind depends upon the mechanical characteristics of the structure: notably, for buildings, the first two vibration modes $x_i(z)$ ($i = 1, 2$), the corresponding generalized masses M_i and associated frequencies n_i, and damping ratios ζ_i. One may visualize the building response, then, as that of a linear, multidegree oscillator to inputs which are random in space and time.

The net formula for the power spectral density of the along-wind displacement x at a point z along the building height can be given as [4]

$$S_x(z, n) \simeq \frac{p^2}{16\pi^4} \sum_i \frac{x_i^2(z)[C_w^2 + 2C_w C_l N(n) + C_l^2]}{n_i^4 M_i^2 \{[1 - (n/n_i)^2]^2 + 4\zeta_i^2 (n/n_i)^2\}}$$

$$\times \int_0^B \int_0^B \int_0^H \int_0^H x_i(z_1) x_i(z_2) U(z_1) U(z_2) S_u^{1/2}(z_1, n) S_u^{1/2}(z_2, n) \quad (25)$$

$$\times \text{Coh}(y_1, y_2, z_1, z_2, n)\, dy_1\, dy_2\, dz_1\, dz_2$$

From this the mean square of the fluctuation x is calculated by

$$\sigma_x^2(z) = \int_0^\infty S_x(z, n)\, dn \tag{26}$$

and the corresponding mean-square acceleration by

$$\sigma_{\ddot{x}}^2(z) = 16\pi^4 \int_0^\infty n^4 S_x(z, n)\, dn \tag{27}$$

The maximum variable excursion $x_{\max}(z)$ is estimated by

$$x_{\max}(z) = K_x(z)\sigma_x(z) \tag{28}$$

where K_x is in the neighborhood of 3.5. For Gaussian processes this is a good estimate, but it can be improved by existing theory [31,32] for more general cases. To this deflection must be added the steady deflection $\bar{x}(z)$ calculated under the steady mean wind $U(z)$.

The concept of a *gust factor* G is conveniently introduced and explained in this context by defining

$$G = 1 + \frac{x_{\max}}{\bar{x}} \tag{29}$$

so the G may be thought of as a design factor by which the standard static wind deflection is multiplied to yield the maximum result.

For a complete discussion of the along-wind buffeting problem of tall buildings and extensive aids to making the necessary calculations in practical cases, the reader is referred to the literature cited, particularly [30].

Buffeting response of suspended-span bridges [33, 38]

The problem of building response, treated briefly above, has several parallels. Space does not permit a full treatment of such cases, but one—that of bridges —will be briefly discussed. It is worthwhile to suggest here the form that solutions to this problem take.

The power spectral density of response in torsional modes $\alpha_i(x)$ of a long suspended-span bridge is given in the form [36]

$$S_\alpha(x, n) = \sum_i \frac{\alpha_i^2(x) \int_0^L \int_0^L \alpha_i(x_1)\alpha_i(x_2) S_{M_1 M_2}^C(x_1, x_2 n)\, dx_1\, dx_2}{16\pi^4 \tilde{n}_{\alpha_i}^4 I_i^2 \{[1 - (n/\tilde{n})^2]^2 + 4\tilde{\zeta}_{\alpha_i}^2 (n/\tilde{n})_{\alpha_i}^2\}} \tag{30}$$

in which L is span length, I_i the generalized mass moment of inertia of the mode relative to the effective rotation axis, $S_{M_1 M_2}^C(x_1, x_2, n)$ the co-spectral density of the buffeting moment on the bridge relative to sections x_1 and x_2, $\tilde{n}_{\alpha_i}$ the natural frequency of mode $\alpha_i(x)$ as modified by aerodynamic forces, and ζ_{α_i} the damping of this mode as influenced by self-exciting aerodynamic effects. The self-excitation effect is a key influence here which the given references treat in a detailed manner. Many of the interesting details are necessarily omitted in the present review.

Buffeting of towers and stacks [39, 40]

These problems are treated in a manner very analogous to that sketched in sections above.

Criticisms of wind-response calculations

The theoretical calculations of responses to wind are all heavily dependent upon experimental information. None can yet proceed from first principles alone. When theoretical approaches are used, their validity is only as great as that of the experimental input. In the case of structures like tall buildings, which are strongly affected by the three-dimensional nature of the flow, purely calculated results can be assumed to have only order-of-magnitude usefulness, as in first approximations for design purposes. The detailed information revealed about building models under tests in turbulent wind makes this abundantly clear. If, further, a structure has complex shape and/or is sited among other structures which particularly affect the local flow, calculation—of the kind available to date, at least—must be restricted to quite modest aims.

Thus, the interesting work done to date on along-wind, torsional, and across-wind response of buildings must be considered as yielding types of *representative* values of building response rather than detailed fully credible information [27–30,41,42]. Some of the assumptions made in the calculations —that the structure is prismatic, that lift and drag, or windward and leeward pressure coefficients, are known with accuracy, that their degrees of correlation are known, that local pressures are proportional to fluctuating wind velocity, that fetches do not create special effects, and so on—are clearly quite sweeping as compared to the detailed information that would be necessary to achieve accurate predictions.

The approaches to calculation of the responses of extended line-like structures, on the other hand, have benefited from the possibility of assuming that the flow locally over a *section* of the structure can be treated as two-dimensional ("strip theory"). This situation has notably enhanced the credibility of calculations made for long, suspended-span bridges, when section aerodynamic information (in the form of "flutter derivatives") for the decks of such structures has been experimentally obtained. Such section information remains, however, in need of substantiation under properly simulated turbulence. While some steps to this end have been taken [38,43], few, if any, bridge section models have to date been tested under correctly scaled turbulence. One effort to this end is currently under way in the joint project between Princeton University and Colorado State University to initiate turbulence mechanically for tests of bridge deck section models.

OVERVIEW

The problems of the assessment of structural loads and responses due to wind have received active treatment in the last quarter century. Most solutions of these problems require strong experimental assistance, which comes principally from wind tunnel model studies. In order for such studies to be realistic, a great deal of background meteorological theory and information are needed. Thereafter, extreme care in simulation is demanded. When calculations, rather than test simulations, are made, these too require a large input of experimental fact. Here, in general, theory still outruns its factual support. While the whole field can be said to be generally in good condition, many areas remain in strong need of detailed criticism and improvement. Much remains to be done in the areas of simulation and in the gathering of full-scale data to guide experimental studies. New and enterprising devices for the simulation of atmospheric turbulence to larger scales could be used. It is worthwhile at the present juncture to make a serious and critical review of the shortcomings of wind engineering as it is presently practiced in order to set the stage for new advances.

REFERENCES

[1] THOM, H. C. S., "New Distributions of Extreme Wind Speed in the United States," *J. Struct. Div., ASCE*, Vol. 94, No. ST7, July 1968, pp. 1787–1801.

[2] BROOKS, C. E. P., CURST, C. S., AND CARRUTHERS, N., "Upper Winds over the World," Part 1: "The Frequency Distribution of Winds at a Point in the Free Air," *Quart. J. Roy. Met. Soc. London*, Vol. 72, 1946, pp. 55–73.

[3] GUMBEL, E. J., *Statistics of Extremes*, Columbia University Press, New York, 1958.

[4] SIMIU, E., AND SCANLAN, R. H., *Wind Effects on Structures*, Wiley, New York, 1978.

[5] DAVENPORT, A. G., VICKERY, B. J., AND MELBOURNE, W. H., "The Structural and Environmental Effects of Wind on Buildings and Structures," Notes, Post-Graduate Course, School of Civil Engineering, The University of Sydney, Sydney, Australia, May 1975.

[6] KAIMAL, J. C., ET AL., "Spectral Characteristics of Surface Layer Turbulence," *Quart. J. Royal Met. Soc. London*, Vol. 98, 1972, pp. 563–589.

[7] SIMIU, E., "Wind Spectra and Dynamic Alongwind Response," *J. Struct. Div. ASCE*, Vol. 100, ST9, Sept. 1974, pp. 1897–1910.

[8] LUMLEY, J. L., AND PANOFSKY, H. A., *The Structure of Atmospheric Turbulence*, Wiley, New York, 1964.

[9] DAVENPORT, A. G., "The Dependence of Wind Loads upon Meteorological Parameters," *Proc. International Seminar on Wind Effects on Buildings and Structures*, University of Toronto Press, Toronto, 1968.

[10] VICKERY, B. J., "On the Reliability of Gust Loading Factors," *Proc. Technical Conference on Wind Loads on Buildings and Structures*, National Bureau of Standards, Building Science Series 30, Washington, D.C., 1970.

[11] SHIOTANI, M., "Structure of Gusts in High Winds," Interim Report, Parts 1–4, The Physical Sciences Laboratory, Nihon University, Funabashi, Chiba, Japan, 1967–1971.

[12] SHIOTANI, M., AND IWATANI, Y., "Correlations of Wind Velocities in Relation to Gust Loading," *Proc. Third International Conference on Wind Effects on Buildings and Structures*, Tokyo, 1971.

[13] BLACKADAR, A. K., PANOFSKY, H. A., AND FIEDLER, F., "Investigation of the Turbulent Wind Field Below 500 Feet Altitude at the Eastern Test Range, Florida," NASA CR-2438, National Aeronautics and Space Administration, Washington, D.C., June 1974.

[14] CERMAK, J. E., "Laboratory Simulation of the Atmospheric Boundary Layer," *AIAA J.*, Vol. 9, No. 9, Sept. 1971, pp. 1746–1754.

[15] DAVENPORT, A. G., AND ISYUMOV, N., "The Application of the Boundary Layer Wind Tunnel to the Prediction of Wind Loading," *Proc. International Research Seminar, Wind Effects on Buildings and Structures*, Ottawa, University of Toronto Press, Toronto, 1968.

[16] HUNT, J. C. R., AND FERNHOLZ, H., "Wind Tunnel Simulation of the Atmospheric Boundary Layer: A Report on Euromach 50," *J. Fluid Mech.*, Vol. 70, Part 3, Aug. 1975, pp. 543–559.

[17] COOK, N. J., "A Boundary Layer Wind Tunnel for Building Aerodynamics," *J. Indust. Aerodyn.*, Vol. 1, 1975, pp. 3–12.

[18] STANDEN, N. M., "A Spire Array for Generating Thick Turbulent Shear Layers for Natural Wind Simulation in Wind Tunnels," Report LTR-LA-94, NAE, National Research Council, Ottawa, Canada, May 1972.

[19] NAGIB, H. M., MORKOVIN, M. V., YUNG, J. T., AND TAN-ATICHAT, J., "On Modeling of Atmospheric Surface Layers by the Counter-Jet Technique" *AIAA J.*, Vol. 14, No. 2, Feb. 1976, pp. 185–190.

[20] BEARMAN, P. W., "An Investigation of the Forces on Flat Plates in Turbulent Flow," *NPL Aero Report 1296*, Teddington, UK, 1969.

[21] VICKERY, B. J., "On the Flow Behind a Coarse Grid and Its Use as a Model of Atmospheric Turbulence in Studies Related to Wind Loads on Buildings," *NPL Aero Report 1143*, Teddington, UK, 1965.

[22] KONISHI, I., SHIRAISHI, N., MATSUMOTO, M., AND OKANAN, H., "Fundamental Consideration on the Aerodynamic Admittance of Long-Spanned Bridges," *Report 18*, Disaster Reseach Center, April 1975 (in Japanese).

[23] HUNT, J. C. R., "Turbulent Velocities Near and Fluctuating Surface Pressures on Structures in Turbulent Winds," *Proc. Fourth International Conference on Wind Effects on Buildings and Structures*, Heathrow, UK, Cambridge University Press, New York, Sept. 1975, pp. 309–320.

[24] TENNEKES, H., "The Logarithmic Wind Profile," *J. Atmosph. Sci.*, Vol. 30, 1973, pp. 234–238.

[25] CSANADY, G. T., "On the Resistance Law of a Turbulent Ekman Layer," *J. Atmosph. Sci.*, Vol. 24, Sept. 1967, pp. 467–471.

[26] DALGLIESH, W. A., "Comparison of Model/Full-Scale Wind Pressures on a High-Rise Building," *J. Indust. Aerodyn.*, Vol. 1, June 1975, pp. 55–66.

[27] DAVENPORT, A. G., "Gust Loading Factors" *J. Struct. Div.*, ASCE, Vol. 93, ST3, June 1967, pp. 11–34.

[28] VELLOZZI, J., AND COHEN, E., "Gust Response Factors," *J. Struct. Div.*, ASCE, Vol. 94, ST6, June 1968, pp. 1295–1313.

[29] SIMIU, E., "Gust Factors and Along-Wind Pressure Correlations," *J. Struct. Div. ASCE*, Vol. 99, ST4, April 1974, pp. 773–783.

[30] SIMIU, E., AND LOZIER, D. W., *The Buffeting of Tall Structures by Strong Winds*, Building Science Series 74, National Bureau of Standards, Washington, D.C., Oct. 1975.

[31] DAVENPORT, A. G., "Note on the Distribution of the Largest Value of a Random Function with Application to Gust Loading," *Proc. Instn. Civil Engrs.*, Vol. 28, 1964, pp. 187–196.

[32] RICE, S. O., "Mathematical Analysis of Random Noise," *Bell Syst. Tech. J.*, Vol. 18, 1944, p. 282, Vol. 19, 1945, p. 46.

[33] DAVENPORT, A. G., "The Response of Slender, Line-like Structures to a Gusty Wind" *Proc. Instn. Civil Engrs.*, London, 1962, pp. 389–407.

[34] DAVENPORT, A. G., "The Buffeting of a Suspension Bridge by Storm Winds," *J. Struct. Div.*, ASCE, June 1962, pp. 233–264.

[35] DAVENPORT, A. G., "The Action of Wind on Suspension Bridges," *Proc. International Symposium on Suspension Bridges*, Lisbon, 1966, pp. 79–100.

[36] SCANLAN, R. H., AND GADE, R. H., "Motion of Suspended Bridge Spans under Gusty Wind," *J. Struct. Div.*, ASCE., Vol. 103, pp. 1867–1883, Sept. 1977.

[37] SCANLAN, R. H., AND TOMKO, J. J., "Airfoil and Bridge Deck Flutter Derivatives" *J. Eng. Mech. Div.*, ASCE, Vol. 97, No. EM6, Dec. 1971, pp. 1717–1737.

[38] SCANLAN, R. H., AND LIN, W. H., "Effects of Turbulence on Bridge Flutter Derivatives" (in review), *J. Eng. Mech. Div.*, ASCE, 1977.

[39] RUSCHEWEYH, H., "Wind Loadings on the Television Tower, Hamburg, Germany," *J. Indust. Aerodyn.*, Vol. 1, No. 4, Aug. 1976, pp. 315–333.

[40] VICKERY, B. J., AND CLARK, A. W., "Lift or Across-Wind Response of Tapered Stacks," *J. Struct. Div., ASCE,* Jan. 1972, pp. 1–20.

[41] VAICAITIS, R., SHINOZUKA, M., AND TAKENO, M., "Response Analysis of Tall Buildings to Wind Loading," *J. Struct. Div., ASCE,* Vol. 101, No. ST3, March 1975, pp. 585–600.

[42] PATRICKSON, C. P., AND FRIEDMANN, P., "A Study of the Coupled Lateral and Torsional Response of Tall Buildings to Wind Loadings," *Report UCLA-ENG-76126,* Dec. 1976, School of Engineering, University of California at Los Angeles.

[43] REINHOLD, T. A., TIELEMAN, H. W., AND MAHER, F. J., "The Torsional Response of a Suspension Bridge Stiffening Truss Model to Turbulence and to an Upstream Obstacle," *Report UPI-E-74-28,* Department of Engineering Science and Mechanics, Virginia Polytechnic Institute and State University, Dec. 1974.

THEORY OF PLASTIC STRUCTURES

*Philip G. Hodge, Jr.**

INTRODUCTION

When a tension specimen is loaded monotonically, it will generally have an initial linear range, followed by a nonlinear one. Figure 1, for example, shows experimental results for mild steel [1] and 24 S–T aluminum [2]. Figure 1(b) also shows two fitted curves for the aluminum data. Although the solid curve is certainly the better fit, the piecewise linear dashed curve has obvious theoretical advantages. For simplicity of exposition we shall assume that it adequately represents the "real" material. Points on the first, steeper, portion of the curve are called *elastic*, and points on the other portion are referred to as *plastic*. A structure is called elastic when all its parts correspond to the elastic portion of the curve. A theory of plastic structures must be used when any part of the structure is on the plastic portion.

It is convenient to define several idealized models of material behavior as indicated in Fig. 2. The salient features of these models can be illustrated

---

*Professor of Mechanics, University of Minnesota.

This research was sponsored by the Office of Naval Research.

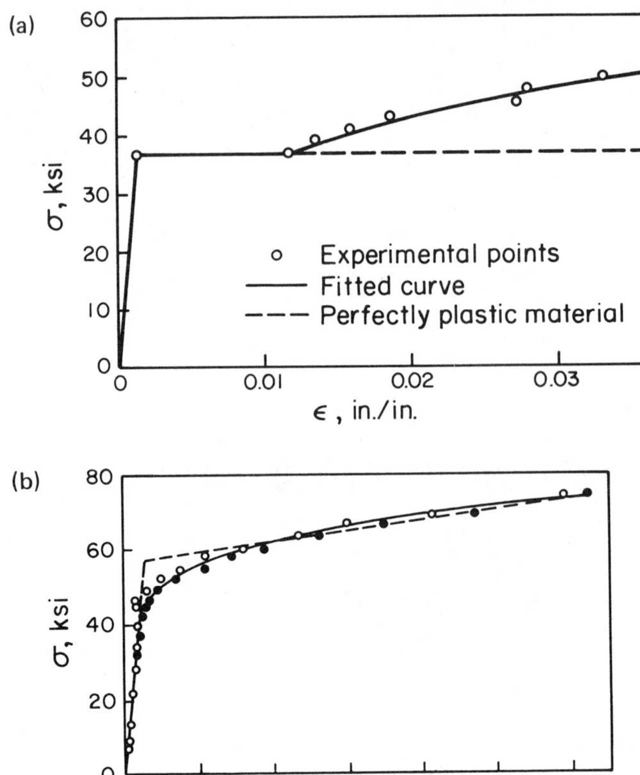

Figure 1 Stress–strain curves: (a) mild steel, (b) 24 S-T aluminum.

by the simple three-bar truss in Fig. 3. The results of loading the truss with a slowly, monotonically increasing load are shown in Fig. 4. The behavior of the "real" truss, curve (a), may be characterized as follows. For $P < P_A$, the elastic limit, the truss is in the *elastic range*—all three bars are elastic. Between P_A and P_B, the yield-point load, the central bar is plastic, but the two diagonal bars are still elastic. Since these bars by themselves would constitute a load-carrying structure, the response of the three-bar truss is still of an elastic order of magnitude and the truss is in the *range of contained plastic deformation*. Along *BC*, all three bars are plastic and a small increment of load produces a much larger increment of deformation, so that we refer to the *range of extensive plastic deformation*.

We consider now, the various idealizations of Fig. 3. The fully elastic model (b) is exact in the elastic range, an increasingly poor approximation in the range of contained plastic deformation, and an absurd approximation for $P > P_B$. The rigid–strain hardening model (c) does not distinguish

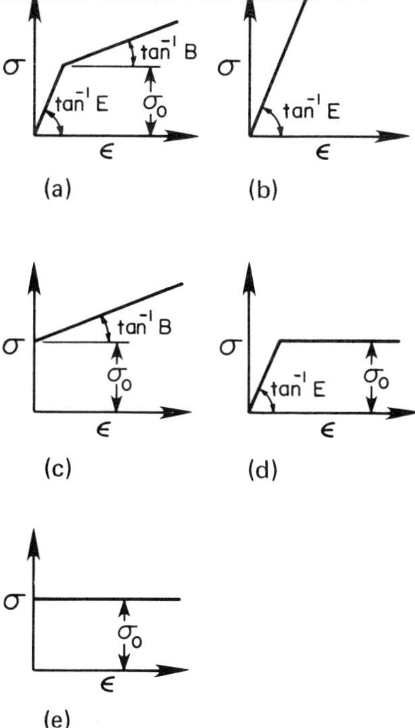

Figure 2 Ideal stress–strain curves. (a) elastic–strain hardening, (b) elastic, (c) rigid–strain hardening, (d) elastic–perfectly plastic, (e) rigid–perfectly plastic.

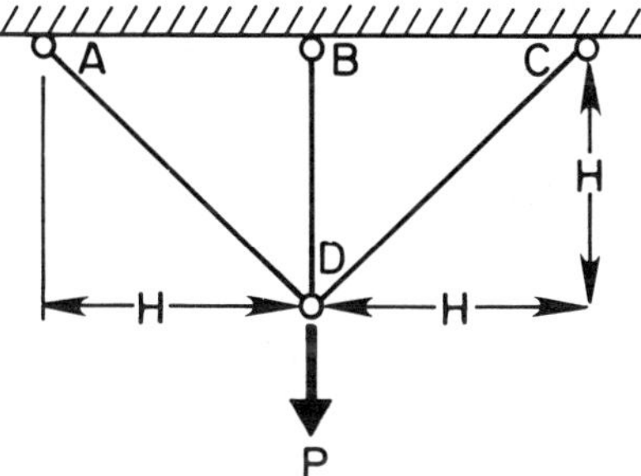

Figure 3 Three-bar truss.

between the elastic and contained ranges but is a good approximation to the extensive plastic deformation range. The elastic–perfectly plastic model (d) is exact in the elastic range and a very good approximation in the range of contained plastic deformation. It is also essentially exact for moderate strains for the mild steel of Fig. 1(a). However, according to this model, the truss cannot support any load greater than the yield-point load

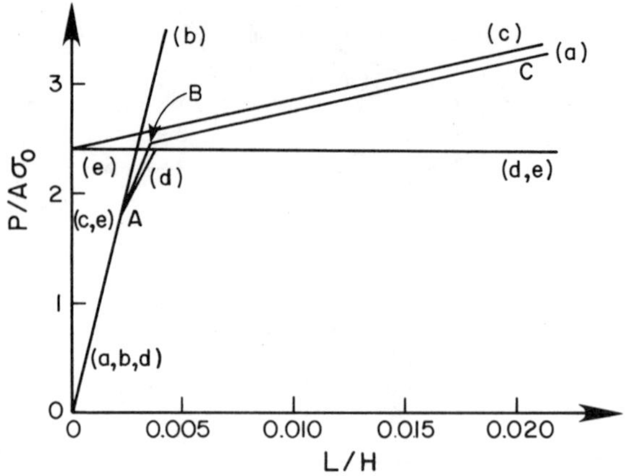

Figure 4 Load-displacement curves for truss (letters in parentheses refer to the materials of Figure 2).

$$P_0 = (1 + \sqrt{2})A\sigma_0 \tag{1}$$

so it is more appropriate to refer to the horizontal segment of the curve as *unrestricted plastic flow*. Any load greater than P_0 could not be equilibrated and hence would produce *accelerated plastic flow*. Finally, the simple rigid–perfectly plastic model (a) does not give a good approximation in any range but does predict the yield-point load (1).

Indeed, all three of the idealized plastic models have exactly the same yield-point load P_0, and this value is a very good approximation to P_B for the real material. Therefore, the simple rigid–perfectly plastic model can be used to predict this significant property of the structure.

Based upon this simple illustration, we may propose the following approach to structural analysis. First, use the rigid–perfectly plastic model to predict P_0. If more detailed information is required concerning deformations under a given load P, use the elastic–perfectly plastic model if $P < P_0$, and use the rigid–strain hardening model if $P > P_0$.

In the remainder of this paper we shall mostly be concerned with the perfectly plastic models, and with their application to structures of various complexity. We shall begin with beams and frames under bending, discuss tension, and then consider torsion and torsion with tension. We shall discuss both the rigid–perfectly plastic model for determining P_0, and the elastic–perfectly plastic model for discussing the response at loads less than P_0. We shall then present more general theories for perfectly plastic and strain-hardening materials, respectively, and proceed to analyze some circular plates for a variety of models. We shall also consider some of the problems that exist in applying finite-element methods to plasticity problems, and the paper will close with a brief concluding section.

BEAMS AND FRAMES

We consider first the pure bending of an elastic–perfectly plastic beam. We make the fundamental assumption of Euler–Bernoulli beam theory that plane sections remain plane and normal, whence the strain across the section will remain linear. For sufficiently small strains the stress will also be linear [Fig. 5(a)],[1] but for a certain moment M_e the stress will reach σ_0 in magnitude

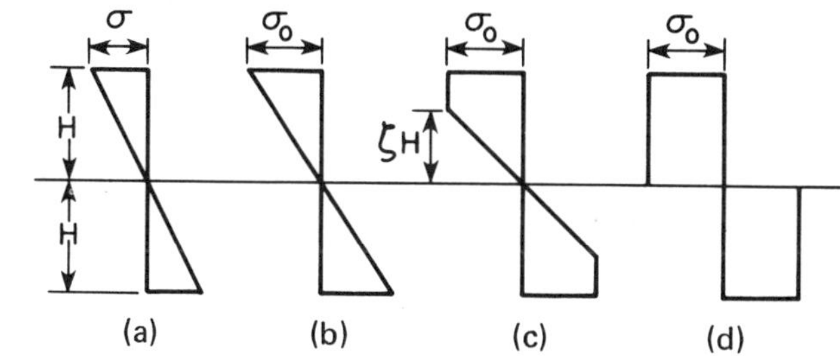

Figure 5 Stress distributions in rectangular beams: (a) elastic, (b) maximum elastic, $M = M_c$, (c) elastic plastic, (d) maximum plastic, $M = M_0$.

at the outer fibers [Fig. 5(b)]. Greater strains in the outer fibers can still be accommodated by $|\sigma| = \sigma_0$, thus leading to the stress distribution in Fig. 5(c). In the limit, the strain will become infinite under the finite moment M_0 obtained from the stress distribution in Fig. 5(d). Figure 6* shows the resulting moment-curvature relation. Curve $OABD$ is the curve for the elastic–perfectly plastic section being considered, for the particular case of a rectangular cross section. Curve $OABC$ shows the diagram for the mild steel of Fig. 1(a). However, in the same spirit in which the nonlinear curve in Fig. 1(b) was replaced by the piecewise linear one, it is usual to solve beam and frame problems according to OED or OFD, depending on whether or not elastic strains are to be included.

The effect of the "knee" AB will depend upon the shape of the section. It will be greater than in Fig. 6 for a circular or cruciform section, less for an I-beam or sandwich section. Indeed, in the limit as the bending resistance of web or core becomes negligible even while the flange or sheet thickness goes to zero, curve OED would become exact.

Figure 7* shows a statically indeterminate propped cantilever beam under increasing load. For $F \le F_1$ the beam is elastic and at $F = F_1$ the moment at A is $-M_0$. As F is increased above F_1, the moment at A must remain at $-M_0$, so that the increment of load above F_1 produces the same response as if a hinge were inserted at A. At $F = F_0$, $M = M_0$ at B. Since the

---

[1]Figure 5 and several other figures are taken from [3] with permission of the publisher. Such figures will be identified by an asterisk following their first reference.

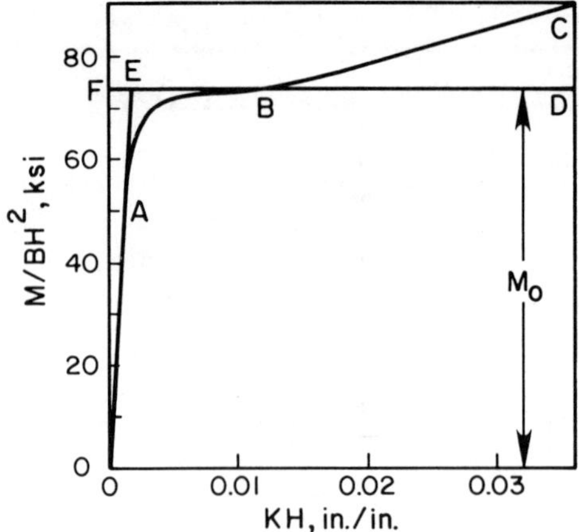

Figure 6 Moment-curvature relations, *OABD*, perfectly plastic material. *OABC*, mild steel. *OAED*, ideal section. *OFED*, rigid plastic material.

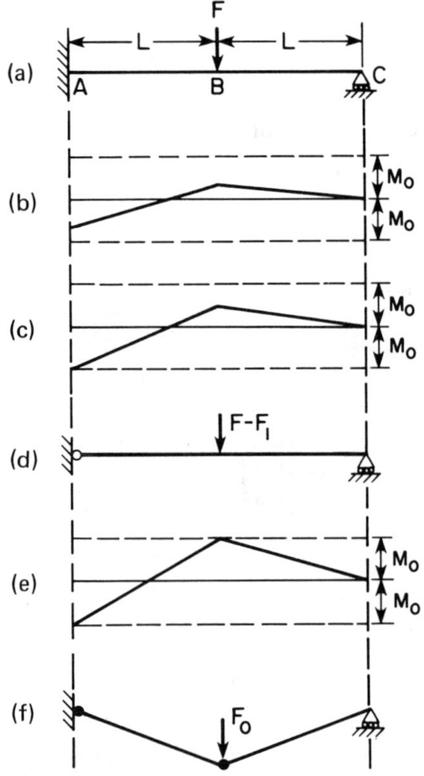

Figure 7 Collapse of statically indeterminate beam: (a) loaded beam, (b) elastic moments, (c) one-hinge moment distribution, (d) beam with one hinge, (e) collapse moments, (f) collapse mechanism.

insertion of a second yield hinge at B would render the beam a mechanism, F_0 is the yield-point load. Notice how closely this example parallels the truss discussed earlier.

Now, if we are interested only in predicting F_0, we may proceed directly to the moment distribution of Fig. 7(e). Indeed, since the moment varies linearly along AB and BC, and since two hinges are necessary to produce a mechanism, it is obvious that we must have hinges at both A and B at the yield-point load. Statics, or even more simply the principle of virtual work, then leads immediately to

$$F_0 = 3M_0/L \tag{2}$$

A less trivial example is furnished by the portal frame in Fig. 8*(a). This frame can be turned into various mechanisms by the insertion of three yield hinges in four possible locations, as illustrated by Fig. 8(b) to (d). The

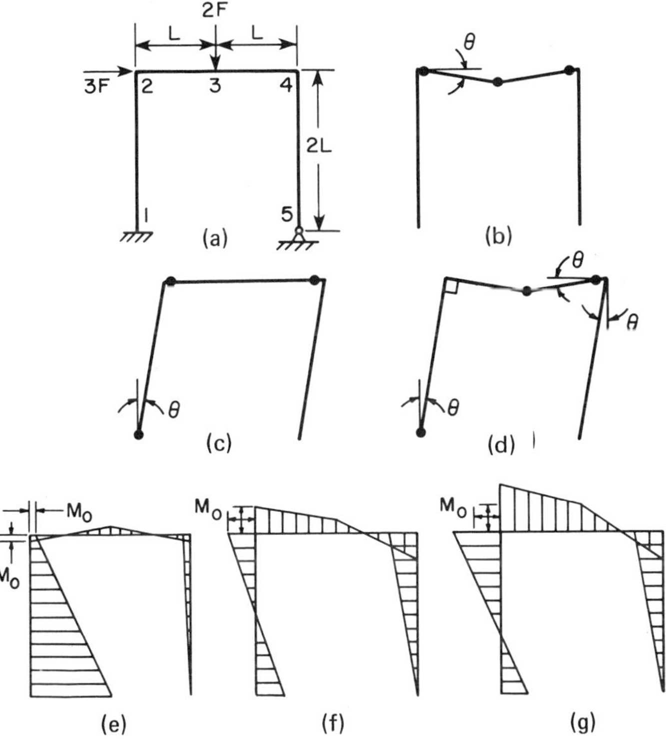

Figure 8 Collapse of simple frame: (a) loaded frame, (b) collapse mode for beam failure only, (c) collapse mode for panel failure only, (d) combined collapse mode, (e) moment distribution for (b) (not admissible), (f) moment distribution for (c) (statically admissible), (g) moment distribution for (d) (not admissible).

corresponding yield-point loads differ:

$$F_b = 2M_0/L \qquad F_c = M_0/2L \qquad F_d = 5M_0/8L \qquad (3)$$

Figure 8(e) to (g) shows the corresponding moment distributions. Evidently, the moments corresponding to F_b and F_d exceed M_0, so F_c must be the correct yield-point load.

We notice that F_c is the smallest of the three mechanism loads, so any of (3) provide an upper bound on the yield-point load. This fact is not just a coincidence. There are two basic theorems of limit analysis, proved independently by Gvozdev [4], Hill [5,6], and Drucker, Greenberg, and Prager [7–9], which enable us to find both upper and lower bounds on the yield-point load.

These theorems are easily stated in a quite general form. We are given a structure with a load distribution fully defined by a single parameter P. We define a statically admissible field as a set of moments that is in internal and external equilibrium with a load P^- and which nowhere violates the yield condition. A kinematically admissible field is defined by a mechanism. At every yield hinge a moment M is assigned equal to $\pm M_0$ in the same sense as the rotation θ. Finally, a load P^+ is defined so that the external work done by P^+ on the mechanism motion is equal to the internal work $\sum M\theta$.

At the actual yield-point load P_0 all conditions of both static and kinematic admissibility will be satisfied. The theorems, which are easily proved [3], state that P_0 is the largest P^- and the smallest P^+:

$$P^- \leq P_0 \leq P^+ \qquad (4)$$

The importance of these theorems is increased by the fact that P_0 in (4) is the same yield-point load for any of the elastic–perfectly plastic, rigid–perfectly plastic, or rigid–strain-hardening models.

Examples of the application of the limit analysis theorems to rectangular and nonrectangular frames, Vierendeel trusses, and grids may be found in [3, Chap. 3].

UNLOADING AND RELOADING

Thus far we have considered only monotonically increasing loads and stresses. If the stress is decreased in an elastic material, the stress–strain curve will simply be retraced. However, for an elastic–perfectly plastic material loaded into the plastic range to point B in Fig. 9, unloading would not retrace BAO, but would be along BC parallel to the initial elastic line OA. Reloading from C would retrace CD and continue along BD as if the unloading excursion had not taken place.

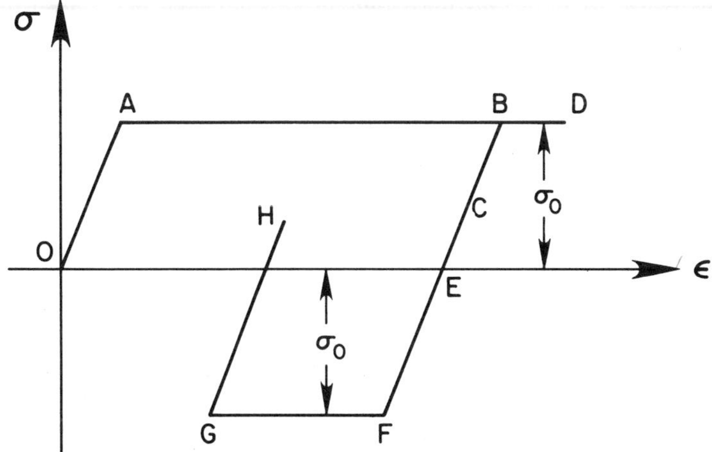

Figure 9 Unloading and reloading of perfectly plastic material.

If, instead of reloading at C, unloading were to continue until the stress were zero, a permanent strain OE would remain in the specimen under zero stress. Further unloading would continue along the same line BCE until the compressive yield stress, $-\sigma_0$, was reached at F. The strain could further decrease to G with no change in stress. Reloading at this point would be along GH parallel to BF and the original elastic line OA.

It is clear from the description above that neither stress nor strain is a unique function of the other, so the constitutive relations must be expressed in rate form. With a need for later generality we may write the law as follows:

$$\sigma^2 - \sigma_0^2 \leq 0 \qquad \dot{\lambda} \geq 0 \qquad \dot{\epsilon} = \dot{\epsilon}^e + \dot{\epsilon}^p \tag{5a,b,c}$$

$$\dot{\epsilon}^e = E^{-1}\dot{\sigma} \qquad \dot{\epsilon}^p = 2\dot{\lambda}\sigma \tag{5d,e}$$

$$\text{If } \sigma^2 - \sigma_0^2 < 0 \qquad \text{or } 2\sigma\dot{\sigma} < 0 \qquad \text{then } \dot{\lambda} = 0 \tag{5f}$$

Observe that when σ remains at $+\sigma_0$, $\dot{\epsilon}$ is not determined but must be nonnegative, whereas if σ remains at $-\sigma_0$, $\dot{\epsilon} \leq 0$. For any given history of strain it will always be clear which branch of (5) is relevant, so the equations may be trivially integrated.

The rigid–perfectly plastic model is a special case of the above in which the elastic portions are vertical in Fig. 9. The equations are given by (5) with $E = \infty$ in (5d), whence $\dot{\epsilon}^e = 0$ in (5c).

Unloading and reloading for strain-hardening materials is more complicated and we delay the subject until later.

Some of the simplified results and techniques used in plastic structural analysis are valid only in the absence of plastic unloading. For this reason, it is important to note that local unloading may take place even under a monotonically increasing external load. A simple example of this phenomenon, first discussed by Drucker [10], is shown in Fig. 10. Three elastic–perfectly plastic bars, with relative yield strengths indicated by the circled numbers, are attached to a fixed rigid bar AB and an otherwise free rigid bar CD. As P is increased, all bars will first be elastic in tension and bar 1 will yield first at $P = P_1$. Bars 2 and 3 will have increasing tensile stresses as P is increased above P_1 to P_2. For $P > P_2$ the load increment must be carried by bars 1 and 2 so that a compressive increment is added to bar 1 (i.e., it unloads). Eventually, at $P = P_0$, bar 1 yields in compression and a mechanism is formed. Thus, even though P is monotonically increased to P_0, the stress in bar 1 is first increased to σ_0, remains there while its strain increases, and then decreases to $-\sigma_0$ when failure occurs. A simple portal frame that exhibits similar behavior is discussed in [3, Chap. 4], and a more complicated truss with several such stress reversals is analyzed in [11].

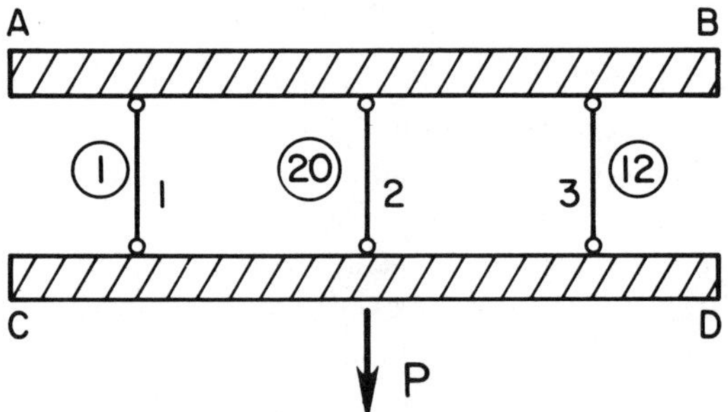

Figure 10   Three-bar truss.

OTHER LIMITATIONS

The yield-point load will always be a limitation on the maximum load which a structure can support. However, there are circumstances in which other considerations may be more restrictive. For example, slender members subject to compressive load may buckle elastically or even plastically before the yield-point load is reached.

Another possibility is excessive deformations. Consider, for example, the beam on four supports in Fig. 11*[12]. The yield-point load is $4M_0/L$, independently of the distance ζL between the outer supports. For very small ζ the configuration is similar to a clamped beam of length $2L$ which has the same yield-point load. However, for large ζ the effective strength added by the outer supports would be expected to decrease. Indeed, in the limit as ζ becomes infinite, the central span should act like a simply supported beam whose yield-point load is only $2M_0/L$.

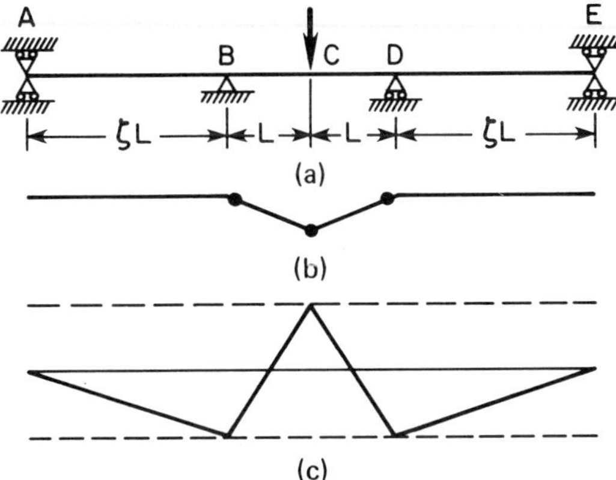

Figure 11 Beam on four supports: (a) loaded beam, (b) collapse mechanism, (c) moment distribution.

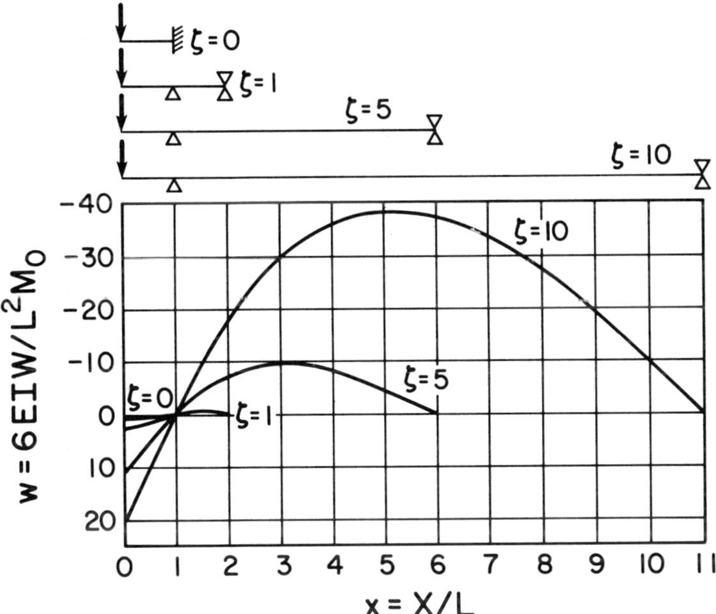

Figure 12 Deformed shape at collapse of beam on four supports.

The solution of this paradox lies in considering the displacements. Figure 12* shows the displacements just prior to the yield-point load for various ζ. For $\zeta = 10$ the displacements would almost certainly render the beam unserviceable before the load reached its yield-point load. For example, if the maximum allowable displacement is 5 times the maximum elastic displacement in the built-in beam, the yield-point load is the limiting criterion only if $\zeta \leq 2$. For $\zeta = 5$ the maximum displacement is reached when $F = 3M_0/L$; for $\zeta = 10$ it occurs when $F = 2.5M_0/L$. In the limit, as ζ tends to infinity, the maximum allowable displacement occurs with the formation of the first hinge at $F = 2M_0/L$ in agreement with the simply supported beam. Details of the solution may be found in [3, Chap. 4] together with a simplified method for estimating displacements.

A less obvious type of limitation can occur when there is more than one load parameter. Consider, for example, the frame in Fig. 13*, under

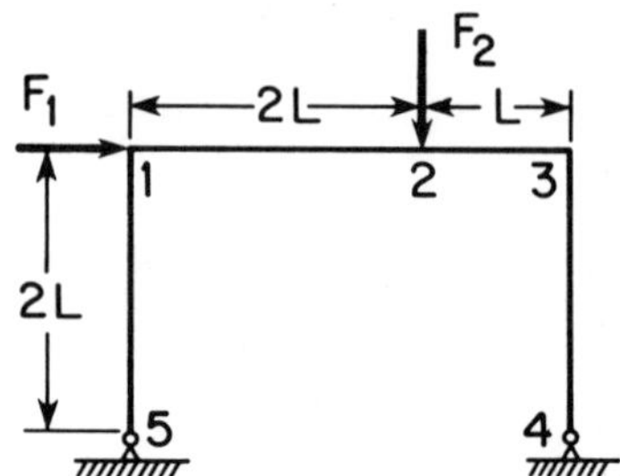

Figure 13   Frame under two time-dependent loads.

independent loads F_1 and F_2. The methods of limit analysis are readily adapted to this two-parameter situation, and the results are shown in Fig. 14*. Load points $f_i = LF_i/M_0$ lying on the hexagon $MNPQRS$ correspond to yield point, with each side representing a different mechanism; points inside the hexagon do not yield, and points outside are unattainable.

Now suppose that f_2 is increased to point A corresponding to a fixed dead-weight load, and f_1 is then alternated between A and B corresponding to a variable wind load. The plastic part of the resultant displacements,

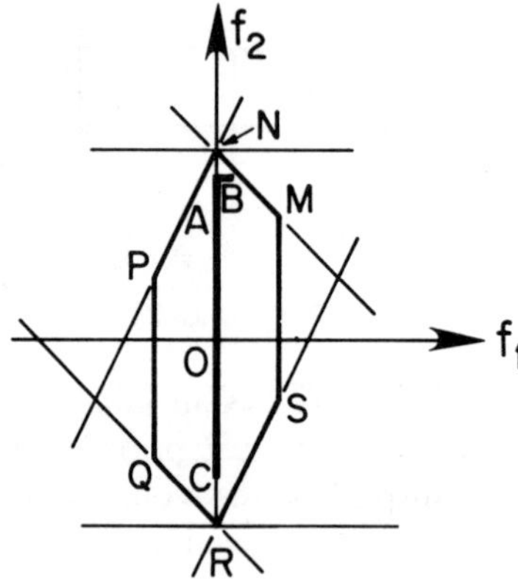

Figure 14   Domain of admissible loads for frame of Fig. 13.

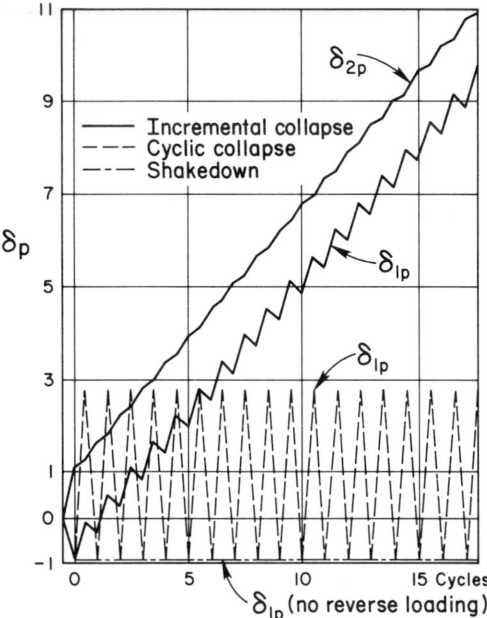

Figure 15 Plastic displacements for various loading cycles for frame of Fig. 13.

computed according to an approximate but complete elastic–plastic analysis, is shown in Fig. 15*. Here δ_{1p} refers to the dimensionless sidesway, δ_{2p} to the dimensionless displacement under the load, and the actual dimensioned quantities are given by

$$\Delta_i = (M_0 L^2/6EI)\delta_i \tag{6}$$

For the rigid–plastic model, E is infinite so that $\Delta_i = 0$ for any number of cycles. However, for any finite E, no matter how large, Δ_i will become arbitrarily large after a sufficient number of cycles, so that the structure can support only a limited number of cycles, even though the load is always strictly within the yield-point hexagon of Fig. 14.

Even one load can cause undesirable behavior if it is alternated. Suppose, for example, that there is no f_1, but that f_2 alternates between points A and C. The resulting plastic displacement is shown by the dashed curve in Fig. 15 and is not in itself alarming since it remains finite. However, each increase in δ_{1p} corresponds to a positive plastic rotation of the yield hinge under the load, whereas each decrease is a negative plastic rotation. Construction of a moment-rotation diagram would show that a large hysteresis loop is present, representing substantial energy input to the hinge. Although details of the phenomenon cannot be discussed within any of the macroscopic theories considered here, it is evident that a relatively few cycles might cause internal damage which would render the frame unserviceable.

The shakedown load of a structure is defined as the largest load for which none of the undesirable features described above occurs. Techniques are available for finding or estimating the incremental collapse load and the cyclic collapse load [3, Chap. 5]. The smallest of these two and the yield-point load is the shakedown load.

BEAMS UNDER TENSION AND BENDING

If a beam is subject to both axial force and bending moments, the strain rate at any distance z from the center axis is

$$\dot{\epsilon} = \dot{e} + z\dot{\kappa} \tag{7}$$

Elastic–plastic stages will be similar to Fig. 5, except that the neutral and center axes will generally be different. For a rigid–perfectly plastic material, we are interested only in a fully plastic cross section. Here, there are four possibilities, as shown in Fig. 16. The corresponding resultants for a rectangular section must satisfy

$$N = -N_0\zeta \qquad M = M_0(1 - \zeta^2) \qquad \dot{\kappa} \geq 0 \qquad -1 < \zeta < 1 \tag{8a}$$

$$N = N_0\zeta \qquad M = -M_0(1 - \zeta^2) \qquad \dot{\kappa} \leq 0 \qquad -1 < \zeta < 1 \tag{8b}$$

$$N = -N_0 \qquad M = 0 \qquad \dot{e} \leq 0 \qquad |\zeta| \geq 1 \tag{8c}$$

$$N = N_0 \qquad M = 0 \qquad \dot{e} \geq 0 \qquad |\zeta| \geq 1 \tag{8d}$$

where ζh is the distance to the neutral axis. In addition, $\dot{e}$ and $\dot{\kappa}$ must be related by

$$\dot{e} + \zeta\dot{\kappa} = 0 \tag{9}$$

since the neutral axis strain rate is zero by definition.

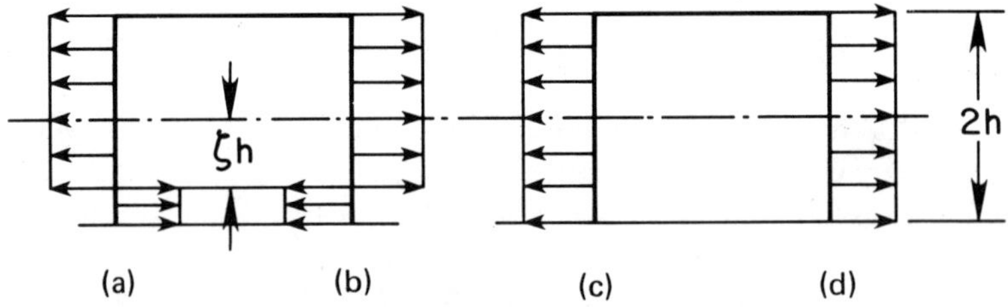

Figure 16 Stress distributions for beam in tension and bending.

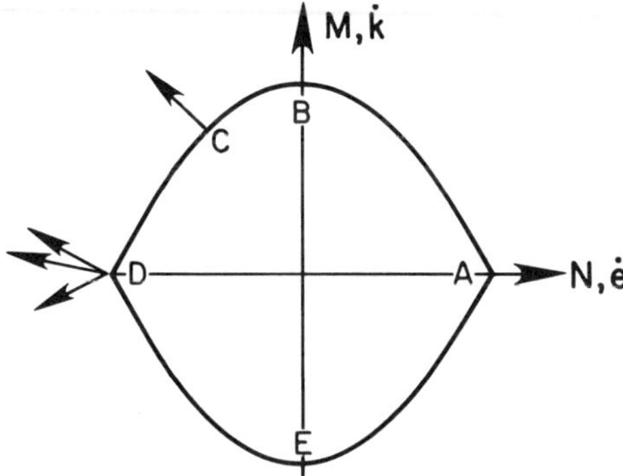

Figure 17 Interaction curve for beam in tension and bending.

Equations (8) define an interaction curve in stress-resultant space, as shown in Fig. 17. Points on the curve correspond to a fully plastic section, points within correspond to a section at least partly elastic, and points outside the curve are unattainable.

Let $\mathbf{Q} = (N, M)$ denote a generalized stress vector, and $\dot{\mathbf{q}} = (\dot{e}, \dot{\kappa})$ a general strain-rate vector. The internal dissipation rate of the beam can then be written

$$D_{\text{int}} = \int_V \sigma\dot{\varepsilon}\, dv = \int_0^L (N\dot{e} + M\dot{\kappa})\, dx = \int_0^L \mathbf{Q} \cdot \dot{\mathbf{q}}\, dx \tag{10}$$

so that Prager's criterion [13] for generalized variables is satisfied.

Now, except for A and D, each point $\mathbf{Q}$ on the yield curve corresponds to a unique ζ and hence, according to (9), to a unique direction for $\dot{\mathbf{q}}$. Further, it is easily shown that $\dot{\mathbf{q}}$ is perpendicular to the yield curve. At points A and D neither ζ nor the normal to the curve are unique. However, all admissible values of ζ correspond to directions between the limiting normals so that in a sense we may state that the generalized strain-rate vector is always normal to the yield curve. The length of $\dot{\mathbf{q}}$ is, of course, undetermined.

To express the preceding geometric results algebraically, we eliminate ζ between (8a) or (8b) to obtain the equations for arcs DBA and DEA, respectively:

$$f_1(N, M) \equiv (N/N_0)^2 + M/M_0 - 1 = 0 \tag{11a}$$

$$f_2(N, M) \equiv (N/N_0)^2 - M/M_0 - 1 = 0 \tag{11b}$$

If $\dot{\mathbf{q}}$ is normal to (11a), then

$$\dot{e} = \dot{\lambda}\partial f/\partial N = 2\dot{\lambda}N/N_0^2$$

$$\dot{\kappa} = \dot{\lambda}\partial f/\partial M = \dot{\lambda}/M_0 \tag{12}$$

where $\dot{\lambda}$ is nonnegative. In vector form, the strain rates for *DBA* and *DEA* are, respectively,

$$\dot{\mathbf{q}} = \dot{\lambda}\nabla f_1(\mathbf{Q}) \qquad \dot{\mathbf{q}} = \dot{\mu}\nabla f_2(\mathbf{Q}) \tag{13}$$

where $\dot{\lambda}$ and $\dot{\mu}$ are each nonnegative. Finally, at points D or A the requirement that $\mathbf{q}$ lie between the limiting normals can be expressed by

$$\dot{\mathbf{q}} = \dot{\lambda}\nabla f_1 + \dot{\mu}\nabla f_2 \qquad \dot{\lambda} \geq 0 \quad \dot{\mu} \geq 0 \tag{14}$$

The theory above may be used to assess the influence of axial forces on the yield-point loads of frames, or to analyze problems of curved beams or arches. Some applications may be found in [3, Chap. 7].

TORSION OF A CIRCULAR CYLINDRICAL BAR

A circular cylinder of length L is fixed at one end and subjected to a pure torque. We make the usual assumption that each cross section undergoes only a rotation proportional to its distance from the fixed end. Then the only displacement is the tangential one,

$$v = \alpha z r \tag{15}$$

where α is the angle of twist per unit length. The only nonzero strain is the shearing strain,

$$\gamma = \gamma_{\theta z} = \alpha r \tag{16}$$

For α sufficiently small, the entire bar will be elastic with

$$\tau = G\alpha r \tag{17}$$

The resultant torque in this case is

$$T = 2\pi \int_0^a r\tau r \, dr = (\pi/2)G\alpha a^4 \tag{18}$$

As α is slowly increased, τ will reach its yield value k in shear at the boundary of the bar for a value α_1 defined by

$$G\alpha_1 a = k \tag{19}$$

so that the maximum elastic torque is

$$T_1 = (\pi/2)ka^3 \tag{20}$$

For larger values of α, an outer annulus of the bar will become plastic while the inner core remains elastic. Thus, for a perfectly plastic material,

$$\tau = G\alpha r \qquad 0 \leq r \leq \rho$$
$$\tau = k \qquad \rho \leq r \leq a \tag{21}$$

Since τ must be continuous at the elastic–plastic boundary ρ,

$$\rho = k/G\alpha \tag{22}$$

The torque in this case is

$$T = \frac{2}{3}\pi k a^3 \left[1 - \frac{1}{4}\left(\frac{k}{G\alpha a}\right)^3\right] \tag{23}$$

Evidently, the limiting fully plastic torque

$$T_0 = \tfrac{2}{3}\pi k a^3 \tag{24}$$

is attained only as α tends to infinity. However, we note that T is within less than 1% of T_0 when α is only three times its maximum elastic value α_1.

COMBINED TENSION AND TORSION

If a beam of circular cross section is subjected to both tension and torsion, it will certainly be limited by the tensile and shear yield stresses σ_0 and k. These values are both experimentally determined constants so that one can be expressed in terms of the other, and there is considerable evidence that $\sigma_0 = \sqrt{3}\,k$. Thus the stress point $\mathbf{Q} = (\sigma, \tau)$ must certainly lie within the rectangle $ABCD$ in Fig. 18. However, if σ and τ are both nonzero, experience suggests that their interaction will pose a more severe limitation and that $\mathbf{Q}$ must lie within some curve within the rectangle. In the next section we will justify taking this curve to be the ellipse

$$f(\mathbf{Q}) \equiv (\sigma^2/3 + \tau^2)/k^2 - 1 = 0 \tag{25}$$

Let us visualize an experiment in which the unit extension e and unit rotation α are controlled. In terms of these parameters, the strain vector is

$$\mathbf{q} \equiv (\epsilon, \gamma) = (e, \alpha r) \tag{26}$$

For e and α sufficiently small, the bar will be everywhere elastic, hence the stress vector is

$$\mathbf{Q} \equiv (\sigma, \tau) = G(3e, \alpha r) \tag{27}$$

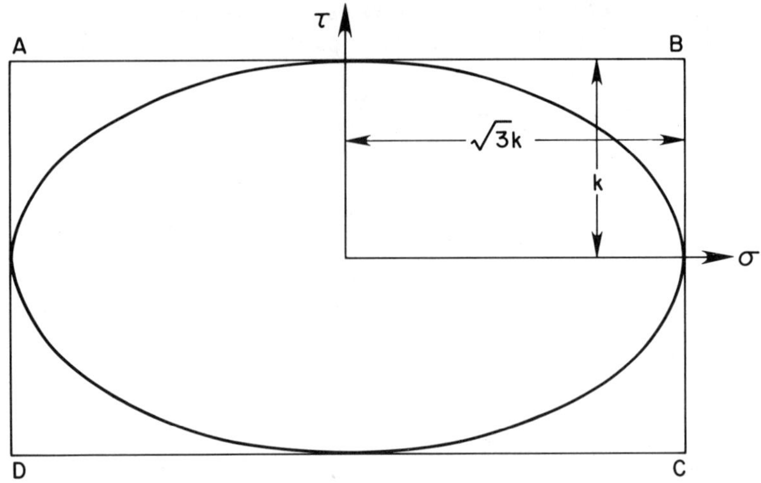

Figure 18   Yield curve for tension and torsion.

where we have, for simplicity, taken $v = 1/2$ so that $E = 3G$. Equation (27) will be valid as long as $\mathbf{Q}$ lies everywhere within the ellipse (25), that is, as long as

$$(\sqrt{3}\,Ge/k)^2 + (G\alpha a/k)^2 - 1 \leq 0 \tag{28}$$

If e and α are such that part of the bar is plastic, the constitutive equations must be expressed in terms of rates. To this end, we first rewrite eqs. (5) in vector form and apply them to the present problem:

$$f(\mathbf{Q}) \equiv (1/k^2)(\sigma^2/3 + \tau^2) - 1 \leq 0 \tag{29a}$$

$$\dot{\lambda} \geq 0 \qquad \dot{\mathbf{q}} = \dot{\mathbf{q}}^e + \dot{\mathbf{q}}^p \tag{29b,c}$$

$$\dot{\mathbf{q}}^e = G^{-1}(\dot{\sigma}/3, \dot{\tau}) \tag{29d}$$

$$\dot{\mathbf{q}}^p = \dot{\lambda}\mathbf{\nabla}f = 2(\dot{\lambda}/k^2)(\sigma/3, \tau) \tag{29e}$$

$$\text{If } f < 0 \qquad \text{or } \dot{f} < 0 \qquad \text{then } \dot{\lambda} = 0 \tag{29f}$$

In generalizing (5e) we have been guided by the results we found in eq. (13).

Let us consider specifically the case where $\dot{\lambda} > 0$. It then follows from (29f) that $f = \dot{f} = 0$, whence the local dissipation rate is

$$\mathbf{Q} \cdot \dot{\mathbf{q}} = \frac{1}{G}\left(\frac{\sigma\dot{\sigma}}{3} + \tau\dot{\tau}\right) + \frac{2\dot{\lambda}}{k^2}\left(\frac{\sigma^2}{3} + \tau^2\right) = 2\dot{\lambda} \tag{30}$$

We can now rewrite eqs. (29) in the more usable form

$$f(\mathbf{Q}) \leq 0 \qquad \mathbf{Q}\cdot\dot{\mathbf{q}} \geq 0 \tag{31a,b}$$

If $f < 0$ or $\dot{f} < 0$, then

$$\dot{\sigma} = 3G\dot{e} \qquad \dot{\tau} = Gr\dot{\alpha} \tag{31c}$$

else

$$\dot{\sigma} = (G/k^2)[(3k^2 - \sigma^2)\dot{e} - \sigma\tau r\dot{\alpha}]$$

$$\dot{\tau} = (G/k^2)[-\sigma\tau\dot{e} + (k^2 - \tau^2)r\dot{\alpha}] \tag{31d}$$

Let us consider a specific loading sequence in which e is slowly increased from 0 to $k/\sqrt{3}\,G$ and is then held there while α is increased to k/Ga. The first stage is fully elastic and is governed by (31c) or, since we start from zero, by (27). At the end of this stage,

$$e = k/\sqrt{3}\,G \qquad \sigma = k/\sqrt{3} \qquad \alpha = \tau = 0 \tag{32}$$

At this point $f = 0$ for all r; hence (31d) are to be integrated with (32) as initial conditions. This process leads to

$$\sigma = k\sqrt{3}\ \text{sech}\ (G\alpha r/k) \qquad \tau = k\ \tanh\ (G\alpha r/k) \tag{33}$$

In particular, the final stress state is

$$\sigma = k\sqrt{3}\ \text{sech}\ (r/a) \qquad \tau = k\ \tanh\ (r/a) \tag{34}$$

Consider now an alternative loading sequence which reaches the same final state by first increasing α to its final value, and then increasing e. As shown in [14, Sec. 14], the final stress state in this case is

$$\sigma = k\sqrt{3}\ \tanh\ \xi \qquad \tau = k\ \text{sech}\ \xi$$

$$\xi = 1 - u + \tanh^{-1} u \qquad u = \sqrt{1 - r^2/a^2} \tag{35}$$

Figure 19, taken from [14], shows how much the stress solution for the same final strains depends upon the history of the loading.

MULTIAXIAL STRESS STATES

The most general state of stress is characterized by the six components of the symmetric tensor $\boldsymbol{\sigma}$. Therefore, general behavior of a perfectly plastic material will be characterized by a yield condition

$$f(\boldsymbol{\sigma}) \leq 0 \tag{36}$$

If the material is isotropic, the form of the yield condition cannot depend upon the orientation of the coordinate axes, so that f can have only three

    Theory of Plastic Structures

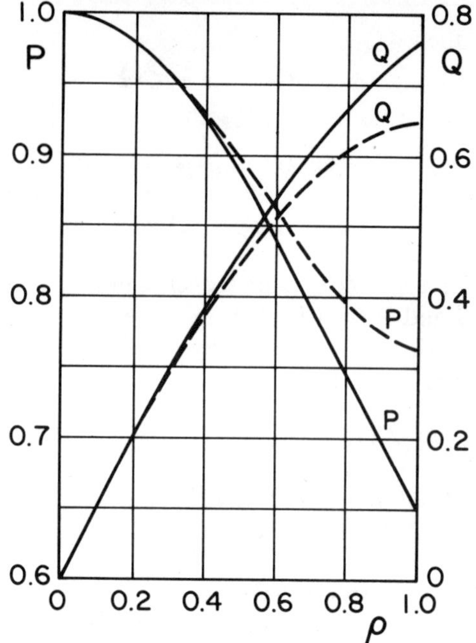

Figure 19   Stress distribution for combined tension and torsion of circular cylinder.

independent arguments. Thus (36) can be replaced by either of

$$f(\bar{\sigma}, J_2, J_3) \leq 0 \qquad f(\sigma_1, \sigma_2, \sigma_3) \leq 0 \tag{37a,b}$$

where $\bar{\sigma}$ is the mean normal stress and J_2 and J_3 are the second and third invariants of the stress deviator tensor, and σ_i are the three principal stresses. Experience indicates that the mean normal stress does not affect the yielding of most structural materials whence (37) can be further simplified to

$$f(J_2, J_3) \leq 0 \qquad f(\sigma_1 - \sigma_3, \sigma_2 - \sigma_3) \leq 0 \tag{38a,b}$$

In 1868, Tresca [15] suggested that yielding was governed by the maximum shearing stress, hence

$$f = \max\left[|\sigma_1 - \sigma_2|, |\sigma_1 - \sigma_3|, |\sigma_2 - \sigma_3|\right] - 2k \tag{39}$$

This condition is generally simple to apply when the principal stress directions are fixed and known but is difficult in more general cases. Therefore, in 1913, Mises [16] proposed approximating (39) by a special case of (38a):

$$\begin{aligned}
f &= 6J_2 - 2\sigma_0^2 \\
&= (\sigma_x - \sigma_y)^2 + (\sigma_x - \sigma_z)^2 + (\sigma_y - \sigma_z)^2 \\
&\quad + 6(\tau_{xy}^2 + \tau_{xz}^2 + \tau_{yz}^2) - 2\sigma_0^2
\end{aligned} \tag{40}$$

If $\sigma_0 = \sqrt{3}\,k$, the two yield conditions will agree in pure shear and will have their maximum discrepancy of $2/\sqrt{3} = 1.155$ in simple tension.

Later experimental evidence has shown that Mises' yield condition is a closer approximation than Tresca's to most real structural materials, but in view of the small difference between them, it is customary to assume whichever one is simplest for a given problem.

We have already derived the constitutive equations for a special case and have assumed them in another special case. In general they are

$$f(\boldsymbol{\sigma}) \leq 0 \qquad \dot{\lambda} \geq 0 \qquad \dot{\boldsymbol{\epsilon}} = \dot{\boldsymbol{\epsilon}}^e + \dot{\boldsymbol{\epsilon}}^p \tag{41a,b,c}$$

$$\dot{\boldsymbol{\epsilon}}^e = \mathbf{E}^{-1}\dot{\boldsymbol{\sigma}} \qquad \dot{\boldsymbol{\epsilon}}^p = \dot{\lambda}\boldsymbol{\nabla}f \tag{41d,e}$$

$$If\, f < 0 \qquad or\, \dot{f} < 0 \qquad then\, \dot{\lambda} = 0 \tag{41f}$$

For the particular case of the Mises yield condition (40), the general results may be simply expressed in terms of the stress and strain deviators,

$$\mathbf{s} = \boldsymbol{\sigma} - \mathbf{I}\bar{\sigma} \qquad \mathbf{e} = \boldsymbol{\epsilon} - \mathbf{I}\bar{\epsilon} \tag{42}$$

where $\bar{\sigma}$ and $\bar{\epsilon}$ are the mean normal stress and strain:

$$f \equiv \mathbf{s}{:}\mathbf{s} - 2k^2 \leq 0 \qquad \dot{\lambda} \geq 0 \tag{43a,b}$$

$$\dot{\mathbf{e}} = \dot{\mathbf{s}}/2G + \dot{\lambda}\mathbf{s} \qquad \bar{\sigma} = 3K\bar{\epsilon} \tag{43c,d}$$

together with (41f). For $\dot{\lambda} > 0$, the requirement $f - \dot{f} - 0$ shows that

$$\dot{\mathbf{e}}{:}\mathbf{s} = \dot{\mathbf{s}}{:}\mathbf{s}/2G + \dot{\lambda}\mathbf{s}{:}\mathbf{s} = 2k^2\dot{\lambda} \tag{44}$$

whence (43) can be rewritten

$$\text{If}\, f < 0 \qquad \text{or}\, \dot{f} < 0 \qquad \text{then}\, \dot{\mathbf{s}} = 2G\dot{\mathbf{e}}$$
$$\text{else}\, \dot{\mathbf{s}} = 2G[\dot{\mathbf{e}} - \mathbf{s}(\dot{\mathbf{e}}{:}\mathbf{s})]/2k^2 \tag{45}$$

Note that although for a perfectly plastic material the strain rate is not uniquely determined by the stress rate, as shown by (45) a given strain rate does uniquely determine the stress rate.

Equations (43), without the elastic terms, were first used by Mises in 1913 [16], following ideas advanced by Saint Venant [17] and Lévy [18] in 1870. The elastic terms were incorporated by Prandtl [19] and Reuss [20] in 1924 and 1930. The more general (41) were first proposed by Mises in 1928 [21]. Later, additional reasons for using them were advanced by Hill [22] and Drucker [23].

For structural problems which are conveniently handled with generalized variables, it is necessary merely to replace the stress matrix $\boldsymbol{\sigma}$ by the generalized stress vector $\mathbf{Q}$, and the strain ϵ by the generalized strain $\mathbf{q}$. The appropriate form of the yield condition f and the elastic matrix $\mathbf{E}$ must, of course, be determined for the particular structure.

STRAIN-HARDENING MATERIAL

If a strain-hardening material is loaded to a point B in the plastic range (Fig. 20) and then unloaded, the unloading will take place along a line BC parallel to the original elastic line OA. If the load is again reversed at C, CB will be retraced and further loading will take place along BD just as if the excursion BCB had not occurred. Actually, BCB will constitute a small hysteresis loop and the reloading may start to behave plastically shortly before B and rejoin line AD shortly after B, but these effects are small compared to the overall elastic and plastic effects and hence are ignored in most models.

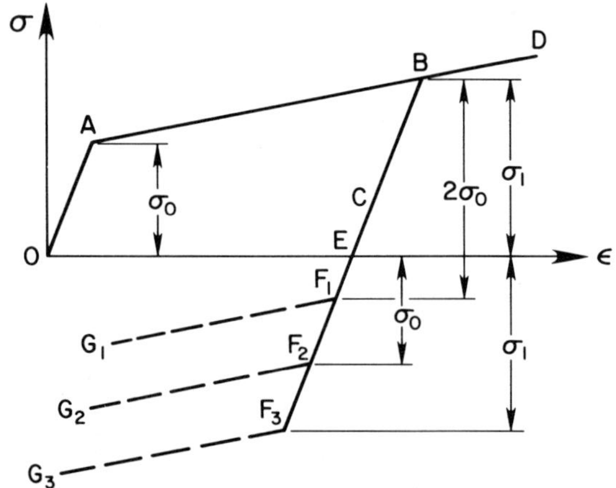

Figure 20   Unloading and reloading of strain-hardening material.

If the bar is fully unloaded to point E, a residual strain OE will remain and, with further unloading, plastic flow in compression will begin at a point F and continue along a line FG.

Thus far the description of behavior is an obvious generalization of Fig. 9 for a perfectly plastic material, but two questions remain to be answered: where is point F, and what is the slope FG? If elastic slopes and initial yield stress are originally symmetric in tension and compression, it seems reasonable to assume that FG is parallel to AB. With this assumption, the future behavior of a tensile specimen at a generic point C and with a given loading history is specified by four numbers: the elastic slope (OA, FB), the plastic slope (AB, GF), the current yield stress in tension (ordinate of B), and the current yield stress in compression (ordinate of F). The two slopes are material constants independent of history, and, for the history shown in Fig. 20, the current tensile yield stress is the value σ_1 at which the most recent plastic flow was reversed. However, various models have been proposed to predict the ordinate of F.

Historically, most of the early work in plasticity which accounted for strain hardening at all assumed that it was *isotropic* (i.e., that the current yield stresses in tension and compression were always equal). Thus, in Fig. 20, compressive plastic flow would take place along $F_3 G_3$.

However, experimental evidence generally suggests that strain hardening in tension will reduce, rather than raise, the magnitude of the compressive yield stress. To account for this "Bauschinger effect," Prager [24,25] proposed a theory that has become known as *kinematic hardening*. In terms of the tension test, kinematic hardening assumes that the elastic range remains constant at $2\sigma_0$, independently of σ_1. Thus unloading would take place along $BF_1 G_1$ in Fig. 20. We note that if the hardening is so great that $\sigma_1 > 2\sigma_0$, then plastic flow in compression would take place under an unloading stress that was still positive.

These two models are generally accepted as the extreme possibilities, and any number of intermediate models can be proposed. For example, $F_2 G_2$ in Fig. 20 shows a compressive yield stress that is independent of σ_1.

For a multiaxial stress, eqs. (41) may still be used with some modifications in the interpretation of $f(\boldsymbol{\sigma})$ and $\dot{\lambda}$. The yield function $f(\boldsymbol{\sigma})$ is no longer a fixed property of the material but will vary with stress history. It will remain constant when the behavior is elastic, but will vary during plastic flow so that $\boldsymbol{\sigma}$ is always on the yield surface. Thus one should write $f = f(\boldsymbol{\sigma}, \mathcal{H})$ where $\mathcal{H}$ stands symbolically for the entire stress–strain history. Then, in eq. (41f), $\dot{f}$ is not a total derivative of the entire function but is defined specifically by $\dot{f} = \boldsymbol{\nabla} f \cdot \dot{\boldsymbol{\sigma}}$.

Since the requirement that $f = 0$ during plastic flow is a partial constraint on the history of f rather than the equation $\dot{f} = 0$, we have one fewer constitutive equation available. The system is balanced by requiring a relation between the magnitude of the plastic strain-rate $\dot{\lambda}$ and the increase in the yield function $\dot{f}$. Thus, when $\dot{\lambda}$ is not equal to zero, it is given by

$$\dot{\lambda} = B\dot{f} \tag{46}$$

For a hardening curve with variable slope, B could be a function of stress; for the linear hardening considered here, B is a material constant proportional to the slope of AB in Fig. 20.

There still remains the problem of how f varies with plastic loading. One reason for the popularity of the isotropic and kinematic models is that they both give easily implemented answers to that question. According to isotropic hardening, the yield surface remains fixed in origin, orientation, and shape but will change in size to accommodate plastic flow. Thus, if the initial yield function is written

$$f(\boldsymbol{\sigma}) = g(\boldsymbol{\sigma}) - \sigma_0 \tag{47}$$

the current yield function after plastic flow will be

$$f(\boldsymbol{\sigma}, \mathcal{K}) = g(\boldsymbol{\sigma}) - \sigma_h \tag{48}$$

where σ_h is the largest value of $g(\boldsymbol{\sigma})$ attained during the preceding history.

In geometric terms, kinematic hardening states that the yield function remains fixed in orientation, shape, and size but will change its origin to accommodate plastic flow. Thus

$$f(\boldsymbol{\sigma}, \mathcal{K}) = f(\boldsymbol{\sigma} - \boldsymbol{\epsilon}^p/B) \tag{49}$$

where $\boldsymbol{\epsilon}^p$ is the current plastic strain.

An alternative method for describing strain hardening is the mechanical sublayer model proposed by White [26] and Besseling [27] and applied by Vaughan [28]. Each material element, infinitesimal or finite, is replaced by a number of subelements connected in parallel. If each subelement is elastic–perfectly plastic but with different yield stresses, the resulting total model has many similarities to kinematic hardening, except that the resulting B is piecewise constant rather than a single number. By including some subelements with isotropic hardening, models may be constructed which are intermediate to overall isotropic or kinematic hardening.

The subject of proper strain-hardening models is one of intense current interest. Indeed, an entire symposium was devoted to the topic at the 1976 ASME meeting and the interested reader is referred to the *Proceedings* of that meeting [29] for some of the current research being done in the area.

CIRCULAR PLATE

For a circular plate under rotationally symmetric loading, appropriate generalized stresses are the dimensionless moments

$$(m_r, m_\theta) = (1/M_0) \int_{-H}^{H} (\sigma_r, \sigma_\theta) z \, dz \tag{50}$$

where $M_0 = \sigma_0 H^2$ is the yield moment and $2H$ the plate thickness. Since the principal directions are known and fixed, it is convenient to use Tresca's yield condition. Setting the third principal stress $\sigma_z = 0$ in (39) and using (50), we obtain

$$f = \max\left[|m_r|,\ |m_\theta|,\ |m_r - m_\theta|\right] - 1 \tag{51}$$

as illustrated in Fig. 21.*

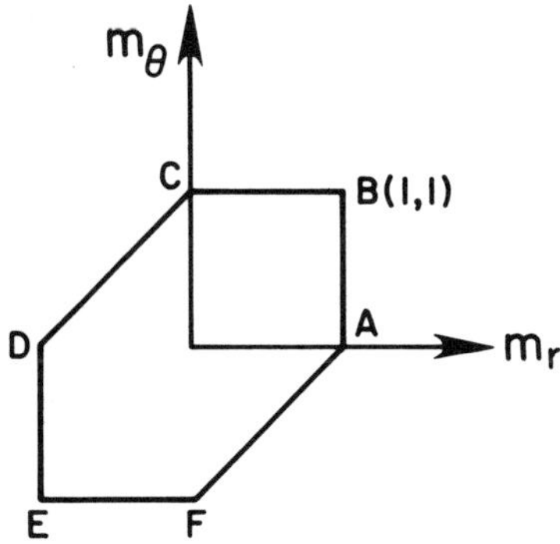

Figure 21   Tresca yield condition for circular plates.

For a constant load the moments must satisfy the equilibrium equation

$$(rm_r)' - m_\theta = -3pr^2 \tag{52}$$

where p is the dimensionless load and r the dimensionless distance. Evidently if a rigid–plastic plate is plastic corresponding to one of the six sides of (51), the problem is statically determinate and easily solved. For example, the stress profile for a simply supported plate of unit radius lies everywhere on side BC,

$$m_\theta = 1 \tag{53a}$$

Substitution of (53a) in (52) and use of the boundary condition $m_r = m_\theta$ at $r = 0$ leads to

$$m_r = 1 - pr^2 \tag{53b}$$

and the boundary condition $m_r(1) = 0$ furnishes the yield-point load

$$p = 1 \tag{53c}$$

Before accepting (53) as the solution, two things must be verified. First, one must show that the stress profile lies on the finite side BC, which is more restrictive than the infinite line (53a). Since (53b,c) show that $0 \leq m_r \leq 1$ for all $0 \leq r \leq 1$, this condition is satisfied.

Second, we must associate a kinematically admissible velocity field with (53). Generalized strains for this problem are

$$\dot{\kappa}_r = -\dot{w}'' \qquad \dot{\kappa}_\theta = -\dot{w}'/r \qquad\qquad (54a,b)$$

On side BC the strain-rate vector must be vertical and directed upward. Thus

$$\dot{w}'' = 0 \qquad \dot{w}' = -C \qquad \dot{w} = C(1 - r) \qquad\qquad (55)$$

where C is any positive constant.

At the center of the plate the slope jumps discontinuously from the symmetry value of 0 to the negative value $-C$. Thus, it follows from (54a) that $\dot{\kappa}_r$ is positively infinite. Since $r = 0$ is at the corner B in Fig. 21, a positive $\dot{\kappa}_r$ is admissible. Thus (55) is a kinematically admissible field and hence (53) is a complete moment solution.

A clamped plate is less simple, since the stress profile lies on BC near the center of the plate and on CD near the edge. The solution in this case is [3, Chap. 10]

$$0 \leq r \leq \eta: \quad m_\theta = 1 \qquad m_r = 1 - pr^2 \qquad \dot{w} = C\eta(1 - \log \eta - r/\eta)$$

$$\eta \leq r \leq 1: \quad m_\theta = \log r + (3p/2)(1 - r^2) \qquad m_r = m_\theta - 1 \qquad (56a)$$

$$\dot{w} = -C\eta \log r$$

where p and η are defined by

$$3p - \log p - 5 = 0 \qquad p\eta^2 = 1 \qquad\qquad (56b)$$

It can be readily verified that the stress profile lies on the finite sides and that the velocity field is kinematically admissible.

As a final example of a rigid–perfectly plastic plate, we consider the annulus in Fig. 22 [30]. The stress profile in Fig. 23(a) is statically admissible, but the associated velocity field shown in Fig. 23(d) is not kinematically

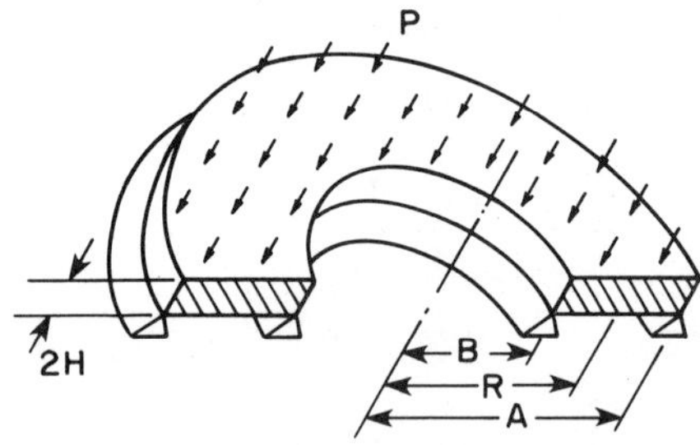

Figure 22   Annular plate, showing load and dimensions.

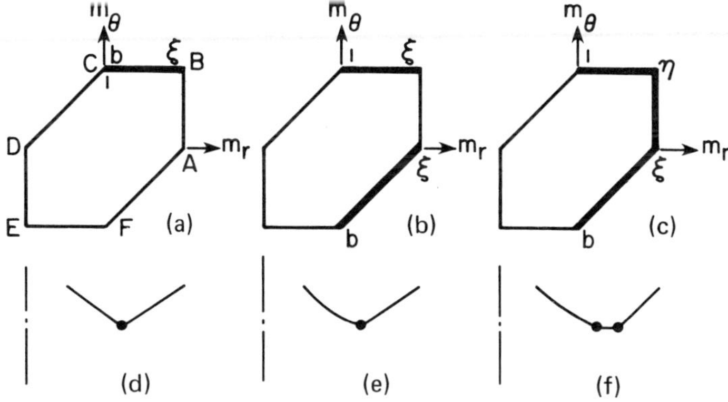

Figure 23 Stress and velocity profiles for annular plate: (a) statically admissible stress profile, (b) statically inadmissible stress profile, (c) correct stress profile, (d) kinematically inadmissible velocity profile, (e) kinematically admissible velocity profile, (f) correct velocity profile.

admissible, since $\dot{\kappa}_\theta < 0$ for $1 < r < \xi$. The velocity profile in Fig. 23(e) would be kinematically admissible if associated with the stress profile in Fig. 23(b), but it turns out that this profile is not statically admissible. The actual solution is shown in Fig. 23(c) and (f), and involves an annulus of the plate remaining rigid. Details of this solution, along with other rigid–perfectly plastic plate problems may be found in [3, Chap. 10] and [31, Chap 4].

We consider next an elastic–perfectly plastic simply supported plate, using the idealized curve OED in Fig. 6. This is equivalent to considering an ideal sandwich plate. The general elastic solution for $v - \frac{1}{3}$ is [32]

$$h^2 \eta w' = 4B/3r + 2Cr/3 + (p/3)r^3$$

$$m_r = B/r^2 - C - (5p/4)r^2 \tag{57}$$

$$m_\theta = B/r^2 - C - (3p/4)r^2$$

where $h = H/A$, $\eta = E/\sigma_0$, and $w = W/H$.

For sufficiently small p the plate is everywhere elastic with

$$h^2 \eta w' = (p/6)(-5r + 2r^3)$$

$$m_r = (5p/4)(1 - r^2) \qquad m_\theta = (p/4)(5 - 3r^2) \tag{58}$$

This solution remains valid until $p = 0.8$, when the plate center reaches point B in Fig. 21. For p slightly larger than 0.8, the center of the plate will become plastic on side BC, with the outer annulus remaining elastic. Thus (53a,b) hold for $0 \le r \le \xi$ and (57) for $\xi \le r \le 1$. Using continuity at $r = \xi$, we obtain the complete moment solution

$$0 \leq r \leq \xi: \quad m_r = 1 + (p/4)(-\xi^4/r^2 + 2\xi^2 - 5r^2)$$

$$m_\theta = 1 + (p/4)(\xi^4/r^2 + 2\xi^2 - 3r^2) \tag{59}$$

$$\xi \leq r \leq 1: \quad m_r = 1 - pr^2 \qquad m_\theta = 1$$

This solution holds until the entire profile lies on BC at the yield-point load $p = 1$.

To find the slope for $p > 0.8$, we observe that in the plastic regime BC, $\dot{\kappa}_r^p$ must vanish, so

$$\eta H \dot{\kappa}_r = \eta H \dot{\kappa}_r^e = \dot{m}_r - \dot{m}_\theta/3 \tag{60}$$

But (60) also holds when the point in question is still elastic, so it can be trivially integrated with respect to time to yield

$$h^2 \eta w'' = -(m_r - m_\theta/3) \tag{61}$$

Therefore, the correct slope solution for $p > 0.8$ is

$$0 \leq r \leq \xi: \quad h^2 \eta w' = 2r/3 + (p/3)(-\xi^4/r - \xi^2 r + r^3)$$

$$\xi \leq r \leq 1: \quad h^2 \eta w' = 2r/3 + (p/3)(-2\xi^3 + r^2) \tag{62}$$

The displacement, of course, can be found by integrating the slope with the boundary condition $w(1) = 0$. Curve (d) in Fig. 24 shows the resulting load versus center deflection for the elastic–perfectly plastic plate. Details of the solution may be found in [32].

This problem can also be solved for an elastic–strain hardening material. The solution for $p < 0.8$ is, of course, the fully elastic one, independent of any potential strain hardening. However, for $p > 0.8$, two plastic regions begin to form simultaneously at the plate center, with $0 \leq r \leq \xi$ in the corner regime B, $\xi \leq r \leq \zeta$ on the side BC, and $\zeta \leq r \leq 1$ still elastic. For isotropic hardening the equations can be solved for each regime and the constants of integration determined by continuity and boundary conditions. For p slightly greater than 1 (the exact amount will depend upon the strain-hardening constant), ζ will equal 1. For larger p, the entire plate will be plastic with $0 \leq r \leq \xi$ in corner B and $\xi \leq r \leq 1$ on side BC. Algebraic details may be found in [32].

Figure 24 shows the load-deflection curves for both the elastic–strain hardening and rigid–strain hardening models. Aside from the fact that the "knee" in the curves, which represents a partially plastic plate, is curved rather than linear, Fig. 24 is qualitatively very similar to Fig. 4. Therefore,

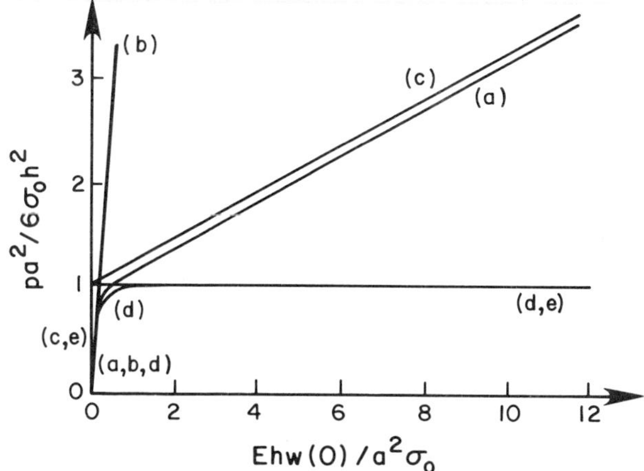

Figure 24   Load-displacement curve for circular plate (letters in parentheses refer to the materials of Fig. 2).

the same conclusions can be drawn from it, and we can gain confidence in our basic approach to plastic structures—that the rigid–perfectly plastic model can be used to find the yield-point load, that strain hardening can be neglected below the yield-point load, and that elastic strains can be ignored above it.

FINITE ELEMENT MODEL

A finite element model has become the accepted way of solving complicated problems in elasticity, and usually the only question is which finite element model to use. Essentially, a trade-off must be made between taking many simple elements with simple internal fields as opposed to fewer complex elements with complex fields.

Almost any finite-element model for elasticity can be easily adapted to include plasticity with or without strain hardening. However, there are certain features of plasticity which are basically different from elasticity and which may well affect the desirability of any given model.

Since plasticity must be formulated as a rate problem, the basic approach is as follows. At a generic time t_0, it is assumed that stress, strain, displacement, and any other quantities of interest are all known, and a finite element model is used to find the rates at time t_0. With all rates known, time is given a finite increment Δt, and quantities at time $t_0 + \Delta t$ are approximated by

$$\phi(t_0 + \Delta t) = \phi(t_0) + \Delta t\,\dot\phi(t_0) \tag{63}$$

The process is then repeated with t_0 replaced by $t_0 + \Delta t$.

There are, then, two distinct repeated stages to the solution: solving the rate problem and applying (63). We consider first three aspects of the rate problem. For definiteness we consider an elastic–perfectly plastic material whose constitutive behavior is governed by eqs. (41). The primary problem is involved with the branch points in (41f) and the inequalities (41a,b).

One of the simplest and most useful finite element models is the constant-strain model, which can be associated with arbitrary triangular or tetrahedral elements in two or three dimensions, respectively. With this element, f and λ will each have a constant value over the entire element, so that the decision concerning branch points must be made on an element-by-element basis.

Since at time t_0 f is known, the first branch point in (41f) is no problem; when $f < 0$, the element is elastic. However, $\dot{f}$ will be obtained as part of the rate solution, so it is not clear which branch should be used when $f = 0$.

A method that appears to work well in practice, at least for one-parameter loads, is to make any reasonable guess as to whether each element with $f = 0$ is elastic or plastic and to solve the rate problem. Each element is then tested. If it was assumed to be elastic with $\dot{\lambda} = 0$, then $\dot{f}$ must be non-positive or else (41a) would be violated; if it was assumed to be plastic with $\dot{f} = 0$, then $\dot{\lambda}$ must be nonnegative or else (41b) would be violated. Each element that fails its test has its assumption reversed and a new rate solution is found and tested. This process is continued until all elements simultaneously pass their test. A simple example has been constructed in which all tests are not passed until the third iteration [11]. Also, to the author's knowledge there is no proof that the iteration process will always converge, although there are no known counterexamples. Further research is needed in this area, and meanwhile it is important that any finite element program continue to test all assumptions until a correct solution is obtained.

An alternative approach to the rate problem is to use the plastic minimum principles [33] and a numerical minimizing technique. The problem is formulated as one in quadratic programming [34,35]. The question of branch points is transformed to one of asking which constraints are active, which is automatically handled by any quadratic programming program.

The second aspect of the rate problem lies in the choice of element. If any element has variable stress, then at time t_0, part of the element has $f < 0$ and part of it $f = 0$. For the $f = 0$ part, the branch-point questions raised above will exist with the further complication that the correct rate solution may be $\dot{f} = 0$ in part of that domain and $\dot{\lambda} = 0$ in the remainder. Even if this problem is solved, numerical integrations across the element will be necessary, substantially increasing the computer time spent in finding each rate solution. Therefore, questions concerning many simple versus few complex elements should be considered directly for elastic–plastic problems, with a strong presumption that simpler elements will be preferable.

The final aspect of the rate problem is that of continuity. Elasticity problems are extremely continuous, and efficient finite element methods will take advantage of that fact. However, problems for elastic–perfectly plastic structures may have discontinuities in stress derivatives (as across the characteristics in a plane strain problem) and even of stresses themselves, as in Fig. 5(d) or 16. Stress discontinuities pose no particular problem; the finite element model merely forces them to occur only at element boundaries, but velocity discontinuities may also occur. For example, we saw earlier that slope discontinuities could occur in beams and plates. It has been shown that finite element models that allow for slope discontinuities are less efficient in the early stages of loading a beam, but become more efficient (i.e., give closer values for a given computation effort) as the yield-point load is approached [36]. A plate model in which slope discontinuities provide the only strains has been developed [37] and has produced some valid results. These ideas have recently been extended [38] to include problems in plane strain.

Even when the rate problem has been solved, there are questions to be answered concerning the application of eq. (63). A plastic element requires that $\dot{f} = 0$. Since the yield surface $f = 0$ is convex [23], if the stress point is not stationary, it must move along this convex surface. However, (63) is motion along a straight line tangent to the surface. Thus at time $t_0 + \Delta t$, the predicted f will be positive, in violation of (41a).

A usual method of resolving this dilemma is to first use (63) to predict a new $\boldsymbol{\sigma}$, which will, in general, have a positive f in each plastic element. We then replace $\boldsymbol{\sigma}$ by $\beta\boldsymbol{\sigma}$, where the scalar β is chosen so that $f(\beta\boldsymbol{\sigma}) = 0$ in each plastic element. However, the stresses $\beta\boldsymbol{\sigma}$ will not precisely satisfy the equilibrium equations, so some type of iteration must be performed.

If the yield function f is piecewise linear, this problem disappears. The stress point may lie on one or more linear faces so that f and λ are replaced by f_i and λ_i throughout (41), and (41e) is replaced by

$$\dot{\boldsymbol{\epsilon}}^p = \sum \dot{\lambda}_i \nabla f_i \tag{64}$$

If the external loading is linear, eq. (63) will now be exact until either an elastic element becomes plastic or until a plastic element encounters a new face.

The yield function is, of course, a physical property of the material. However, it can only be measured with limited accuracy, so it is reasonable to approximate it by any convenient function. For finite element models, a piecewise linear approximation has the obvious advantage of eliminating the problem associated with (63) and any strictly convex yield function. A systematic procedure for constructing a piecewise linear approximation to a given nonlinear f is given in [39].

A third possible solution to this difficulty is to introduce a small plastic viscosity [40,41]. The resulting viscoplastic model permits expansion of the yield surface, but for small v the expansion would be severely limited, so that the solution will be very close to that for a perfectly plastic material. Indeed, for some real materials the viscoplastic model may be more realistic than the perfectly plastic one.

The preceding discussion of (63) has been implicitly based upon a model with constant stress in each element. The use of a more complex element would not eliminate this problem, and all the suggested remedies would be more difficult to apply. For this stage too, then, there are strong arguments in favor of using a large number of simple elements in the solution of plasticity problems [43].

CONCLUSIONS

It is evident that the preceding sections make no claim to completeness either in content or in references. We have tried merely to sample the vast field of plastic structures, beginning with the simplest truss and beam problems, in the hope of rousing the reader's interest. Many interesting and important topics have had to be omitted entirely. Among these omissions are plastic buckling, the effect of geometric changes, and dynamic effects, and we have not even mentioned the vast literature on plastic shells [42].

The reader interested in more details concerning the material covered will find most of it and more in [3]. For the material not covered, almost every new issue of most mechanics journals contains articles extending the theory of plasticity and the list of problems which it has solved.

REFERENCES

[1] D. VASARHELYI, AND R. A. HECHTMAN, Welded Reinforcement of Openings in Structural Steel Members, *Welding J. Research*, Suppl., **31**, 169s–180s (1952).

[2] J. MARIN, Stress–Strain Relations in the Plastic Range for Biaxial Stresses, *J. Franklin Inst.*, **248**, 231–249 (1949).

[3] P. G. HODGE, JR., *Plastic Analysis of Structures*, McGraw-Hill Book Company, New York, 1959.

[4] A. A. GVOZDEV, The Determination of the Value of the Collapse Load for Statically Indeterminate Systems Undergoing Plastic Deformation, *Proceedings of the Conference on Plastic Deformations*, Akademiia Nauk SSSR, Moscow–Leningrad, 1938, pp. 19–33; translation into English by R. M. Haythornthwaite, *Int. J. Mech. Sci.*, **1**, 322–335 (1960).

[5] R. HILL, On the State of Stress in a Plastic–Rigid Body at the Yield Point, *Phil. Mag.*, **42** (7), 868–875 (1951).

[6] R. HILL, A Note on Estimating the Yield-Point Loads in a Plastic–Rigid Body, *Phil. Mag.*, **43** (7), 353–355 (1952).

[7] H. J. GREENBERG, AND W. PRAGER, Limit Design of Beams and Frames, *Proc. ASCE*, **77** (Sec. 59), 1951.

[8] D. C. DRUCKER, H. J. GREENBERG, AND W. PRAGER, The Safety Factor of an Elastic–Plastic Body in Plane Strain, *J. Appl. Mech.*, **18**, 371–378 (1951).

[9] D. C. DRUCKER, W. PRAGER, AND H. J. GREENBERG, Extended Limit Design Theorems for Continuous Media, *Q. Appl. Math.*, **9**, 381–389 (1952).

[10] D. C. DRUCKER, Plasticity of Metals—Mathematical Theory and Structural Applications, *Trans. ASCE*, **116**, 1059–1072 (1959).

[11] P. G. HODGE, JR., Complete Solutions for Elastic–Plastic Trusses, *SIAM J. Appl. Math.*, **25**, 435–447 (1973).

[12] F. STÜSSI, AND C. F. KOLLBRUNNER, Beitrag zum Traglastverfahren, *Bautechnik*, **13**, 264–267 (1935).

[13] W. PRAGER, The General Theory of Limit Design, *Proc. 8th Intern. Congr. Appl. Mech.*, Istanbul, 1952, **2**, 65–72 (1956).

[14] W. PRAGER, AND P. G. HODGE, JR., "Theory of Perfectly Plastic Solids," University Microfilms International, Ann Arbor, Mich. ("Books on Demand" OP69234), originally John Wiley & Sons, Inc., New York, 1948.

[15] H. TRESCA, Mémoire sur l'écoulement des corps solides, *Mém. Pres. Par Div. Sav.*, **18**, 733–799 (1868).

[16] R. V. MISES, Mechanik der festen Koerper im plastisch deformablen Zustand, *Goettinger Nachr., Math.-Phys. Kl.*, **1913**, 582–592 (1913).

[17] B. DE SAINT VENANT, Mémoire sur l'établissement des équations différentielles des mouvements intérieurs opérés dans les corps solides ductiles au delà des limites ou l'elasticité pourrait les ramener à leur premier etat, *C. R. Ac. Sci. (Paris)*, **70**, 473–480 (1870).

[18] M. LÉVY, Mémoire sur les équations générales des mouvements intérieurs des corps solides ductiles au delà des limites où l'élasticité pourrait les ramener à leur premier état, *C. R. Ac. Sci (Paris)*, **70**, 1323–1325 (1870).

[19] L. PRANDTL, Spannungsverteilung in plastischen Koerpern, *Proc. 1st Internat. Congr. Appl. Mech.*, Delft, 1924, pp. 43–54.

[20] E. REUSS, Beruecksichtigung der elastischen Formaenderungen in der Plastizitaetstheorie, *Z. Angew. Math. Mech.*, **10**, 266–274 (1930).

[21] R. V. MISES, Mechanik der plastischen Formaenderung von Kristallen, *Z. Angew. Math. Mech.*, **8**, 161–185 (1928).

[22] R. HILL, Plastic Distortion of Non-uniform Sheets, *Phil. Mag.*, (7), **40**, 971–983 (1949).

[23] D. C. DRUCKER, Some Implications of Work Hardening and Ideal Plasticity, *Q. Appl. Math.*, **7**, 411–418 (1950).

[24] W. Prager, A New Method of Analyzing Stresses and Strains in Work-Hardening Plastic Solids, *J. Appl. Mech. Trans. ASME*, **78**, 493–496 (1956).

[25] W. Prager, The Theory of Plasticity—A Survey of Recent Achievements, *Proc. Inst. Mech. Eng.*, **169**, 41–67 (1955).

[26] G. N. White, Jr., Application of the Theory of Perfectly Plastic Solids to Stress Analysis of Strain Hardening Solids," *GDAM Rept. 51*, Brown University, 1950.

[27] J. F. Besseling, A Theory of Plastic Flow for Anisotropic Hardening of an Initially Isotropic Material, *Rept. S410*, Nat. Aero. Res. Inst., Amsterdam, 1953.

[28] D. K. Vaughan, A Comparison of Current Work-Hardening Models Used in the Analysis of Plastic Deformations, M.S. thesis, Texas A & M University, 1973.

[29] "Constitutive Equations in Viscoplasticity—Computational and Engineering Aspects," J. Stricklin and K. J. Saczalaski, eds., American Society of Mechanical Engineers, Winter Annual Meeting, New York, 1976.

[30] P. G. Hodge, Jr., Yield Point Load of an Annular Plate, *J. Appl. Mech.*, **26**, 454–455 (1959).

[31] P. G. Hodge, Jr., "Limit Analysis of Rotationally Symmetric Plates and Shells," University Microfilms International, Ann Arbor, Mich. ("Books on Demand OP63418), originally Prentice-Hall, Inc., Englewood Cliffs, N.J., 1963.

[32] P. G. Hodge, Jr., Boundary Value Problems in Plasticity, *Plasticity*, E. H. Lee and P. S. Symonds, eds., Pergamon Press, New York, 1960, pp. 297–337.

[33] H. J. Greenberg, Complementary Minimum Principles for an Elastic-Plastic Material, *Q. Appl. Math.*, **7**, 85–95 (1949).

[34] P. G. Hodge, Jr., T. Belytschko, and C. T. Herakovich, Quadratic Programming and Plasticity, *Computation Approaches in Applied Mechanics*, E. Sevin, ed., American Society of Mechanical Engineers, New York, 1969, pp. 73–84.

[35] P. G. Hodge, Jr., V. K. Garg, and S. C. Anand, A Finite Element Method for Plasticity Problems, *Developments in Theoretical and Applied Mechanics*, Vol. 7, M. Chi, ed., Catholic University, Washington, D.C., 1974, pp. 369–383.

[36] P. G. Hodge, Jr., Finite Element Methods in Plasticity, *Proc. 7th U.S. Nat. Congr. Appl. Mech.*, S. K. Datta, ed., American Society of Mechanical Engineers, New York, 1974, pp. 114–119.

[37] P. G. Hodge, Jr., and A. A. McMahon, A Simple Finite Element Model for Elastic-Plastic Plate Bending, *Comp & Struct.*, **2**, 841–854 (1972).

[38] H. M. v. Rij and P. G. Hodge, Jr., A Slip Model for Finite Element Plasticity, *J. Appl. Mech.*, **45**, 527–532 (1978).

[39] P. G. Hodge, Jr., Automatic Piecewise Linearization in Ideal Plasticity, *Comp. Methods Appl. Mech. Eng.*, **10**, 249–272 (1977).

[40] Q. S. NGUYEN, AND J. ZARKA, Quelques méthodes de résolution numérique en élastoplasticité classique et en élasto-viscoplasticité, Séminaire Plasticité et Viscoplasticité 1972, Ecole Polytechnique, Paris; see also discussion of [43].

[41] O. C. ZIENKIEWICZ, Discussion of [43].

[42] P. G. HODGE, JR., Plastic Analysis and Pressure Vessel Safety, *Appl. Mech. Rev.*, **24**, 741–747 (1971).

[43] P. G. HODGE, JR., Computer Solutions of Plasticity Problems, *Problems of Plasticity*, A. Sawczuk, ed., Noordhoff International Publishing, Leyden, 1974, pp. 261–286.

MECHANICS OF SHELLS IN THE AGE OF COMPUTATION

G. A. Wempner*

INTRODUCTION

Nearly a century has elapsed since the inception of Love's theory, yet most current theories of elastic shells retain certain basic features of that first approximation. Meanwhile the theory has undergone refinements; in particular, the kinematic relations have been refined and extended to include finite rotations and deflections. With the capabilities of modern computers, the nonlinear theories can now be utilized to treat important practical problems, notably the buckling and postbuckling of imperfect shells. Also, with the aid of computers, theories and computational procedures can be devised to treat problems of inelastic shells.

Here, we restrict our attention to questions of particular relevance in practice, and we address these questions from the physical or mathematical viewpoints that reveal the essential physical and geometric features. First,

*Georgia Institute of Technology, Atlanta, Georgia.

The support of the National Science Foundation and the Alexander von Humboldt Foundation of the German Federal Republic; the assistance of Mr. C.-M. Hwang, Ms. S. McCombs, and the cooperation of Professor W. Wunderlich are most gratefully acknowledged.

we identify the salient features of the first approximation, which has proven to be an adequate and effective tool for the analyses of most thin elastic shells. Briefly, we recount a few simplifications and formulations which have rendered theories amenable to formal solution, but we make no attempt to review the multitude of theories and solutions that have been obtained. Also, we comment only briefly upon the alternative approaches toward the development of shell theories, first approximations, and higher approximations. Our few remarks on higher approximations are limited to questions of layered and anisotropic shells, which are emerging as important forms in modern technology. Two theories of inelastic shells are described; one is obtained by a direct approach and the other by a derivation from a theory of three dimensions. The distinctive features of some discrete approximations are outlined. We conclude our review of the subject with a discussion of the limitations of shell theories and of discrete approximations.

Many topics are beyond the scope of this work; thermal and temporal effects, vibrations, and limit analyses are excluded. We attempt only to sketch enough of the history to indicate the past and current trends. Of necessity, past efforts have been concerned primarily with attempts to achieve formal solutions to boundary-value problems of elastic shells. Presently, we retain our interest in such solutions as a means to describe localized phenomena, at edges, cutouts, attachments, and so on, but direct our attention to discrete approximations, to the utilization of computers, as an effective means to describe the overall response of actual shells, linear and nonlinear, elastic and inelastic.

Numerous references are appended so that the interested reader can pursue any aspect in greater detail. These references reflect not only the author's cognizance and his ignorance, but also the limitations of space and the accessibility of these works. Hopefully, the listing will provide sufficient access to the vast literature on the theories and analyses of shells.

THE NATURE OF SHELL THEORY

A shell is a thin body; the thickness is small compared to the radii of curvature and small compared to the distances between edges. Some shells, such as thin plates, are quite flexible, but others are surprisingly stiff, like the shell of an egg. This stiffness can be attributed to the curvature which requires that the shell stretch to deform. Any stretching is accompanied by forces that support loads as a membrane. In contrast, the plate must support lateral pressure by bending and by the actions of transverse shear forces and bending moments.

A theory of shells is an approximation that describes the essential features of the response, deformation, and stresses in terms of surface variables and the properties of a surface. In other words, a theory of shells is an approximation of a thin three-dimensional body in terms of two dimensions. The inherent approximations preclude any further description of distributions through the thickness.

According to our earlier remarks, a theory of shells should account for the stretching and the attendant forces that are perceived in membranes, as well as for the changes of curvature, transverse shear, and bending moments perceived in a plate. The simplest theory that embodies the essential features is the *first approximation*.

THE FIRST APPROXIMATION

Strains and stresses

The salient features of the first approximation follow:
The position of the same particle of a reference surface is given by vectors $\vec{r}$ and $\vec{R}$ in an initial and subsequent (undeformed and deformed) state, respectively. Associated with each position is a triad of vectors, $\vec{a}_i$ and $\vec{A}_i$, respectively: $\vec{a}_\alpha$ and $\vec{A}_\alpha$ are tangential to the same, but convected, coordinate lines of the surfaces; $\hat{a}_3$ and $\hat{A}_3$ are the unit normal vectors. Two surface tensors *fully* describe the differential geometry of the surfaces in the initial and subsequent states. The *metric* tensors of the two states are

$$a_{\alpha\beta} \equiv \vec{a}_\alpha \cdot \vec{a}_\beta, \qquad A_{\alpha\beta} = \vec{A}_\alpha \cdot \vec{A}_\beta \tag{1a,b}$$

The *curvature* tensors of the two states are

$$b_{\alpha\beta} \equiv -\hat{a}_{3,\alpha} \cdot \vec{a}_\beta, \qquad B_{\alpha\beta} \equiv -\hat{A}_{3,\alpha} \cdot \vec{A}_\beta \tag{2a,b}$$

These tensors play key roles, akin to the metric tensor of the three-dimensional space. Therefore, it seems natural to introduce two strain tensors:

$$\gamma_{\alpha\beta} \equiv \tfrac{1}{2}(A_{\alpha\beta} - a_{\alpha\beta}), \qquad \kappa_{\alpha\beta} \equiv (B_{\alpha\beta} - b_{\alpha\beta}) \tag{3a,b}$$

The strains $\gamma_{\alpha\beta}$ determine the extensions and shears of the surface, and the strains $\kappa_{\alpha\beta}$ determine the changes of curvature and twist.

Associated with each component of strain is a component of stress such that the power of the stresses assumes the form

$$\dot{w} = n^{\alpha\beta}\dot{\gamma}_{\alpha\beta} + m^{\alpha\beta}\dot{\kappa}_{\alpha\beta} \tag{4}$$

The position and orientation of the deformed surface are defined by the vectors $\vec{R}$ and $\hat{A}_3$, respectively. The velocity $\dot{\vec{R}}$, an angular velocity $\dot{\vec{\phi}}$, and the strain rate $\dot{\gamma}_{\alpha\beta}$ determine the motion of the triad [21]:

$$\dot{\hat{A}_3} = \dot{\vec{\phi}} \times \hat{A}_3, \qquad \dot{\hat{A}_\alpha} = \dot{\gamma}_{\alpha\beta}\vec{A}^\beta + \dot{\vec{\phi}} \times \vec{A}_\alpha \qquad (5a,b)$$

To determine the physical identity of the stresses, we consider the work done by forces $\vec{N}^\alpha$ and couples $\vec{M}^\alpha$ in effecting a motion of an element bounded by edges at (θ^α) and $(\theta^\alpha + d\theta^\alpha)$ as depicted in Fig. 1. The forces $\vec{N}^\alpha$ and body force $\vec{F}$ work upon the velocity $\dot{\vec{R}}$, and the couples $\vec{M}^\alpha$ and body couple $\vec{M}$ work upon the angular velocity $\dot{\vec{\phi}}$. The work done upon the element, after dividing by the area $(\sqrt{A}\,d\theta^1\,d\theta^2)$ and passing to the limit, follows:

$$\dot{w} = \frac{1}{\sqrt{A}}[(\vec{N}^\alpha \cdot \dot{\vec{R}})_{,\alpha} + \vec{F} \cdot \dot{\vec{R}} + (\vec{M}^\alpha \cdot \dot{\vec{\phi}})_{,\alpha} + \vec{M} \cdot \dot{\vec{\phi}}] \qquad (6)$$

The vectors $\dot{\vec{R}}_{,\alpha} \equiv \dot{\vec{A}}_\alpha$, and $\dot{\vec{\phi}}_{,\alpha}$ can be expressed in terms of the spin $\dot{\vec{\phi}}$ and strain rates, $\dot{\gamma}_{\alpha\beta}$ and $\dot{\kappa}_{\alpha\beta}$, in accordance with eqs. (3a,b) and (5a,b). For simplicity, tensorial components of the forces and couples are introduced:

$$\frac{\vec{N}^\alpha}{\sqrt{A}} \equiv N^{\alpha\beta}\vec{A}_\beta + Q^\alpha\hat{A}_3, \qquad \frac{\vec{M}^\alpha}{\sqrt{A}} \equiv \hat{A}_3 \times \vec{R}^\alpha, \qquad \frac{\vec{M}}{\sqrt{A}} = \hat{A}_3 \times \vec{C}$$

$$(7a,b,c)$$

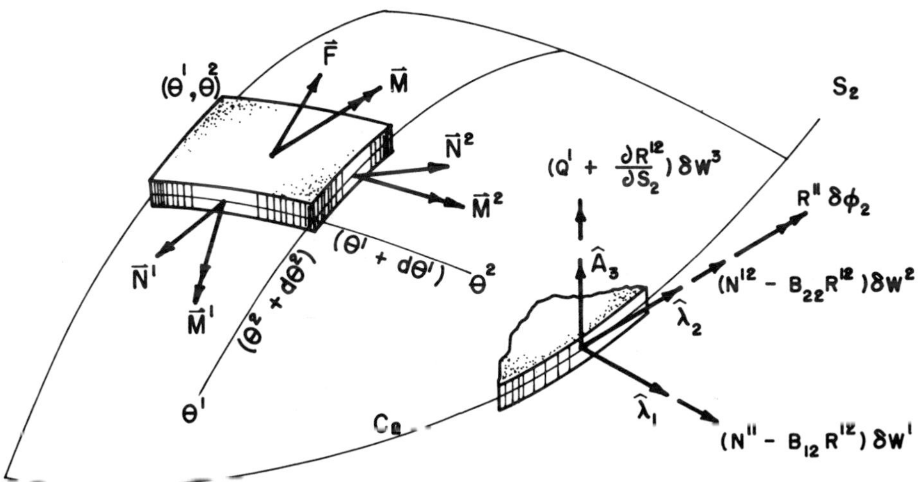

Figure 1

After some intermediate steps, eq. (6) assumes the form

$$\dot{w} = \frac{1}{\sqrt{A}}(\vec{N}^{\alpha}_{,\alpha} + \vec{F})\cdot\dot{\vec{R}} + \hat{A}_3 \times \left[\frac{1}{\sqrt{A}}(\sqrt{A}\,\vec{R}^{\alpha})_{,\alpha} - Q^{\alpha}\vec{A}_{\alpha} + \vec{C}\right]\cdot\dot{\vec{\phi}}$$

$$+ (N^{\alpha\beta} - B^{\alpha}_{\mu}R^{\mu\beta})(\vec{A}_{\alpha} \times \vec{A}_{\beta})\cdot\dot{\vec{\phi}} + (N^{\alpha\beta} + B^{\beta}_{\mu}R^{\mu\alpha})\dot{\gamma}_{\alpha\beta} - R^{\beta\alpha}\dot{\kappa}_{\alpha\beta} \quad (8)$$

The coefficients of the velocity vectors $\dot{\vec{R}}$ and $\dot{\vec{\phi}}$ must vanish independently; these provide the equations of equilibrium (or motion). Corresponding to spin about the normal, we obtain

$$(N^{\alpha\beta} - B^{\alpha}_{\mu}R^{\mu\beta})\epsilon_{\alpha\beta} = 0 \quad (9)$$

In other words, the parenthetical stress must be symmetrical. Following Sanders [31] and Leonard [32], we achieve consistency between eqs. (4) and (9) and the final terms of (8), if we define the symmetrical stress tensors

$$n^{\alpha\beta} \equiv N^{\alpha\beta} + B^{\beta}_{\mu}R^{\mu\alpha}, \qquad m^{\alpha\beta} \equiv -\tfrac{1}{2}(R^{\alpha\beta} + R^{\beta\alpha}) \quad (10a,b)$$

The strains $\gamma_{\alpha\beta}$ and $\kappa_{\alpha\beta}$ of eqs. (3a,b) are not the only measures suited to the first approximation. Others have been employed for purposes of simplifying, or otherwise modifying, the form of the governing equations. In particular, Sanders [31] and Koiter [33] introduced modifications of the strain $\kappa_{\alpha\beta}$, which is augmented by terms of the type $b^{\gamma}_{\alpha}\gamma_{\gamma\beta}$. The introduction of a modified strain $\tilde{\kappa}_{\alpha\beta}$ renders an exact formulation [35,36] of the static geometric analogy of Lur'e [28] and Gol'denweizer [29]. The analogy between equilibrium and compatibility equations, and between stress components $(n^{\alpha\beta}, m^{\alpha\beta})$ and strain components $(\bar{\epsilon}^{\lambda\alpha}\bar{\epsilon}^{\mu\beta}\tilde{\kappa}_{\lambda\mu}, \bar{\epsilon}^{\lambda\alpha}\bar{\epsilon}^{\beta\mu}\dot{\gamma}_{\lambda\mu})$ has a practical consequence: the stresses are given by three *stress functions* as the analogous strains are given by the components of displacement.

It is noteworthy that the foregoing development rests only on the geometric properties of the reference surface and the concept of work. In reality, the stress $\vec{R}^{\alpha}$ (couple) does work upon rotation of the normal; because these couples act upon edges of the real shell, the hypothesis of Kirchhoff–Love is implicit [37,26]. Otherwise, the results might apply to any shell, of any material. If the shell were simply elastic, an energy potential Ψ would depend upon the strains $\Psi = \Psi(\gamma_{\alpha\beta}, \kappa_{\alpha\beta})$. Again, the balance of work and energy provides the general forms:

$$n^{\alpha\beta} = \frac{\partial\Psi}{\partial\gamma_{\alpha\beta}}, \qquad m^{\alpha\beta} = \frac{\partial\Psi}{\partial\kappa^{\alpha\beta}} \quad (11a,b)$$

Also, the work principles of classical plasticity may apply, whereupon a yield condition Y depends upon the stresses [21], $Y = Y(n^{\alpha\beta}, m^{\alpha\beta})$. Thus, the normality criterion provides the general forms:

$$\dot{\gamma}^P_{\alpha\beta} = \dot{\lambda}\frac{\partial Y}{\partial n^{\alpha\beta}}, \qquad \dot{\kappa}^P_{\alpha\beta} = \dot{\lambda}\frac{\partial Y}{\partial m^{\alpha\beta}} \qquad (12\text{a,b})$$

Edge conditions

An edge of the reference surface is defined by curve C_2 of Fig. 1; S_2 denotes the arc length along C_2. The unit tangent vectors $\hat{\lambda}_1$ and $\hat{\lambda}_2 \equiv \partial\vec{R}/\partial S_2$ and the normal $\hat{A}_3$ form an orthogonal triad. A virtual motion of the triad is defined by the virtual displacement along the curve, $\delta\vec{W}(S_2)$, and by a virtual rotation $\delta\phi_2$ about the tangent $\hat{\lambda}_2$. Three components of force and a couple are associated with the three components of $\delta\vec{W}$ and the rotation $\delta\phi_2$. The virtual work of each is given in Fig. 1; the parenthetical factors are the actions associated with each motion. It is significant that only these four independent actions (or motions) can be prescribed upon the edge of a shell, according to the first approximation. In particular, the twisting couple R^{12} cannot be prescribed. This essential feature was established by Kirchhoff [37] and a physical interpretation was given by Thompson and Tait [38]. Further discussion of these boundary conditions may be found in the works of Friedrichs [39], Reissner [40], Koiter [41], and Wempner [42]. The Kirchhoff conditions have been modified by Heijden, who presented a comprehensive discussion of their effects upon the elastic shell and gave numerous references to practical questions of stress distributions, and concentrations, at edges [43].

Geometric nonlinearities

Equations governing important practical problems often involve nonlinearities, which stem from relatively large rotations and changes of curvature, although the strains ($\gamma_{\alpha\beta}$) remain small. Large rotations and twisting are apparent in springs (cylindrical, helicoidal, conical), while the interactions between forces and changes of curvature account for the buckling of shells. Since these nonlinearities are geometric in origin, it seems appropriate to view them independently of the constitutive equations.

Nonlinear terms are evident in the strain-displacement equations and the equations of equilibrium (or motion). The prominent examples appear in the equations of the surface strains and the equation governing equilibrium of the normal force:

$$\gamma_{\alpha\beta} = \tfrac{1}{2}(\vec{A}_{\alpha} \cdot \vec{W}_{,\beta} + \vec{A}_{\beta} \cdot \vec{W}_{,\alpha}) + \tfrac{1}{2}\vec{W}_{,\alpha} \cdot \vec{W}_{,\beta} \tag{13}$$

$$\sqrt{A}(N^{\alpha\beta}b_{\alpha\beta} + \underline{N^{\alpha\beta}\kappa_{\alpha\beta}}) + (\sqrt{A}Q^{\alpha})_{,\alpha} + F^3 = 0 \tag{14}$$

Here, the offensive terms are underlined. In most cases the final term of eq. (13) can be replaced by the product of moderate rotations; discussions of such approximations are found in the works of Sanders [34], Koiter [36], and Wempner [139]. The product of force and change of curvature in eq. (14) is vital to analyses of buckling, that is, to examinations of adjacent states of equilibrium. If a nonlinear problem is formulated via the principle of virtual work, then the nonlinear terms of eq. (13) produce those of eq. (14).

Linear constitutive equations

Love's first approximation for the strain energy of the *isotropic* elastic shell follows [1]:

$$\psi = \frac{h}{2}C^{\alpha\beta\gamma\eta}\left(\gamma_{\alpha\beta}\gamma_{\gamma\eta} + \frac{h^2}{12}\kappa_{\alpha\beta}\kappa_{\gamma\eta}\right) \tag{15}$$

$$C^{\alpha\beta\gamma\eta} = \frac{E}{1+v}\left(a^{\alpha\gamma}a^{\beta\eta} + \frac{v}{1-v}a^{\alpha\beta}a^{\gamma\eta}\right) \tag{16}$$

Here, the reference surface is at the middle of the shell and h denotes the thickness. The approximation is founded upon the Kirchhoff–Love hypothesis and the neglect of transverse normal stress. The validity of Love's approximation in eq. (15) and the corresponding stress–strain relations of eqs. (11a,b) has been supported by the works of Koiter [33], Naghdi [13], and Budiansky and Sanders [45]. As noted earlier by Reissner [44], a generally valid higher approximation may require consideration of transverse shear strain and normal stress. Estimates of the inherent errors of Love's approximation (15) are given by John [46], Niordson [47,55], and Koiter and Simmonds [48]. The latter cite an error (root mean square) in the three-dimensional stress of order $\epsilon^2 = h^2/L^2 + h/R$, where R and L are the smallest principal curvature and shortest wavelength of the deflection pattern.

The distinctive feature of the approximation (15) is the separation of the energies attributed to extensional and flexural deformations. Consequently, the stress–strain relations of extension and flexure are uncoupled.

The kinematic and dynamic description of the first approximation may be adequate for the approximation of thin heterogeneous and anisotropic shells (e.g., laminated shells), but the extensional and flexural variables are generally coupled in the constitutive equations of anisotropic shells. Details may be found in the works of Ambartsumyan [49], Librescu [50,51], Reissner [52,53], and Noor [54].

METHODS OF DERIVATION

Several distinct methods have been employed to obtain theories of shells. From a practical viewpoint, the approach of Kirchhoff, Aron, and Love has been the most fruitful [37,25,26,1]. The cornerstone is the hypothesis that normals to the surface remain normal. Hence, dynamic and kinematic equations follow precisely; however, additional approximations are needed in the kinematic and constitutive equations to achieve a workable theory.

Another approach involves the deduction of the two-dimensional theory from the equations of three dimensions. This approach has been taken by Goodier [57], Johnson and Reissner [58], Green [59], Gol'denweizer [60], Rutten [61], and Reissner [62]. A common feature is the expansion of dependent variables in powers of a thinness parameter (e.g., h/R). The procedure is systematic and mathematical, following a priori scaling of the dominant terms.

In the direct approach of the Cosserats, the two-dimensional theory of a continuous surface is developed by analogy with the mechanics of three-dimensional continua. Accordingly, the mechanical properties of the surface must also be obtained by direct experimentation upon the shell. The latter is difficult, if not impossible.

Variational theorems offer a systematic approach to approximations in the manner of the Rayleigh-Ritz method. The principle of virtual work provides the two-dimensional conditions of equilibrium, consistent with the Kirchhoff–Love hypothesis; likewise, the stationary theorem of Hu-Washizu [63,64] also provides the kinematic and constitutive equations, and the theorem of Hellinger–Reissner [65,66] provides constitutive equations that express the displacements in terms of stresses. Versions of these theorems are also applicable to the incremental theory of plasticity [68]. Conjugate variables and their governing equations are consistent with the underlying approximation; for example, the stresses ($n^{\alpha\beta}$, $m^{\alpha\beta}$) and strains ($\gamma_{\alpha\beta}$, $\kappa_{\alpha\beta}$) are consistent with the Kirchhoff–Love hypotheses, according to the principle of virtual work. A variational procedure also provides the two-dimensional functional, which, in turn, provides a mechanism for subsequent approximations of the two-dimensional fields.

Our treatments of strain, stress, and edge conditions are presented in the spirit of an "intrinsic theory" as introduced by Synge and Chien [30]; the developments rest upon differential geometry, without explicit reference to displacement. Also, the results are presented in the spirit of the "direct method" as employed by the Cosserats [27]; from the onset, all results pertain to a surface, without explicit reference to the three-dimensional body. Of course, the shell is a three-dimensional body with distributions of properties, stress, and strain throughout. From a practical viewpoint it becomes essential to correlate the two-dimensional description with the actual properties of the shell.

THEORIES OF THIN ELASTIC SHELLS

A complete theory, in primitive form, consists of dynamic, kinematic, and constitutive equations. Alternatively, the theory can be given as the stationary, or minimum, condition of a functional.

The general nonlinear equations of the first approximation are given in the primitive form by Sanders [34] and Koiter [36].

Mathematical formulations of the theory may assume various forms: the principle of stationary potential leads to a system of three partial differential equations (equilibrium conditions) in the displacements (W^i) of the middle surface. Alternatively, the three equations of equilibrium may be augmented by the three equations of compatibility, all expressed in terms of the stresses $(n^{\alpha\beta}, m^{\alpha\beta})$. Linear versions of these equations are given by Novozhilov [5] and Gol'denweizer [6].

Until the recent development of digital computers, attention necessarily focused upon the formal solution of the governing equations, that is, solutions expressed by known functions. Consequently, much effort was directed toward simplifications and reductions that were amenable to solution.

Perhaps the greatest simplification is the membrane theory, which neglects entirely the flexural stiffness $(m^{\alpha\beta} = 0)$. Membrane theory provides estimates of the stresses $(n^{\alpha\beta})$ in certain thin shells under distributed loads; however, such estimates may be erroneous near edges. The theory and numerous solutions are given by Gol'denweizer [6], Novozhilov [5], and Flügge [11]. The applicability of membrane solutions was examined recently by Steele [78].

Simplifications of the equations of the first approximation fall in two broad categories: first, approximations of a geometrical character, which always impose certain limitations; and second, simplifications, or reductions, of a mathematical character (e.g., a judicious choice of variables). Space allows us to mention but a few important examples.

Perhaps the most important kinematic approximation pertains to shallow shells; tangential components of displacement are neglected in the approximations of rotation and changes of curvature. Such an approximation was employed in the analyses of the stability of cylinders by Donnell

[71] and in the nonlinear theories of Marguerre [72], and Mushtari and Vlasov [9,4]. The general theory is expressed by two fourth-order differential equations in terms of a displacement and stress function [36,74]. In the usual sense, a shallow shell is one that differs little from a plate; however, in practice it is enough that patterns of deflection occur within regions that are relatively small compared to the radii of curvature [36].

The linear equations governing small axisymmetric deformations of shells of revolution were reduced to a system of two second-order differential equations by Meissner [70] following the work of H. Reissner [69]. Further discussion and references are presented by Simmonds [81]. Equations governing finite axisymmetric deformations were achieved by E. Reissner [73].

The linear equations governing shells of specific forms have been reduced to a system of two coupled fourth-order equations. Formulations for the cylinder, catenoidal, helicoidal, and minimal surfaces are given by Sanders [75], Wan [76], Simmonds [77], and Latta and Simmonds [80].

BUCKLING OF THIN ELASTIC SHELLS

The great practical advantage of thin shells is their capacity to sustain large distributed loads by virtue of curvature and attendant membrane stresses. A *perfect thin* sphere is *theoretically* in equilibrium under extreme external pressures, with no appreciable bending; likewise, the perfect thin cylinder is theoretically in equilibrium under extreme axial thrust. The *extreme* loads are not realized, because the equilibrium states become unstable. The critical (membrane) stresses at the point of bifurcation (from stable to unstable states) are often much less than the yield stress, approximately $\sigma_c \equiv \sqrt{3(1 - v^2)}\, E(h/R)$. Unfortunately, shells are never perfect and small imperfections cause buckling at stresses much lower than these critical values, as low as $0.1\sigma_c$. Moreover, the buckling is a violent snap-through to a severely deformed state.

In 1941, von Kármán and Tsien employed an approximation of the buckled form to examine the *postbuckled* states of the cylindrical shell under axial load [83]. Their results showed a load-deflection path, which descended precipitiously through unstable states to a minimum, and then ascended through stable states (severely deformed). In 1945, Koiter examined the energy criteria for stability at the critical load, the conditions for instability and the effects of imperfections [84]. Briefly, two schools emerged. As early as 1934, Donnell had suggested that the actual buckling occurred at a (limit) point, much below the bifurcation point, because of imperfection and premature bending [82]. This view was supported by the work of Koiter. Yet, others believed that an accurate calculation of the minimum postbuckled load could serve as a practical limit. With the advent of the digital computer, refined analyses of the postbuckled states revealed that the minimum post-buckled load is much less than the actual buckling load [85,88]. An explanation is provided, in part, by Yoshimura's kinematic model, in which the cylinder is severely, but inextensionally, deformed [86]. In such modes, the cylinder acts like an accordion, and sustains the minimum load via flexural stresses. It appears now that accurate calculations of *actual* buckling loads require the nonlinear equations that govern the equilibrium of the *imperfect* shell. It appears, too, that the theory (the first approximation) is adequate for most situations *provided* that the kinematic equations are sufficiently precise.

The interested reader will find further information and numerous references in the works of Koiter [91], Almroth and Starnes [89], Brush and Almroth [90], and Tvergaard [92].

HIGHER APPROXIMATIONS

Theories based upon the Kirchhoff–Love hypothesis have the same basic kinematic and dynamic equations. However, some constitutive equations retain terms of higher degree in the thickness parameter (h/R) in order to accommodate thick shells. Notable examples are the equations of Flügge [2], Lur'e [94], and Byrne [95]; these are discussed by Naghdi [13].

Numerous authors have abandoned initial constraints, like the Kirchhoff–Love hypothesis, in order to obtain theories that apply to thick shells. Examples are the works of Neuber [97], Hildebrand, Reissner, and Thomas [98], Green and Zerna [100], and Zerna [103]. Neuber expanded the stresses in power series; Zerna expanded the displacements in powers. More recently, Soler expressed the displacements and stresses by series of Legendre polynomials [111].

Variational theorems have been employed to derive theories of arbitrary order: included are the membrane theory (zero order) and first approximation (first order). The principles of virtual displacements and virtual forces were employed by Wempner [114] and Oliveira [107,108], respectively.

Some comparisons of various shell theories and three-dimensional theories are found in the works of Klosner and Levine [104]. Sensenig [110], and Wunderlich [112].

Briefly, let us consider the simplest refinements that are needed to accommodate certain shells of practical importance. Perhaps the simplest is the sandwich, which consists of a soft core and very thin, but stiff, facings. The essential modification of the first approximation is the admission of transverse shear strain $(\gamma_{\alpha 3})$ and then the transverse extensional strain (γ_{33}); these three strain components define the relative displacements of the facings.

The essential modification is accomplished by the addition of the corresponding terms in eqs. (5a), (4), and (8):

$$\dot{\vec{A_3}} = \dot{\gamma}_{i3}\vec{A^i} + \dot{\vec{\phi}} \times \vec{A}_3 \tag{17}$$

$$\dot{w} = n^{\alpha\beta}\dot{\gamma}_{\alpha\beta} + m^{\alpha\beta}\dot{\kappa}_{\alpha\beta} + 2q^{\alpha}\dot{\gamma}_{\alpha 3} + t\dot{\gamma}_{33} \tag{18}$$

The additional stresses (q^{α} and t) are transverse shear and normal stresses. All theories of transverse shear deformations contain the counterparts of the strains $\gamma_{\alpha 3}$ and stress q^{α} appearing in eq. (18). Appropriate constitutive equations can be obtained by energy methods [21,109]. Equations were derived by Reissner with the theorem of stationary complementary energy [96]; these equations were subsequently recast by Reissner [101] and by Naghdi [102] with the theorem of Hellinger–Reissner [65,66].

In most cases a theory for thin multilayered shells can be developed upon the hypothesis that the displacement is piecewise linear between interfaces; then the theory for N layers contains $(6 + 3N)$ strains (and stresses). If transverse extensions are negligible, the number is reduced to $(6 + 2N)$ and, if the transverse shear strains are negligible, the number is again (6) (i.e., the Kirchhoff–Love hypothesis prevails). Curiously, a higher approximation for the homogeneous thick shells can be achieved by the same mechanism (i.e., a piecewise linear approximation through a sufficient number of hypothetical layers).

Theories of the conventional weak-core sandwich are given by Reissner [99] and Wempner [105,106]. More elaborate theories and additional references are presented by Librescu [50,51] and by Malcolm and Glockner [113].

ELASTIC–PLASTIC BEHAVIOR

Direct theory

As noted earlier, the choice of strains ($\gamma_{\alpha\beta}$, $\kappa_{\alpha\beta}$) and the identification of stresses ($n^{\alpha\beta}$, $m^{\alpha\beta}$) impose no restrictions on the material. A first approximation of the elastic–plastic shell is conceivable, by the direct method or by a derivation from three-dimensional theories.

According to classical plasticity, a yield condition is required in stress space ($n^{\alpha\beta}$, $m^{\alpha\beta}$). In 1948, Ilyushin proposed a yield condition in accord with the Mises criterion and a deformation theory based upon the Kirchhoff–Love hypothesis [117]. Specifically, he introduced three stress invariants:

$$Q_{NM} \equiv \tfrac{3}{2}m_{N\alpha\beta}m_M^{\alpha\beta} - \tfrac{1}{2}m_{N\alpha}^{\alpha}m_{M\beta}^{\beta} \tag{19}$$

where suffixes (N, M) represent the numbers 0 or 1, and[1]

$$m_0^{\alpha\beta} \equiv \frac{n^{\alpha\beta}}{2h\sigma_0}, \qquad m_1^{\alpha\beta} \equiv \frac{6}{\sqrt{3}}\frac{m^{\alpha\beta}}{h^2\sigma_0} \qquad (20a,b)$$

In our notations, Ilyushin's yield condition takes the form

$$F = Q_{00} + A|Q_{01}| + BQ_{11} = 1 \qquad (21)$$

The invariants Q_{NM} depend upon two invariants of the strains $(\gamma_{\alpha\beta}, \kappa_{\alpha\beta})$ while the coefficients A and B are constants. Rozenblum developed a particular form of (21) wherein $A = 0$ [118]. A useful comparison of these forms was given by Robinson [121]. Modifications of Ilyushin's theory were developed by Crisfield for steel shells [122].

A workable theory of elastic–plastic behavior has been conceived by Bieniek [124]. The essential features follow: the initial yielding occurs at the top $(+)$ or bottom $(-)$ surfaces under the condition:

$$F_0 \equiv Q_{00} + 2\sqrt{3}|Q_{01}| + 3Q_{11} = 1 \qquad (22)$$

If the material is ideally plastic, then there exists a limit condition $F_L = 1$. Bieniek also assumed a limit condition of the form given in eq. (21). To satisfy the limit condition under simple bending, $B = \frac{4}{3}$. One additional limiting state of stress is required to determine A; the state $(m_0^{11} = m_0^{22}, m_0^{12} = 0;$ $m_1^{11} = m_1^{22}, m_1^{12} = 0)$ of maximum Q_{01} provides the value $A = \frac{2}{3}$. As loading progresses, the yield condition changes; the surface $F_0(m_N^{\alpha\beta})$ approaches the limit surface $F_L(m_N^{\alpha\beta})$. A key step in Bieniek's development is his recognition that the surface translates in the hyperspace of moment $(m_1^{\alpha\beta})$; for example, in simple bending the yield condition shifts from the initial value $(1/\sqrt{3})$ to the limit value $(\sqrt{3}/2)$, but the elastic range is unchanged so that yielding occurs subsequently at the value $(-1/2\sqrt{3})$. Accordingly, the yield condition is taken in the form

$$F = Q_{00} + \tfrac{2}{3}|Q_{01}| + 3Q_{11}^* = 1 \qquad (23)$$

where

$$Q_{11}^* = Q_{11}(m_1^{\alpha\beta} - {}^*m_1^{\alpha\beta}) \qquad (24)$$

The quantities ${}^*m_1^{\alpha\beta}$ are called "hardening parameters" and account for the translation. The parameters evolve with plastic strain $(\dot{\kappa}_{\alpha\beta}^P)$; if the loading criteria are satisfied, as in classical plasticity,

$$F = 1, \qquad \frac{\partial F}{\partial m_N^{\alpha\beta}}\dot{m}_N^{\alpha\beta} > 0 \qquad (25a,b)$$

---

[1]To correlate Bieniek's direct theory and our derived theory, we adopt the definitions (20a,b). His dimensionless stress $(m_1^{\alpha\beta})$ and strain $(\kappa_{\alpha\beta})$ are altered by the factor $(1/\sqrt{3})$.

then,

$$*\dot{m}_{1\alpha\beta} = 2(1 - F_L)\frac{I_0}{I_1}\,\dot{\kappa}^P_{\alpha\beta} \tag{26}$$

where I_0 and I_1 are invariants of plastic strain rates. In the absence of plastic bending ($\dot{\kappa}^P_{\alpha\beta} = 0$), $I_1 = 0$; in the absence of plastic extension ($\dot{\gamma}^P_{\alpha\beta} = 0$), $I_0 = I_1$. The plastic strains follow the normality law and augment the elastic strains:

$$\dot{\gamma}_{\alpha\beta} = \dot{\gamma}^E_{\alpha\beta} + \dot{\lambda}\frac{\partial F}{\partial m_0^{\alpha\beta}}, \qquad \dot{\kappa}_{\alpha\beta} = \dot{\kappa}^E_{\alpha\beta} + \dot{\lambda}\frac{\partial F}{\partial m_1^{\alpha\beta}} \tag{27a,b}$$

The parameter $\dot{\lambda}$ is eliminated by means of eq. (26) and the following condition:

$$\dot{F} \equiv \frac{\partial F}{\partial m_N^{\alpha\beta}}\dot{m}_N^{\alpha\beta} + \frac{\partial F}{\partial *m_1^{\alpha\beta}}\,*\dot{m}_1^{\alpha\beta} = 0 \tag{28}$$

Then, in the usual manner the stresses and strains can be related by a tangent modulus.

Derived theory

A derivation from the three-dimensional equations of classical elasticity and plasticity entails several difficulties. Parts of the shell yield while adjoining parts remain elastic; interfaces between elastic and inelastic regions move. Upon unloading, elastic behavior again prevails; upon reloading, new regions of inelasticity are initiated and the existing regions are altered. With the initiation and evolution of inelastic regions, the distribution of stress through a section can become exceedingly complicated. The problems are to monitor the interfaces and to describe the stress distribution. We circumvent the former problem by employing an endochronic theory which entails no initial yield condition, hence no abrupt transition from the elastic to inelastic regions [120]. We treat the latter problem with a higher approximation of the stress distribution.

Following Ilyushin [117] and Pipkin and Rivlin [119], evolution of plastic strain is measured by an arc length ζ in the space of the strain deviator:

$$(\dot{\zeta})^2 = \dot{\bar{\gamma}}^{\alpha\beta}\dot{\bar{\gamma}}_{\alpha\beta} + \dot{\bar{\gamma}}^{\alpha}_{\alpha}\dot{\bar{\gamma}}^{\beta}_{\beta} \tag{29}$$

Following Valanis we introduce an alternative measure λ, such that $d\lambda/d\zeta > 0$ ($\zeta \geq 0$) [120]. We adopt a form that is particularly useful for our initial formulations:

$$\dot{\lambda} = \sqrt{\tfrac{3}{2}\sigma^n}\,\dot{\zeta}, \qquad \sigma \equiv \tfrac{3}{2}(\sigma^{\alpha\beta}\sigma_{\alpha\beta} - \tfrac{1}{3}\sigma^\eta_\eta\sigma^\mu_\mu) \tag{30),(31}$$

wherein σ is a second invariant of the stress deviator in plane stress ($\sigma^{33} = 0$). Our σ and σ^{ij} are dimensionless such that the yield condition of Mises corresponds to $\sigma = 1$. The plane stress–strain relation follows Prandtl–Reuss [115,116]:

$$\dot{\bar{\gamma}}_{\alpha\beta} = \dot{\bar{\gamma}}_{\alpha\beta}^{E} + \dot{\bar{\gamma}}_{\alpha\beta}^{P} = D_{\alpha\beta\gamma\eta}\dot{\sigma}^{\gamma\eta} + (\sigma_{\alpha\beta} - \tfrac{1}{3}\sigma_{\mu}^{\mu}G_{\alpha\beta})\dot{\lambda} \qquad (32\text{a,b})$$

wherein $D_{\alpha\beta\gamma\eta}$ is a dimensionless component of the flexibility tensor. The final term of eq. (32) is the plastic strain $\dot{\bar{\gamma}}_{\alpha\beta}^{P}$ which vanishes under the unloading condition:

$$\dot{\sigma}^{\alpha\beta}\dot{\bar{\gamma}}_{\alpha\beta}^{P} < 0 \qquad (33)$$

If n in eq. (30) is sufficiently large, the behavior approaches ideal plasticity ($\sigma \rightarrow 1$), so that the formulation admits comparison with the classical theory of the ideal material.

To represent the stress and strain distributions, we use the Legendre polynomials to benefit from the orthogonality:

$$\sigma^{\alpha\beta} \doteq \sqrt{1 + 2N}\, m_{N}^{\alpha\beta}P_{N}(\theta^{3}) \qquad (N = 0, 1, \ldots) \qquad (34)$$

As a first approximation we accept the linear distribution of strain:

$$\dot{\bar{\gamma}}_{\alpha\beta} = \dot{\epsilon}_{\alpha\beta}P_{0} + \sqrt{3}\,\dot{\kappa}_{\alpha\beta}P_{1} \qquad (35)$$

The strains $\epsilon_{\alpha\beta}$ and $\kappa_{\alpha\beta}$ are the dimensionless counterparts of the extensional and flexural strains of the usual first approximation.

In accordance with (34) and (35), the power of the stresses has the dimensionless form

$$\dot{w} = \tfrac{1}{2}\int_{-1}^{1} \sigma^{\alpha\beta}\dot{\bar{\gamma}}_{\alpha\beta}\sqrt{\frac{g}{a}}\, d\theta^{3} = m_{0}^{\alpha\beta}\dot{\epsilon}_{\alpha\beta} + m_{1}^{\alpha\beta}\dot{\kappa}_{\alpha\beta} \qquad (36\text{a,b})$$

According to eqs. (34) and (32b), the power has the alternative form:

$$\dot{w} \doteq D_{\alpha\beta\gamma\eta}m_{N}^{\alpha\beta}\dot{m}_{N}^{\gamma\eta} + m_{M}^{\alpha\beta}(m_{N\alpha\beta} - \tfrac{1}{3}m_{N\mu}^{\mu}a_{\alpha\beta})\dot{f}_{NM} \qquad (36\text{c})$$

where[2]

$$\dot{f}_{NM} = \frac{\sqrt{1 + 2N}\sqrt{1 + 2M}}{2}\int_{-1}^{1} P_{N}P_{M}\dot{\lambda}\, dz \qquad (37)$$

The balance of eqs. (36b) and (36c) is satisfied by the stress–strain relations

$$\dot{\epsilon}_{\alpha\beta} = \dot{e}_{\alpha\beta}^{0} + \dot{p}_{\alpha\beta}^{0}, \qquad \dot{\kappa}_{\alpha\beta} = \dot{e}_{\alpha\beta}^{1} + \dot{p}_{\alpha\beta}^{1}, \qquad 0 = \dot{e}_{\alpha\beta}^{N} + \dot{p}_{\alpha\beta}^{N} \qquad (38_{0,1,N})$$

[2]Terms of higher order in thinness (h/R) are omitted from the integrals of eqs. (36) and (37).

wherein

$$\dot{e}^N_{\alpha\beta} \equiv D_{\alpha\beta\gamma\eta}\dot{m}^{\gamma\eta}_N, \qquad \dot{p}^N_{\alpha\beta} \equiv (m_{M\alpha\beta} - \tfrac{1}{3}m^\mu_{M\mu}a_{\alpha\beta})\dot{f}_{MN} \tag{39$_{\text{E, P}}$}$$

In accordance with eq. (33), $\dot{f}_{MN} = \dot{p}^N_{\alpha\beta} = 0$ if

$$\dot{m}^{\alpha\beta}_N\dot{p}^N_{\alpha\beta} < 0 \tag{40}$$

The plots of Figs. 2 and 3 are the results of the derived theory, according to Wempner [125].

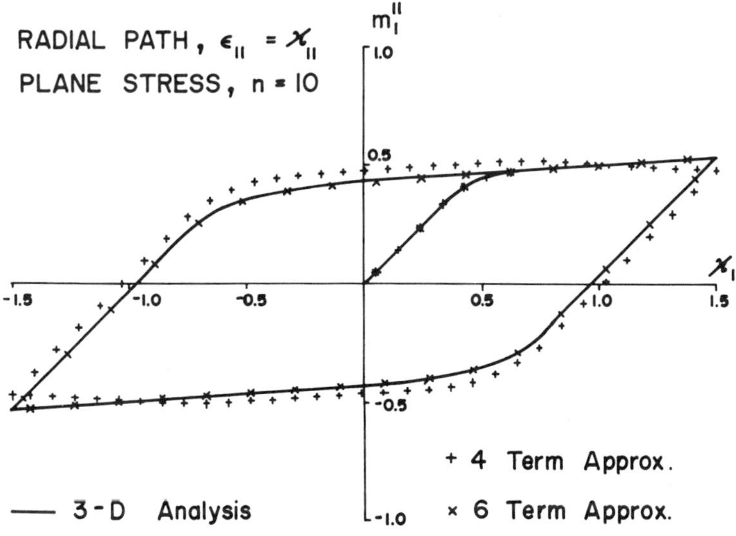

Figure 2

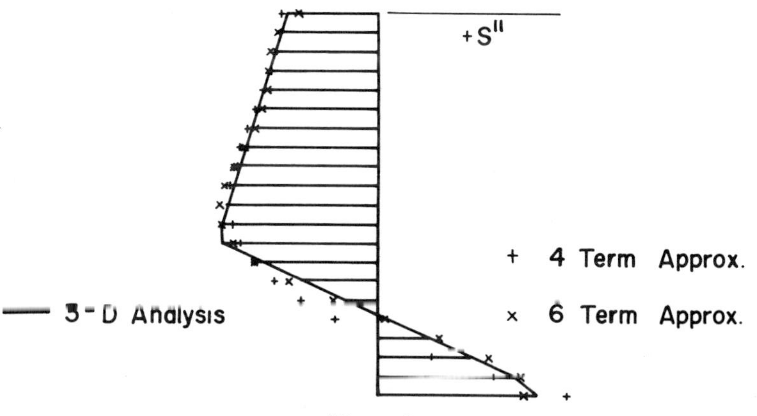

Figure 3

DISCRETE APPROXIMATIONS

Forms of approximation

A theory of shells expresses the problems of these thin bodies in terms of two-dimensional fields of displacements, strains, and stress, which are governed by partial differential equations and edge conditions, or by the stationary conditions of functionals. Formal solutions exist for simple problems, but many problems of real shells (e.g., the reinforced roofs of buildings or the hulls of space and marine vehicles) defy such solution. The alternative is an approximation. Three forms of approximation are widely used.

One form is the truncated Fourier series; such approximation of a function $W(\theta)$ of one variable θ has the form

$$W \doteq \sum_{K=0}^{M} A_K F_K(\theta) \tag{41}$$

where the support of the function $F_K(\theta)$ is generally the entire domain in question. Examples are the trignometric, Bessel, and Legendre functions.

A second form is Lagrange interpolation, which has the form

$$W \doteq \sum_{K=0}^{M} W_K L_K^M(\theta) \tag{42}$$

where $W_K \equiv W(\theta_K)$, L_K^M is a polynomial of degree M; as M increases, the support of L_K^M increases; for example, the interpolation ($M = 3$) spans four nodal values ($\theta_0 \leq \theta \leq \theta_3$).

A third form is Hermite interpolation, which takes the form

$$W \doteq \sum_{K=0}^{M} \sum_{N=0}^{Q} W_K^N H_{NK}^P(\theta), \qquad P = 2Q + 1 \tag{43}$$

where

$$W_K^N \equiv \left. \frac{\partial^N W}{\partial \theta^N} \right|_{\theta=\theta_K}, \qquad K = 0, \ldots, \dot{M}$$

The support of H_{NK}^P is confined to the adjoining nodes ($\theta_{K-1} < \theta < \theta_{K+1}$), *irrespective* of the order (P) of the approximation.

Forms such as (41), (42), and (43) are utilized as well in the approximations of our two-dimensional variables [131,133], but basic features are evident in these one-dimensional forms.

Approximations by Fourier expansions (41) have been employed via the methods of Rayleigh–Ritz [126,127] and Galerkin [128]; these have proven most effective in problems of vibrations and buckling [17,10,90].

Lagrange interpolation (42) is most often employed to obtain approximations via difference equations [131,133]. Higher-order interpolation and the corresponding difference equations involve more nodal values; the matrices of linear problems become denser and the equations at edges may be cumbersome.

The forms (41), (42), and (43) are distinguished by the character of the functions F_K, L_K^M, and H_{NK}^M. In general, the function F_K has global support; a variation of A_K produces variations throughout the shell. Consequently, the success of such approximations (41) often hinges upon the ability (or luck) of the user in preselecting suitable functions. Also, the form (41) is often ill-suited to describe phenomena of a local character. By contrast, the functions H_{NK}^M of (43), or L_K^2 of (42), are hill-like functions which vanish beyond the adjoining nodes $(\theta < \theta_{K-1}, \theta_{K+1} > \theta)$. Algebraic equations governing the nodal values can be obtained by the methods of Rayleigh–Ritz or Galerkin. Because these difference equations involve only values (W_K^N) at adjoining nodes, the procedure is well suited to phenomena of a local character, e.g., plastic versus elastic behavior in different parts of the shell.

The application of eq. (43) can be demonstrated by the problem of a beam. The displacement $W(\theta)$ provides a stationary value of the functional

$$U = \int_0^L [\tfrac{1}{2}(W_{,11})^2 + pW]\, d\theta \tag{44a}$$

Employing the second-order approximation $(N = 0, 1)$ and the method of Rayleigh–Ritz, we obtain two stationary conditions at each node $(\partial U/\partial W_N^0 = 0, \partial U/\partial W_N^1 = 0)$. The results are two difference equations; the first is a counterpart of the Euler equation $(W_{,1111} = -p)$.

As noted earlier, the hypothesis of Kirchhoff–Love, and the continuity of slope, can be replaced by a discrete counterpart [137,139]. The concept can be demonstrated again with the beam: if ϕ denotes the small rotation and Q a Lagrange multiplier, then the displacement W, rotation ϕ, and multiplier Q must render a stationary value of the modified functional

$$U = \int_0^L [\tfrac{1}{2}(\phi_{,1})^2 - Q(\phi - W_{,1}) + pW]\, d\theta \tag{44b}$$

Continuity of slope $(W_{,1})$ is no longer required, and the functions (W, ϕ, Q) can be approximated by linear interpolation $(L_K^0 = H_{0K}^0)$. Three stationary conditions hold at each node $(\partial U/\partial W_N = 0, \partial U/\partial \phi_N = 0, \partial U/\partial Q_N = 0)$. In the limit $(\theta_{K+1} \to \theta_K \leftarrow \theta_{K-1})$ the difference equations approach the familiar differential equations:

$$\frac{dQ}{d\theta} = p, \qquad \frac{d^2\phi}{d\theta^2} = -Q, \qquad \frac{dW}{d\theta} = \phi$$

Approximations of the form (43) can be employed in conjunction with the principles of virtual work and minimum potential or with the theorems of Hu, Washizu, Hellinger and Reissner. Accordingly, the displacements, stresses, or strains may be variables.

Finite elements

In the procedure of finite elements the shell is subdivided into discrete elements, quadrilateral or triangular. The displacements and/or stresses and/or strains are approximated within elements; usually, the approximation is accomplished by interpolation between nodes of the element that are usually at corners or edges. By means of a variational theorem and the methods of Rayleigh–Ritz, the displacements, stresses, and strains are expressed in terms of the nodal values. The shell is viewed as the assemblage of elements, wherein sufficient conditions of continuity must be imposed upon the displacements, rotations, and interactions in order to ensure convergence to the solution of the underlying theory ($\theta^{\alpha}_{N+1} \rightarrow \theta^{\alpha}_{N}$). With some exceptions, the procedure is another approach that leads to difference equations governing approximations of the form (43); exceptions include relaxation of continuity requirements along interelement edges.

In the approximations of elastic shells of revolution, the form (41) is often used for the circumferential coordinate, while a form of interpolation is used for the meridional coordinate. Similarly, the finite element may be a narrow frustrum of the shell, small only in the meridional dimension.

Several general features of the element approximations by polynomials are evident: (a) the element must be small compared to the pattern of deformations (unless the pattern can be anticipated); (b) the motion of the element can be decomposed into a rigid-body motion and a deformation; (c) in view of (a), the kinematic approximations of the shallow shell are applicable to the deformation of the element; and (d) the roles of the surface tensor in eqs. (3a) and (3b) dictate similar approximations of the undeformed and deformed surfaces.

Whatever the procedure, the discrete approximation must be accomplished by the solution of large systems of algebraic equations, linear or nonlinear. The latter are achieved by various techniques of prediction and correction via successive stepwise linear solutions [164–168].

LIMITATIONS

In practical applications, shells are subject to local conditions of loading and discontinuities; attachments, cutouts, and intersections produce stress concentrations, yielding, or fracture, which are not always predictable by a theory of surfaces. Still some local phenomena can be predicted by the theories of two dimensions. Certainly, parts of a shell experience nearly plane stress and the effect of an elliptical hole may be predicted by the Kolosoff solution for plane stress. The overall deflections, stress resultants, and buckling are usually predictable by the first approximation and often by a discrete approximation.

The first approximation may be viewed as a partial approximation of the three-dimensional body. Subsequent subdivision into finite elements produces three-dimensional elements of thickness h. In circumstances of local phenomena, more elements are needed through the thickness and/or across the surface, and in circumstances of very localized conditions, only a three-dimensional theory can suffice.

ROLES OF ANALYSIS AND COMPUTATION

From the inception of shell theory to the present, research has focused upon the formulation of theories and the formal solution of boundary-value problems. During the past two decades, the rapid development of the digital computer has allowed the numerical approximation of the equations and their solution. Moreover, the capacities of computers now admit the approximation of actual shells, reinforced and imperfect, elastic and inelastic.

Now, in addition to our concerns about the theories, their validity and accuracy, we must also question the discrete approximations, their validity and accuracy. As Budiansky and Sanders proposed a "best" first-order shell theory, we must seek the best discrete approximations and the best computational procedures. Perhaps this task will prove still more challenging as there are evidently many approximations that are valid and good if enough time and effort is expended in computation. It is hoped that the best approximations can be identified without excessive computational experimentation and attendant costs.

The traditional theories and analyses of three- and two-dimensional continua play essential roles; they often provide the only means to describe certain local phenomena. They provide the formal solutions to specific problems that are needed to assess the theories and approximations. The approximations are certainly needed for the host of problems not amenable to formal solutions.

Finally, the ultimate test of a theory or approximation is a confirmation by physical experimentation. The art of engineering involves the selection of the proper tool to predict a particular phenomenon.

REFERENCES

General

[1] LOVE, A. E. H., *The Mathematical Theory of Elasticity*, 1927, Dover, New York, 1944.

[2] FLÜGGE, W., *Statics and Dynamics of Shells*, Springer, Berlin, 1934.

[3] LUR'E, A. I., *Statics of Thin-Walled Shells*, Moscow, 1947; translation AEC-tr-3798, 1959.

[4] VLASOV, V. Z., *General Theory of Shells and Its Application in Engineering*, Moscow, 1949; translation NASA-TT-F-99, 1964.

[5] NOVOZHILOV, V. V., *The Theory of Thin Shells*, Moscow, 1951; translation P. Noordhoff, Leyden, The Netherlands, 1953.

[6] GOL'DENWEIZER, A. L., *Theory of Elastic Thin Shells*, Moscow, 1953; translation, Pergamon, New York, 1961.

[7] NASH, W. A., *Bibliography on Shells and Shell-Like Structures*, David Taylor Model Basin, Report 863, 1954.

[8] GREEN, A. E., AND ZERNA, W., *Theoretical Elasticity*, Oxford University Press, London, 1954; 2nd ed., 1968.

[9] MUSHTARI, KH. M., AND GALIMOV, K. Z., *Nonlinear Theory of Thin Elastic Shells*, USSR, 1957; translation NASA-TT-F-62, 1961.

[10] TIMOSHENKO, S., AND WOINOWSKY-KRIEGER, S., *Theory of Plates and Shells*, McGraw-Hill, New York, 1959.

[11] FLÜGGE, W., *Stresses in Shells*, Springer, Berlin, 1960.

[12] KOITER, W. T. (Editor), *The Theory of Thin Elastic Shells*, Proc. IUTAM Symp., Delft, North-Holland, Amsterdam, 1960.

[13] NAGHDI, P. M., *Foundations of Elastic Shell Theory* (in Progress in Solid Mechanics), Wiley, New York, 1963.

[14] OLZAK, W., AND SAWCZUK, A. (Editors), *Non-classical Shell Problems*, North-Holland, Amsterdam, 1964.

[15] DURGAR'YAN, S. M. (Editor), *Theory of Shells and Plates*, Proc. 4th All-Union Conf., 1964; translation NASA-TT-F-341, 1966.

[16] VOL'MIR, A. S., *Flexible Plates and Shells*, USSR; translation AFFDL-TR-66-216, 1967.

[17] KRAUS, H., *Thin Elastic Shells*, Wiley, New York, 1967.

[18] NIORDSON, F. I. (Editor), *Theory of Thin Shells*, Proc. IUTAM Symp., Copenhagen, Springer, Berlin, 1969.

[19] NAGHDI, P. M., *The Theory of Shells and Plates* (in Encyclopedia of Physics, Vol. 6.2), 1972.

[20] KOITER, W. T., AND SIMMONDS, J. G., *Foundations of Shell Theory*, Proc. 13th Int. Cong. Theo. Appl. Mech., 1972; Springer, Berlin, 1975.

[21] WEMPNER, G. A., *Mechanics of Solids with Applications to Thin Bodies*, McGraw-Hill, New York, 1973.

[22] FUNG, Y. C., AND SECHLER, E. E. (Editors), *Thin-Shell Structures*, Prentice-Hall, Englewood Cliffs, N.J., 1974.

[23] SIMMONDS, J. G., *Recent Advances in Shell Theory*, pp. 617–626, in Advances in Eng. Sci., NASA CP-2001, 1976.

[24] GOULD, P. L., *Static Analysis of Shells*, Lexington Books, Lexington, Mass., 1977.

The first approximation

[25] ARON, H., *Das gleichgewicht und Bewegung einer unendlich dünnen beliebig gekrümmten elastischen Schale*, Zeit. reine u. ang. Math., Vol. 78, 1874.

[26] LOVE, A. E. H., *On the Small Free Vibrations and Deformations of Thin Elastic Shells*, Phil. Trans. Roy. Soc. London, Ser. A., Vol. 179, pp. 491–546, 1888.

[27] COSSERAT, E., and COSSERAT, F., *Theorie des Corps Déformables*, Paris, 1909.

[28] LUR'E, A. I., *The General Theory of Thin Elastic Shells*, Prikl. Mat. Mekh., Vol. 4, No. 2, 1940.

[29] GOL'DENWEIZER, A. L., *The Equations of the Theory of Shells*, Prikl. Mat. Mekh., Vol. 4, No. 2, 1940.

[30] SYNGE, J. L., AND CHIEN, W. Z., *The Intrinsic Theory of Elastic Shells and Plates*, Th. von Kármán Anniv. Vol., pp. 103–120, 1941.

[31] SANDERS, J. L., *An Improved First-Approximation Theory for Thin Shells*, NASA Report 24, 1959.

[32] LEONARD, R. W., *Nonlinear First Approximation Thin Shell and Membrane Theory*, Thesis, Virginia Polytechnic Institute, 1961.

[33] KOITER, W. T., *A Consistent First Approximation in the General Theory of Thin Elastic Shells*, Proc. IUTAM Sym., Delft, 1959; North-Holland, Amsterdam, pp. 12–13, 1960.

[34] SANDERS, J. L., *Nonlinear Theories for Thin Shells*, Quart. Appl. Math., Vol. 21, pp. 21–36, 1963.

[35] NAGHDI, P. M., *A New Derivation of the General Equations of Elastic Shells*, Int. J. Eng. Sci., Vol. 1, pp. 509–522, 1963.

[36] KOITER, W. T., *On the Nonlinear Theory of Thin Elastic Shells*, Koninkl. Nederl. Akad. Wetenschap., Vol. 69, No. 1, pp. 1–54, 1966.

Edge conditions

[37] KIRCHHOFF, G., Über das gleichgewicht und die Bewegung einer elastichen Scheibe, J. Math. (Crelle), Vol. 40, pp. 51–58, 1850.

[38] THOMPSON, W., AND TAIT, P. G., *Treatise on Natural Philosophy*, Vol. 1, pt. 2, art. 645–648, 1867.

[39] FRIEDRICHS, K. O., *Kirchhoff's Boundary Conditions and the Edge Effect for Elastic Plates*, Proc. Sym. Appl. Math., Vol. 3, 1950.

[40] REISSNER, E., *Variational Considerations for Elastic Beams and Shells*, J. Eng. Mech. Div., ASCE, Vol. 88, No. 1, pp. 23–57, 1962.

[41] KOITER, W. T., *On the Dynamic Boundary Conditions in the Theory of Thin Shells*, Koninkl. Nederl. Akad. Wetenschap., Vol. 67, 1964.

[42] WEMPNER, G. A., *The Boundary Conditions for Thin Shells and Their Physical Meaning*, ZAMM, Vol. 47, No. 2, p. 136, 1967.

[43] HEIJDEN, A. M. A. VAN DER, *On Modified Boundary Conditions for the Free Edge of a Shell*, Doctoral Dissertation, Delft, 1976.

Constitutive equations of elastic shells

[44] REISSNER, E., *Stress Strain Relations in the Theory of Thin Elastic Shells*, J. Math. Phys., Vol. 31, pp. 109–119, 1952.

[45] BUDIANSKY, B., AND SANDERS, J. L., *On the "Best" First-Order Linear Shell Theory*, Progress in Applied Mechanics, Macmillan, New York, 1963.

[46] JOHN, F., *Estimates for the Derivatives of the Stresses in Thin Shell and Interior Shell Equations*, Comm. Pure Appl. Math., Vol. 18, pp. 235–267, 1965.

[47] NIORDSON, F. I., *A Note on the Strain Energy of Elastic Shells*, DCAMM, Rept. 9, Lyngby, Denmark, 1970.

[48] KOITER, W. T., AND SIMMONDS, J. G., *Foundations of Shell Theory*, Proc. 13th Intl. Cong. Theor. Appl. Mech., Springer, Berlin, pp. 150–176, 1972.

[49] AMBARTSUMYAN, S. A., *Theory of Anisotropic Shells*, Moscow, 1961; translation NASA-TT-F-118, 1964.

[50] LIBRESCU, L. I., *The Elasto-Statics and Kinetics of Anisotropic and Hetrogeneous Shell-Type Structures*, Bucarest, 1969; Noordhoff, 1975.

[51] LIBRESCU, L. I., *Some Results Concerning the Refined Theory of Elastic Multilayered Shells*, Mech. Appl., Vol. 20, No. 1,2,3, and 4, 1975.

[52] REISSNER, E., AND TSAI, W. T., *Pure Bending, Stretching, and Twisting of Anisotropic Cylindrical Shells*, J. Appl. Mech., 1972.

[53] REISSNER, E., AND TSAI, W. T., *On Pure Bending and Stretching of Orthotropic Laminated Cylindrical Shells*, J. Appl. Mech., Vol. 41, No. 1, pp. 168–172, 1974.

[54] NOOR, A. K., AND CAMIN, R. A., *Symmetry Considerations for Anisotropic Shells*, Comp. Methods Appl. Mech. Eng., Vol. 9, North-Holland, Amsterdam, pp. 317–335, 1976.

[55] NIORDSON, F. I., *A Consistent Refined Shell Theory*, DCAMM, Rept. 104, Lyngby, Denmark, 1976.

[56] TEICHMANN, G., AND SCHUMANN, W., *Some Remarks Concerning Linear Theory of Heterogeneous Orthotropic Shells*, Acta Mech., Vol. 25, pp. 291–299, 1977.

Methods of derivation

[57] GOODIER, J. N., *On the Problems of the Beam and the Plate in the Theory of Elasticity*, Trans. Roy. Soc. Canada, Vol. 32, pp. 65–88, 1938.

[58] JOHNSON, M. W., AND REISSNER, E., *On the Foundations of the Theory of Thin Elastic Shells*, J. Math. Phys., Vol. 37, pp. 371–392, 1958.

[59] GREEN, A. E., *On the Linear Theory of Thin Elastic Shells*, Proc. Roy. Soc., Ser. A, Vol. 266, pp. 143–160, 1962.

[60] GOL'DENWEIZER, A. L., *Construction of an Approximate Thin Shell Theory by Means of Asymptotic Integration of the Equations of Elasticity*, Prikl. Mat. Mekh., Vol. 27, pp. 593–608, 1963.

[61] RUTTEN, H. S., *Asymptotic Approximation in the Three-Dimensional Theory of Thin and Thick Elastic Shells*, Proc. 2nd IUTAM Symp. Shell Theory, pp. 115–134, Copenhagen, 1969.

[62] REISSNER, E., *On Consistent First Approximations in the General Linear Theory of Thin Elastic Shells*, Ing.-Arch., Vol. 40, pp. 402–419, 1971.

[63] HU, H. C., *On Some Variational Principles in the Theory of Elasticity and Plasticity*, Scientia Sinica, Vol. 4, 1955.

[64] WASHIZU, K., *On the Variational Principles of Elasticity and Plasticity*, Tech. Report 25-18, MIT, 1955.

[65] HELLINGER, E., *Die Allgemeinen Ansätze der Mechanik der Kontinua*, Encykl. Math. Wissen., Vol. 4, No. 4, 1914.

[66] REISSNER, E., *On a Variational Theorem in Elasticity*, J. Math. Phys., Vol. 29, 1950.

[67] REISSNER, E., *On the Foundations of Generalized Linear Shell Theory*, IUTAM Symp., Copenhagen, 1967; Theory of Thin Shells, Springer, Berlin, 1967.

[68] WEMPNER, G. A., *Discrete Approximations of Elastic-Plastic Bodies by Variational Methods* (Variational Methods in Engineering, edited by C. A. Brebbia and H. Tottenham, Southampton University Press), 1973.

Theories of thin elastic shells

[69] REISSNER, H., *Spannungen in Kugelschalen (Kuppeln)*, Festschrift Müller-Breslau, p. 181, 1912.

[70] MEISSNER, E., *Das Elastizitätsgesetz für dünne Schalen von Ringflächen Kugel, oder Kegelform*, Phys. Zeitschr., Vol. 14, p. 343, 1913.

[71] DONNELL, L. H., *Stability of Thin-Walled Tubes Under Torsion*, N.A.C.A., TR 479, 1934.

[72] MARGUERRE, K., *Zur Theorie der Gekrümmten Platte grosser Formänderung*, Proc. 5th Int. Cong. Appl. Mech., pp. 93–101, 1938.

[73] REISSNER, E., *On Axisymmetrical Deformations of Thin Shells of Revolution*, Proc. Symp. Appl. Math., Vol. 3, McGraw-Hill, New York, pp. 27–52, 1950.

[74] LUKASIEWICZ, S., *On the Technical Theory of Non-shallow Shells*, Arch. Budowiy Maszyn, Vol. 15, 1968.

[75] SANDERS, J. L., JR., *On the Shell Equations in Complex Form*, Proc. 2nd Symp. Thin Elastic Shells, Springer, Berlin, pp. 135–136, 1969.

[76] WAN, F. Y. M., *The Exact Reduction of Equations of Elastic Shells of Revolution*, Studies Appl. Math., Vol. 48, pp. 361–375, 1969.

[77] SIMMONDS, J. G., *Simplification and Reduction of the Sanders-Koiter Linear Shell Equations for Various Midsurface Geometries*, Quart. Appl. Math., Vol. 28, pp. 259–275, 1970.

[78] STEELE, C. R., *Membrane Solutions for Shells with Edge Constraint*, J. Eng. Mech. Div., ASCE Vol. 100, pp. 497–510, 1974.

[79] PIETRASZKIEWICZ, W., *Lagrangean Non-Linear Theory of Shells*, Arch. Mech. (Polish), Vol. 26, No. 2, pp. 221–228, 1974.

[80] LATTA, G. E., AND SIMMONDS, J. G., *The Sanders-Koiter Shell Equations Can Be Reduced to Two Coupled Equations for All Minimal Midsurfaces*, Quart. Appl. Math., Vol. 33, No. 2, 1975.

[81] SIMMONDS, J. G., *Rigorous Expunction of Poisson's Ratio from the Reissner–Meissner Equations*, Int. J. Solids Structures, Vol. 11, pp. 1051–1056, 1975.

Buckling of thin elastic shells

[82] DONNELL, L. H., *A New Theory for the Buckling of Thin Cylinders Under Axial Compression and Bending*, Trans. Amer. Soc. Mech. Eng., Vol. 56, p. 795, 1934.

[83] KÁRMÁN, TH. VON, AND TSIEN, H. S., *The Buckling of Thin Cylindrical Shells Under Axial Compression*, J. Aero. Sci., Vol. 8, p. 303, 1941.

[84] KOITER, W. T., *On the Stability of Elastic Equilibrium*, Thesis, Delft, 1945; translation NASA TTF-10, 1967.

[85] KEMPNER, J., *Postbuckling Behavior of Axially Compressed Circular Cylindrical Shells*, J. Aero. Sci., Vol. 21, p. 329, 1954.

[86] YOSHIMURA, Y., *On The Mechanism of Buckling of a Circular Cylindrical Shell Under Axial Compression*, NACA TM-1390, 1955.

[87] DONNELL, L. H., *Effect of Imperfections on Buckling of Thin Cylinders Under External Pressure*, J. Appl. Mech., 1956.

[88] ALMROTH, B. O., *Postbuckling Behavior of Axially Compressed Circular Cylinders*, AIAA, Vol. 1, p. 630, 1963.

[89] ALMROTH, B. O., AND STARNES, J. H., JR., *The Computer in Shell Stability Analysis*, J. Eng. Mech. Div., ASCE, Vol. 101, No. EM6, pp. 873–888, 1975.

[90] BRUSH, D. O., AND ALMROTH, B. O., *Buckling of Bars, Plates, and Shells*, McGraw-Hill, New York, 1975.

[91] KOITER, W. T., *Current Trends in the Theory of Buckling*, IUTAM Symp., Cambridge, 1974; Buckling of Structures, Springer, Berlin, 1976.

[92] TVERGAARD, V., *Buckling Behavior of Plate and Shell Structures*, IUTAM Cong., Delft, 1976; Theoretical and Applied Mechanics (edited by W. T. Koiter), North-Holland, Amsterdam, 1976.

[93] BUDIANSKY, B. (Editor), *Buckling of Structures*, IUTAM Symp., Cambridge, 1974; Springer, Berlin, 1976.

Higher approximations

[94] LUR'E, A. I., *The General Theory of Thin Elastic Shells*, Prikl. Mat. Mekh., Moscow, Vol. 4, No. 7, 1940.

[95] BYRNE, R., *Theory of Small Deformations of Thin Elastic Shell*, Seminar in Mathematics, N.S. 2, University of California, Los Angeles, pp. 103–152, 1944.

[96] REISSNER, E., *The Effects of Transverse Shear Deformation on the Bending of Elastic Plates*, J. Appl. Mech., Vol. 12, 1945.

[97] NEUBER, H., *Algemeine Schalentheorie I, II*, Zeit. Ang. Math. Mech., Vol. 29, No. 4,5, pp. 97–108, 142–146, 1949.

[98] HILDEBRAND, F. B., REISSNER, E., AND THOMAS, G. B., *Notes on the Foundations of the Theory of Small Displacements of Orthotropic Shells*, NACA-TN-1833, 1949.

[99] REISSNER, E., *Small Bending and Stretching of Sandwich-Type Shells*, NACA 975, Washington, D.C., 1950.

[100] GREEN, A. E., AND ZERNA, W., *The Equilibrium of Thin Elastic Shells*, Quart. J. Mech. Appl. Math., Vol. 3, pp. 9–22, 1950.

[101] REISSNER, E., *Stress-Strain Relations in the Theory of Thin Elastic Shells*, J. Math. Phys., Vol. 31, p. 109, 1952.

[102] NAGHDI, P. M., *On the Theory of Thin Elastic Shells*, Quart. Appl. Math., Vol. 14, p. 369, 1957.

[103] ZERNA, W., *Mathematisch strenge Theorie elasticher Schalen*, Zeit. Ang. Math. Mech., Vol. 7, pp. 333–341, 1962.

[104] KLOSNER, J. M., AND LEVINE, H. S., *Comparison of Elasticity and Shell Theory Solutions*, Proc. 2nd Aerospace Sciences Meeting, AIAA, 1965.

[105] WEMPNER, G. A., AND BAYLOR, J. L., *A Theory of Sandwich Shells*, Proc. 8th Midwestern Mech. Conf., 1963; Developments in Mechanics, Pergamon, London, 1965.

[106] WEMPNER, G. A., *Theory of Moderately Large Deflections of Sandwich Shells with Dissimilar Facings*, Int. J. Solids Struct., Vol. 3, pp. 367–382, 1967.

[107] OLIVEIRA, E. R. DE ARANTES E, *A Theory of Shells Involving Moments of Arbitrary Order*, Nat. Lab. Civil Eng., Lisbon, 1967.

[108] OLIVEIRA, E. R. DE ARANTES E, *A Further Research on the Analysis of Shells Using Moments of Arbitrary Order*, Nat. Lab. Civil Eng., Lisbon, 1967.

[109] WEMPNER, G. A., *Plate Theories via Energy Potentials*, Zeit. Ang. Math. Mech., Vol. 48, pp. 463–470, 1968.

[110] SENSENIG, C. B., *A Shell Theory Compared with the Exact Three Dimensional Theory of Elasticity*, Int. J. Eng. Sci., Vol. 6, No. 8, p. 435, 1968.

[111] SOLER, A. I., *Higher-Order Theories for Structural Analysis Using Legendre Polynomial Expansions*, J. Appl. Mech., Vol. 36, No. 4, pp. 757–762, 1969.

[112] WUNDERLICH, W., *A Method for the Analysis of Thick Elastic Shells and Comparisons Between Various Approximations for Thin Shells*, Proc. Int. Colloquium, IASS, Madrid, 1969.

[113] MALCOLM, D. J., AND GLOCKNER, P. G., *Nonlinear Sandwich Shell Theory and Cosserat Surface Theory*, J. Eng. Mech. Div., ASCE, Vol. 98, pp. 1183–1203, 1972.

[114] WEMPNER, G. A., *Invariant Multicouple Theory of Shells*, J. Eng. Mech. Div., ASCE, pp. 1397–1415, 1972.

Elastic-plastic behavior

[115] PRANDTL, L., *Spannungsverteilung in Plastischen Körper*, Proc. 1st Int. Cong. Appl. Mech., Delft, Vol. 34, 1924.

[116] REUSS, A., *Berücksichtigung der elastischen Formänderungen in der plastizitäts Theorie*, Zeit. Ang. Math. Mech., Vol. 10, 1930.

[117] ILYUSHIN, A. A., Plasticity (in Russian), Gostekhizdat, Moscow, 1948; Plasticity (in French), Eyrolles, Paris, 1956.

[118] ROZENBLUM, V. I., *An Approximate Theory of the Equilibrium of Plastic Shells*, Prikl. Mat. Mekh., Vol. 18, 1954.

[119] PIPKIN, A. C., AND RIVLIN, R. S., *Mechanics of Rate Independent Materials*, Zeit. Ang. Math. Mech., Vol. 16, 1965.

[120] VALANIS, K. C., *A Theory of Plasticity Without a Yield Surface*, Arch. Mech., Warsaw, Vol. 23, No. 4, 1971.

[121] ROBINSON, M., *A Comparison of Yield Surfaces for Thin Shells*, Int. J. Mech. Sci., Vol. 13, pp. 345–354, 1971.

[122] CRISFIELD, M. A., *On an Approximate Yield Criterion for Thin Steel Shells*, TRRL Lab., Rep. 658, Berkshire, Great Britain, 1974.

[123] WEMPNER, G., AND ABERSON, J. A., *A Formulation of Inelasticity from Thermal and Mechanical Concepts*, Int. J. Solids Structures, Vol. 12, pp. 705–721, 1976.

[124] BIENIEK, M. P., AND FUNARO, J. R., *Elastic-Plastic Behavior of Plates and Shells*, Tech. Rep. DNA 3954T, Weidlinger Associates, New York, 1976.

[125] WEMPNER, G., AND HWANG, C. M., *A Derived Theory of Elastic-Plastic Shells*, Int. J. Solids Structures, Vol. 13, pp. 1123–1132, 1977.

Discrete approximations

[126] RAYLEIGH, LORD, *Theory of Sound*, 1894; Dover, New York, 1945.

[127] RITZ, W., *Über eine neue Methode zur Lösung gewisser Variations Probleme der Mathematischen Physik*, Zeit. Reine Ang. Math., Vol. 135, pp. 1–61, 1908.

[128] GALERKIN, B. G., *Series Expansions for Some Cases of Equilibrium of Plates and Beams* (in Russian), Wjestnik Ing., Petrograd, 1915.

[129] ARGYRIS, J. H., *Energy Theorems and Structural Analysis*, Aircraft Eng., Vol. 26, pp. 347–356, 1954; Vol. 27, pp. 42–58, 8–94, 125–134, 145–158, 1955.

[130] ARGYRIS, J. H., AND KELSEY, S., *Structural Analysis by the Matrix Force Method with Applications to Aircraft Wings*, Wiss. Ges. Luftfahrt Jahrb., pp. 78–98, 1956.

[131] HILDEBRAND, F. B., *Introduction to Numerical Analysis*, McGraw-Hill, New York, 1956.

[132] TURNER, M. J., CLOUGH, R. W., MARTIN, H. C., AND TOPP, L. J., *Stiffness and Deflection Analysis of Complex Structures*, J. Aero. Sci., Vol. 23, pp. 805–823, 1956.

[133] COLLATZ, L., *The Numerical Treatment of Differential Equations*, 3rd ed., Springer, Berlin, 1960.

[134] Clough, R. W., *The Finite Element Method in Plane Stress Analysis*, J. Struct. Div., ASCE; Proc. 2nd Conf. Elect. Comp., pp. 345–378, 1960.

[135] Argyris, J. H., and Kelsey, S., *Energy Theorems and Structural Analysis*, Butterworth, London, 1961.

[136] Goldberg, J. E., *Computer Analysis of Shells*, Proc. Sym. Theory Shells (to honor L. H. Donnell, edited by D. Muster); pp. 3–22, 1967.

[137] Wempner, G. A., *New Concepts for Finite Elements of Shells*, Zeit. Ang. Math. Mech., Vol. 48, pp. 174–176, 1968.

[138] Johnson, M. W., Jr., and Mclay, R. W., *Convergence of the Finite Element Method in the Theory of Elasticity*, J. Appl. Mech., Vol. 35, No. 2, pp. 274–278, 1968.

[139] Wempner, G., *Finite Elements, Finite Rotations and Small Strains of Flexible Shells*, Int. J. Solids Structures, Vol. 5, pp. 117–153, 1969.

[140] Strang, G., *The Finite Element Method and Approximation Theory*, Symp. Num. Sol. Part. Diff. Eq., University of Maryland, 1970.

[141] Argyris, J. H., *The Impact of the Digital Computer on the Engineering Sciences*, Aero. J., Vol. 74, No. 709, No. 710, pp. 13–41, pp. 111–127, 1970.

[142] Zienkiewicz, O. C., *The Finite Element Method: From Intuition to Generality*, App. Mech. Rev., Vol. 23, No. 3, pp. 249–256, 1970.

[143] Fraeijs de Veubeke, B. (Editor), *High Speed Computing of Elastic Structures*, Proc. IUTAM Symp., Liege Press, 1971.

[144] Forsberg, K., and Hartung, R., *An Evaluation of Finite Difference and Finite Element Techniques for General Shells*, Proc. IUTAM Symp. (edited by B. Fraeijs de Veubeke), Vol. 2, pp. 837–859, 1971.

[145] Key, S. W., and Beisinger, Z. E., *The Analysis of Thin Shells by the Finite Element Method*, Proc. IUTAM Symp., Liege Press, 1971.

[146] Argyris, J. H., and Scharpf, D. W., *Finite Element Theory of Plates and Shells Including Shear Strain Effects*, Proc. IUTAM Symp., Liege Press, 1971.

[147] Bushnell, D., Almroth, B. O., and Brogan, F., *Finite-Difference Energy Method for Nonlinear Shell Analysis*, J. Computers Structures, Vol. 1, pp. 361–387, 1971.

[148] Hartung, R. F. (Editor), *Computer Oriented Analysis of Shell Structures*, AFFDL-TR-71-79, 1971.

[149] Atluri, S., and Pian, T. H. H., *Theoretical Formulation of Finite Element Methods in Linear-Elastic Analysis of General Shells*, J. Struct. Mech., Vol. 1, No. 1, pp. 1–41, 1971.

[150] Taylor, R. L., *On Completeness of Shape Functions for Finite Element Analysis*, Int. J. Num. Meth. Eng., Vol. 4, No. 1, pp. 17–22, 1972.

[151] Argyris, J. H., and Lochner, N., *On the Application of the SHEBA Shell Element*, Comp. Methods Appl. Mech. Eng., Vol. 1, pp. 317–347, 1972.

[152] Wilson, E. L., Taylor, R. L., Doherty, W. P., and Ghaboussi, J., *Incompatible Displacement Models* (in Numerical and Computer Methods in Structural Mechanics, edited by Fenves et al.), Academic Press, New York, 1973.

[153] ARGYRIS, J. H., HASSE, M., AND MALEJANNAKIS, G. A., *Natural Geometry of Surfaces with Specific Reference to the Matrix Displacement Analysis of Shells, I, II and III*, Koninkl. Nederl. Akad. Wetenschap., Vol. 76, 1973.

[154] LEONARD, J. W., AND LI, C-T., *Strongly Curved Finite Element for Shell Analysis*, J. Eng. Mech. Div., ASCE, pp. 515–535, 1973.

[155] TONG, P., AND PIAN, T. H. H., *Postbuckling Analysis of Shells of Revolution by the Finite-Element Method; Thin Shell Structures*, Fung, Y. C., and Sechler, E. E., editors, pp. 436–452, Prentice-Hall, Englewood Cliffs, N.J., 1974.

[156] BUSHNELL, D., *Thin Shells*, Proc. Symp. Struct. Mech. Comp. Prog. (edited by Pilkey, W., Saczalski, K., and Schaeffer, H.), University of Virginia Press, Charlottesville, Va., 1974.

[157] FRIED, I., *Finite-Element Analysis of Thin Elastic Shells with Residual Energy Balancing and the Role of the Rigid Body Modes*, J. Appl. Mech., 1975.

[158] GALLAGHER, R. H., *Shell Elements* (in Finite Element Methods in Structural Mechanics), Proc. World Congress on Finite Elements, 1975.

[159] LEVINE, H. S., WINTER, R., ARMEN, H., JR., AND PIFKO, A., *Application of The Finite Element Method to Inelastic Shell Analysis*, Proc. N.S.F. Symp. Approx. and Num. Meth. Inelastic Shells, Georgia Tech., 1975.

[160] NOOR, A. K., AND MATHERS, M. D., *Shear-Flexible Finite-Element Models of Laminated Composite Plates and Shells*, NASA TN D-8044, 1975.

[161] MATSUI, T., AND MATSUOKA, O., *A New Finite Element Scheme for Instability Analysis of Thin Shells*, Int. J. Num. Meth. Eng., Vol. 10, pp. 145–170, 1976.

[162] GALLAGHER, R. H., *Finite Element Structural Analysis: An Overview*, ASCE Annual Convention, Preprint 2765, 1976.

[163] ASHWELL, D. G., AND GALLAGHER, R. H. (Editors), *Finite Elements for Thin Shells and Curved Members*, Wiley, New York, 1976.

Computational methods

[164] YAGHMAI, S., AND POPOV, E., *Incremental Analysis of Large Deflections of Shells of Revolution*, Int. J. Solids Structures, Vol. 7, pp. 1375–1393, 1971.

[165] BELYTSCHKO, T., AND HSIEH, B. J., *Nonlinear Transient Analysis of Shells and Solids of Revolution by Convected Elements*, AIAA/ASME/SAE Structures Conf., AIAA Paper 73-359, 1973.

[166] BIENIEK, M. P., FUNARO, J., AND BARON, M. L., *Numerical Analysis of the Dynamic Response of Elastic-Plastic Shells*, Tech. Rep. No. 20, Weidlinger Associates, New York, 1976.

[167] ALMROTH, B. O., MELLER, E., AND BROGAN, F. A., *Computer Solutions for Static and Dynamic Buckling of Shells*, Proc. IUTAM Symp., Cambridge, 1974; Springer, Berlin, 1976.

[168] BUSHNELL, D., *A Strategy for the Solution of Problems Involving Large Deflections, Plasticity and Creep*, Int. J. Num. Meth. Eng., Vol. 11, pp. 683–708, 1977.

STRUCTURAL DESIGN WITH WOOD— AN OVERVIEW

*James R. Goodman**

INTRODUCTION

The objective of this paper is to provide an overview of the current and projected state-of-the-art for structural design with wood. Background information concerning the future availability of wood and its renewable nature will be provided. Special concepts relating to the biological origin of wood and the effects of variability on structural behavior will be presented. Structural systems of wood will be discussed along with the special role wood plays in housing and light-frame construction. Finally, a few brief comments will be provided concerning educational problems faced by the wood industry and those faced by society regarding efficient use of this critical resource.

A quote by Baxter from a recent presentation at an international conference on World Resources serves to set the stage for the subsequent discussions of the structural use of wood [3].

In contrast with most human material needs timber is not scarce but it is clear that mankind must behave sensibly in managing its timber resources to

*Professor of Civil Engineering, Colorado State University, Fort Collins, Colorado.

avoid unfortunate consequences in the future. Perhaps a better way to put it is that, in the case of timber, there is still time to plan its development and use in a way that will provide for man's foreseeable needs without robbing posterity of its rightful inheritance. As a material, timber is full of virtues. Its development and use causes little pollution; it is versatile, easy to handle and work, light in weight, resilient, durable and beautiful.

WOOD AS A RENEWABLE RESOURCE

Timber is perhaps the nation's most abundant renewable construction material resource. The actual tonnage of wood produced today in the United States exceeds the total tonnage of all other construction materials [31]. Renewable resources in the form of forest and agricultural products are used in large quantities (4.4 billion tons in 1972) for a variety of industrial uses [6]. Their uses for housing and other structural purposes, paper and paperboard, textiles, chemical feedstocks and fuel constitute one of America's largest industrial sectors, and one that has continuously grown [31].

Since the United States operates essentially in the context of the world wood supply rather than the national situation, at least for the long term, it is worthwhile to take a brief view of the world situation. Softwood species are preferred worldwide for many uses. They are the principal sources of renewable structural materials in the form of lumber, plywood, and wood composites, such as particle board. The principal softwood-producing areas are the United States, Europe, Canada, and the USSR. Some supplies are available in areas such as Australia, New Zealand, Chile, and Brazil. To be more specific, the following quote by Baxter provides some revealing statistics concerning total timber (hardwood and softwood) supply and consumption [3]:

> About one-third of the land area of the world is under forests and these contain some 330×10^9 m³ of timber. Slightly more than half of this timber is in America where Latin America has the greater share. The USSR has about a fifth, Asia nearly 15%, Africa 10% and Europe less than 4%. In 1971 total removals were about 2.4×10^9 m³, so the resources represent about 140 years' supply at the 1971 rate of usage. It has been estimated that the rate will rise 50% by 1985.

> Growing times vary enormously with species and climate but an overall average of 70 years is probably of the right order and it could be argued that man can double his rate of usage and still keep within the capacity of the earth to keep up the supply indefinitely.

However, before rejoicing over this seemingly happy state of affairs, one should consider several aspects of the situation from the U.S. viewpoint. A number of factors cloud the simple numbers that can be estimated using current measures. Of serious concern is the amount of land being removed from usable production and the future political decisions that may cause this trend to continue. Unlike most other resource-based industries, the forest products industry owns only a small percentage of its raw material [26]. As shown in Fig. 1, more than half the total inventory of standing softwood sawtimber is federally owned. The uncertainty concerning the management and control of these vital timber-growing lands is a serious problem affecting the future availability of wood for structural and other uses.

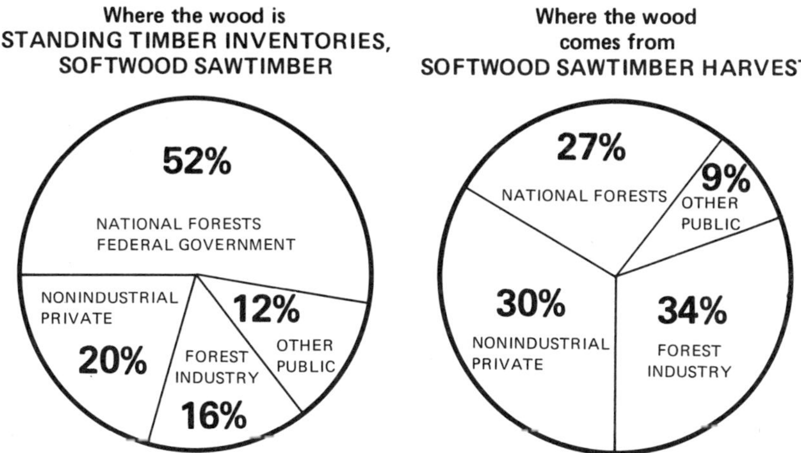

Figure 1 Softwood timber inventory and production in the United States (after Hodges [26]).

Other factors include insect attack and fires, which threaten large areas of timber. In addition, much of the growing timber used in making world projections is found in regions of Canada and the USSR that are virtually inaccessible, at least with today's technology and cost structure. Further, the use of fertilizer, projected as part of the increased growth rate potential of tomorrow's forests, may not be feasible in view of the dwindling non-renewable petroleum resources. Thus, while we possess much of the needed technical competence to manage our forests more intensively (*the role of foresters*), many factors, such as land ownership, costs of reforestation, thinning, fire protection, and other unknowns, may preclude application of this technology.

In addition, from the U.S. viewpoint, our needs are clearly growing at an ever-increasing rate. If the United States were to look to Canada and the USSR for meeting these needs rather than to develop its own unrealized softwood-growing potential, it would be forced to compete with the growing number of other importing industrial nations for its supply. Obviously, this situation would create serious price dislocations.

The commodity basis for lumber prices is clearly evident by referring to Fig. 2, which traces the recent decade, showing housing starts versus lumber prices [26]. Housing is clearly of prime concern in any discussion of timber use, as can be seen in Fig. 3 [26]. More discussion of the importance of wood structural design associated with housing will be presented later.

Engineering solutions can provide at least a partial answer to a possible

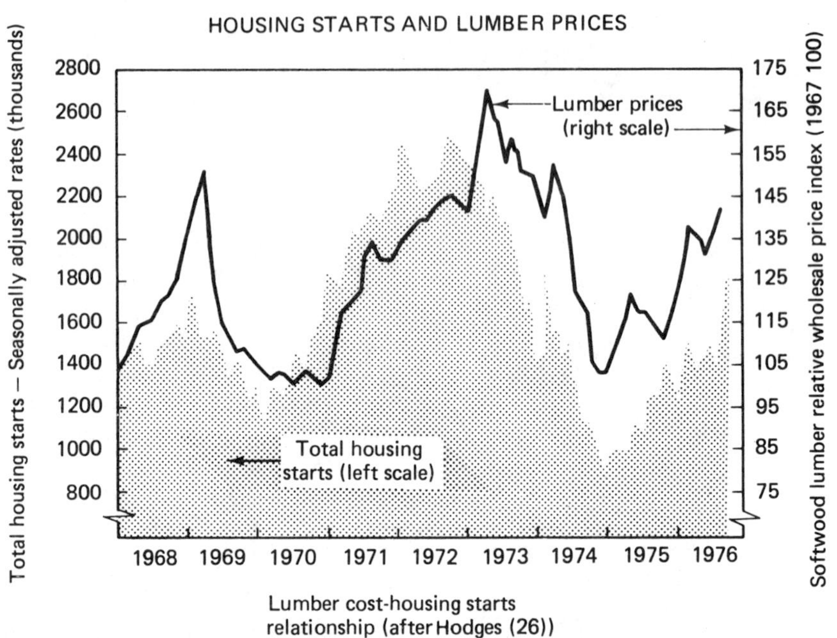

Figure 2 Lumber company–housing starts relationship (after Hodges [26]).

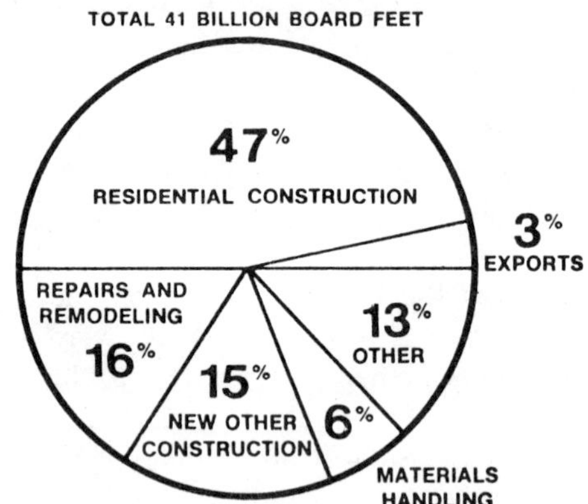

Figure 3 Softwood lumber consumption pattern (after Hodges [26]).

long-range shortage of wood for structural use. Estimates of the potential for saving wood material through proper design range upward of one-third without sacrificing structural performance [6]. Translated to an equivalent gain in overall forest productivity, this amounts to as much as 15%. It should be noted that this gain is immediately available while gains due to forest management may take at least 70 years in some growing areas. The need to utilize efficiently every fiber of wood which can be grown with proper environmental and safety constraints (*the role of the engineer*) cannot be overemphasized. This aspect of structural design with wood will be stressed in this overview.

WOOD AS A STRUCTURAL MATERIAL

To understand the behavior of wood as a structural material it is important to keep in mind the basic characteristics of the wood substance and its biological nature. Varying conditions of geographical location, soil type, precipitation, and so on, all contribute to the variation in material properties important to structural design. These facts, coupled with species differences and a variety of available wood products, create a diverse array of material considerations for designers using wood. Table 1 provides a summary of some of the broad categories of advantages and disadvantages for wood and wood-based materials use in structures. Not included are nonstructural considerations, such as aesthetics, energy consideration of material production, and renewability of the timber resource, as discussed previously.

As an example illustrating the concepts listed in Table 1, consider the strength-and-stiffness-to-weight ratios for wood as compared with some other

Table 1

BASIC CONCEPTS RELATING TO USE OF WOOD IN STRUCTURES

Advantages	*Disadvantages*
1. High load-carrying capacity (along the grain) per unit weight	1. Affected by moisture changes
2. Low cost per unit weight	2. Susceptible to decay and insect attack if not treated
3. Raw material easily shaped	3. Strength properties are time dependent
4. Easy workability with simple tools	4. Joints, although easy to construct, are difficult to make fully effective
5. Wide variety of products available	5. Large inherent variation
6. Heavy timber construction has good fire resistance	6. Orthotropic material properties
7. High stiffness to weight ratios	7. Tendency to brittle behavior in tension
8. Low energy for construction and in end use	8. Low absolute stiffness compared to metals

materials. For selected structural materials, a comparison is made in Table 2 of stiffness parameters as well as bending and tension load parameters. These results illustrate the strength-to-weight ratio relationships which place wood in perspective with other structural materials. While absolute strength and stiffness are considerably lower than for the metals, ratios with weight compare very favorably as shown.

Table 2

STIFFNESS AND STRENGTH PARAMETERS FOR STRUCTURAL MATERIALS

	Density,	Bending Stiffness		Bending Strength		Tensile Strength	
	D	E		MOR	$MOR/$	σ_u	
Material	(g/cm^3)	$(kips/cm^3)$	E/D^3	$(kips/cm^2)$	D^2	$(kips/cm^2)$	σ_u/D
Mild steel	7.80	2,110,000	4,425	4,650	76.4	4,650	596
Duraluminium (95% Al)	2.80	703,000	32,020	3,860	492.3	3,860	1,378
Concrete	2.30	175,000	14,380	50	9.5	28	12.2
Nylon	1.10	1,750	1,315	915	756.2	1,050	955
Douglas-fir	0.48	137,000	1,238,000	860	3,733	860	1,792
Spruce (Engelmann)	0.34	91,500	2,328,000	610	5,277	660	1,941

Designers with wood must also be aware of the orthotropic nature of wood. This complex material constitutive law gives rise to many phenomena in structural behavior which set wood apart in its design and resulting performance. An extensive study of the orthotropic behavior of wood and and its three-dimensional action under compressive loading has been made by Goodman and Bodig [12], which was also summarized in *Wood Structures* [16] more recently. A method of predicting elastic parameters for wood using only measurements of along-the-grain stiffness was given by Bodig and Goodman [4,5] and the results are reproduced here as Table 3. Poisson's ratios were found to be reasonably constant for most softwood species and are also shown in Table 3. Use of the predicted values provides the analyst with a reasonable method of evaluating the required elastic parameters for use in finite element studies or for other analysis methods. Greater emphasis on three-dimensional structures, with complex connection details, makes it mandatory to consider more sophisticated analyses than for the simple post-and-beam structures of the past.

Another inherent concern regarding use of wood for structural applications relates to its variability. Designers familiar with other materials, but not with wood, may find this feature of wood structural design confusing. Typical variability is represented in Fig. 4, which illustrates a histogram for

Table 3

PREDICTED ELASTIC PARAMETERS AS FUNCTIONS OF MODULUS
OF ELASTICITY ALONG THE GRAIN FOR SOFTWOODS (10^6 psi)

E_L	E_R	E_T	G_{LR}	G_{LT}	G_{RT}
1.0	0.1073	0.0584	0.0978	0.0921	0.00814
1.2	0.1155	0.0661	0.1018	0.0957	0.00905
1.4	0.1230	0.0733	0.1057	0.0993	0.00990
1.5	0.1265	0.0768	0.1077	0.1011	0.01030
1.6	0.1298	0.0802	0.1097	0.1029	0.01069
1.8	0.1362	0.0869	0.1137	0.1065	0.01145
2.0	0.1421	0.0933	0.1177	0.1101	0.01216
2.2	0.1477	0.0995	0.1216	0.1137	0.01285
2.4	0.1530	0.1055	0.1256	0.1173	0.01352
2.6	0.1581	0.1113	0.1296	0.1209	0.01416

Average Poisson's Ratio Values

Poisson's Ratio	Softwoods
v_{LR}	0.37
v_{LT}	0.42
v_{RT}	0.47
v_{TR}	0.35
v_{RL}	0.041
v_{TL}	0.033

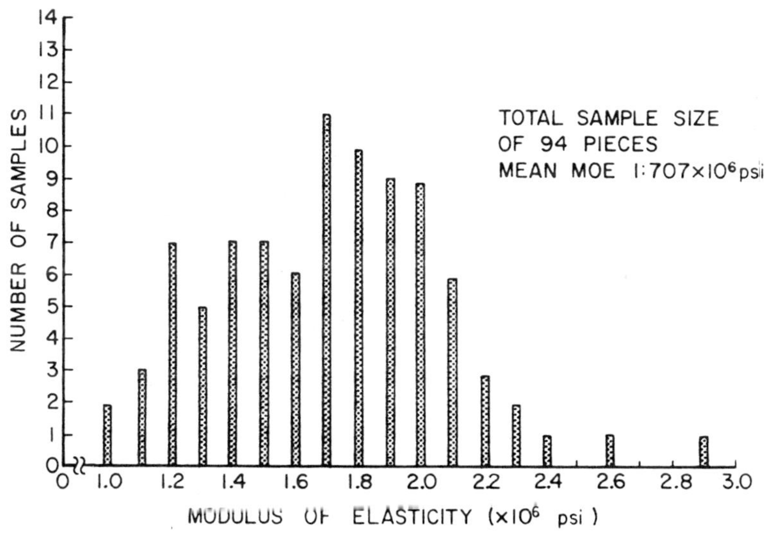

Figure 4 Histogram of modulus of elasticity of visually graded lumber (after Dawson [7]).

127

modulus of elasticity (MOE) taken from a sample of visually graded lumber [7]. Extensive studies of the variability of lumber properties, of clear wood properties, and even of glulam beam behavior, have been conducted over the years. Generally, design values have been set for strength at the lower 5% exclusion limit while stiffness properties have been taken to be the mean values. Designers must be aware of the consequences of these methods of assigning design strength and stiffness. The proper method of recognizing the full import of the variability of wood on structural safety and performance is through reliability-based design methods. Extensive research is currently under way in this area of endeavor [2,7,48] and more discussion of these methods will be presented later.

Other basic characteristics of wood should be considered more extensively, such as the tendency of wood, especially in the perpendicular-to-grain direction, to exhibit brittle tensile behavior. Some additional references to these properties will be made in relation to a subsequent section on structural systems with wood.

This brief overview of wood as a structural material serves only to introduce a complex subject area. Designers with wood should make every attempt to develop a "feel" for the material and its unique behavior. Most of the structural problems that arise with wood are a result of a lack of understanding of variability, moisture effects, orthotropic behavior, or other basic concepts which are a part of the biological nature of the material.

STRUCTURAL DESIGN CONSIDERATIONS WITH WOOD

Fundamental design problems

Basic concepts of wood as a structural material presented in Table 1 give rise to certain fundamental design problems for structures of wood. These design problems and subsequent specification and construction concerns are summarized in Table 4. Examples will be chosen to illustrate certain key features of design with wood which are not common to other structural materials. It is possible in this paper to present only selected examples of special design considerations. For more details concerning wood structural design the reader is referred to *Wood Structures* [16] and the *Timber Construction Manual* [42], or to a design textbook such as those of Hoyle [27], Gurfinkel [23], or Andersen [1].

Nonrigid connections for wood members–an example design problem

An area of structural design with wood which often is most difficult is that of connections. Connections in wood structures may be rigid or semirigid. Examples of rigid connections are the rigid adhesives, used in glulam beams

Table 4

DESIGN, SPECIFICATIONS, AND CONSTRUCTION CONSIDERATIONS
FOR WOOD STRUCTURES

Design	*Specifications*	*Construction*
1. Duration of load effects	1. Large number of products, species, grades, and sizes	1. Lightweight material easily erected
2. Orthotropic properties	2. Limited technical data on some products	2. Lightweight creates special problems during partially erected construction stage
3. Semirigid nature of connections	3. Diversity of wood industry requires knowledge of many agencies and products groups	3. Prefabrication is easily done and resulting components are lightweight
4. Exposure conditions	4. Complex material behavior in structural systems may not be adequately covered by codes	4. Large structural components may create shipping problems
5. Variety of products, species, sizes, and grades		5. On-site changes require only simple tools
6. Joint details		
7. Variability		
8. Size effects		

and for glued plywood gussets in some truss systems. Semirigid joints usually result from mechanical connectors and/or elastomeric adhesives. Semirigid joints lead to complex structural analysis problems even for rather simple residential roof trusses which are produced by the thousands each day [39]. For layered systems fastened with nails, for example typical housing wood joist floor systems, an incomplete composite behavior results which requires an extensive finite-element analysis for complete description of its structural behavior [40]. Studies to account for the incomplete composite behavior due to semirigid connections in wood systems have been conducted by many investigators, as will be described subsequently. An extensive treatment of composite behavior in wood structural systems is given by Goodman and Gutkowski [17] as part of a book on the composite behavior of structural systems.

To illustrate the concept of incomplete composite action in layered wood systems, consider a T-beam as shown in Fig. 5. The behavior of this beam depends on the stiffness of the interlayer connection used in its construction. As shown in Fig. 4(c), the strain distribution reflects the existence of interlayer slip. Analysis of this type of problem has been studied by several researchers [22,32,33] and was continued and expanded by Goodman [9] under the direction of the honoree of this symposium, Dr. E. P. Popov, with whom a paper was subsequently published [10]. Since that time, an extensive series of studies have been conducted for beams, columns, floor systems, walls, roofs, and so on [11,15,28–30,34–36,40,41,45–47].

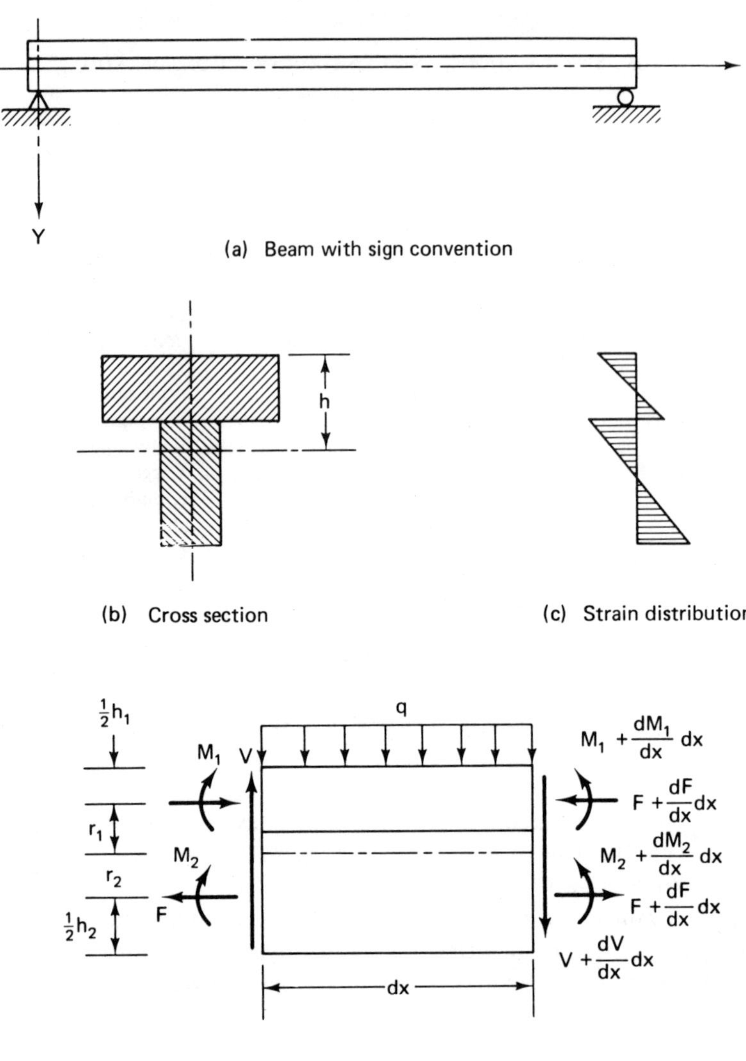

(a) Beam with sign convention

(b) Cross section

(c) Strain distribution

(d) Beam element

Figure 5 Layered wood beam system with incomplete composite behavior.

Beams such as shown in Fig. 5 with nailed or elastomeric connections behave as illustrated in Fig. 6. Analysis, including the effects of gaps in sheathing, is done using finite-element analysis [40].

One of the primary design areas where the concepts of composite action in wood systems may be applied is that of residential housing. Current design methods for houses constructed of wood generally lag behind those for other materials. For example, the beneficial aspects of the composite T-

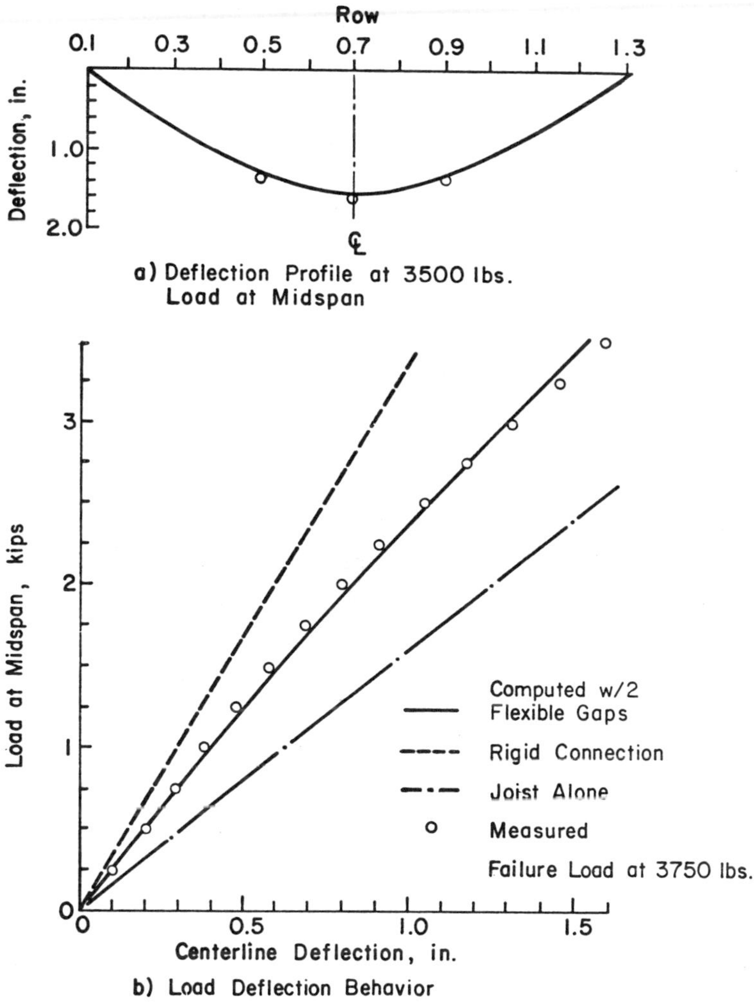

Figure 6   Example T-beam behavior for incomplete composite action.

beam and two-way action (load sharing) of wood joist systems which results from the presence of sheathing are neglected. Thus, floors designed using these assumptions usually have excessive strength and their stiffness cannot be accurately predicted. For nailed wood joist system construction, as well as for elastomerically glued construction, the assumption of either rigid connections or no connection is unrealistic for evaluating composite action. Only a recognition of the partial degree of interaction that results between each of the layers (sheathing and joist layers) will give correct results for system behavior.

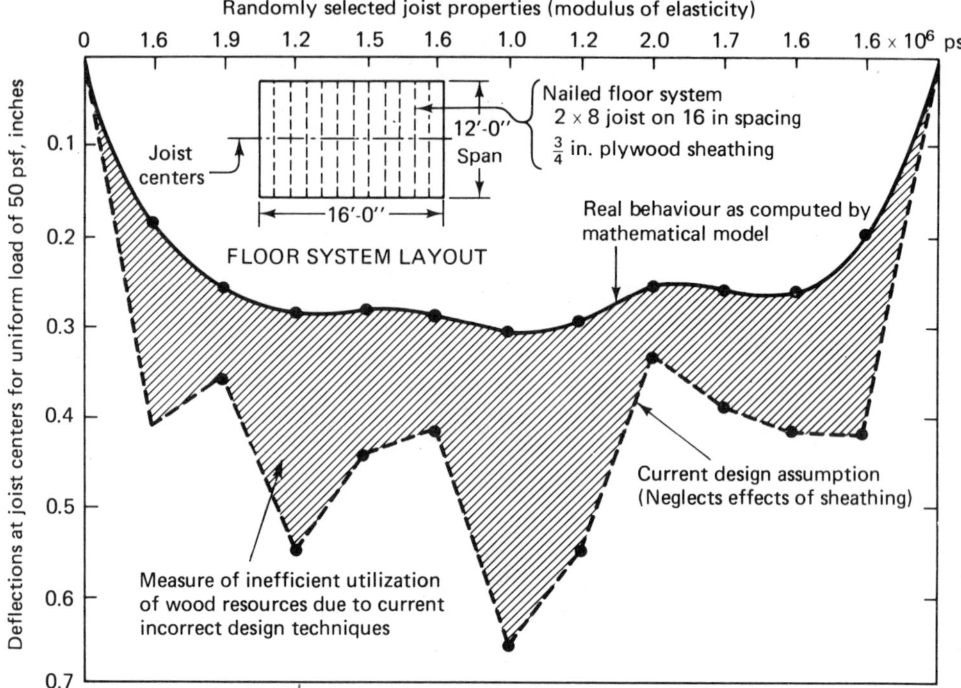

Figure 7 Design versus real behavior of wood joist floor system with varying joist properties.

As an example of the type of behavior in wood joist systems, consider floor analysis, shown in Fig. 7. Using a finite-element model developed at Colorado State University, an analysis was made of a typical 2 × 8 joist system with $\frac{3}{4}$-in. plywood sheathing [41]. To illustrate the difference between the current design approach and the results of applying the rational analysis procedure incorporating composite and two-way behavior, the example given in Fig. 7 is offered. This example shows the degree of difference possible between the two concepts for design in terms of deflection behavior. In current design each joist is required to carry its individual load, calculated by the assumptions of neglecting composite and two-way behavior. The predicted deflection computed using the rational analysis procedure is shown to be as small as one-half that given by the present simplified method. Numerous studies have been conducted which verify the mathematical model [15,29,30,44–47]. Design procedures are now being developed to incorporate the true behavior of complete composite systems. Similar work is being conducted at Oregon State University on wall design [34] and for glulam bridges at Colorado State University [24].

Performance of wood structures in wind, earthquakes, and fire

This symposium is especially concerned about structural performance under conditions of wind and earthquake as well as fire. An aspect of the behavior of wood systems and their connectors is a generally excellent performance under earthquake and wind loadings. The same action which leads to loss of composite action and rigidity, that of semirigid connections, tends to allow the wood-frame system to distort without tearing apart. The interlayer motions and the joint rotations keep the stresses low enough in the frame to avoid stress-actuated failures. A recent article by Toumi has a summary assessment of light wood-frame buildings under the 1964 Alaska earthquake, the May 6, 1975, Omaha tornadoes, and other potentially damaging phenomena, such as hurricane winds [43]. His general conclusions were that, with proper attention to roof-to-wall connections and wall-to-floor connections, conventionally designed wood light-frame structures will withstand all but the most severe situations.

A series of investigations and recommendations for proper storm-resistant construction were made following hurricane Camille. Results of these studies are well documented in references by Stern [38], by publications of the Southern Pine Association [37], and by Zornig and Sherwood [49]. Research should be conducted concerning the many types of light-frame structures built of wood, however, to ensure that such systems have adequate diaphragm action and other types of lateral stability. The light weight of wood structures can also lead to serious problems during the erection sequence if proper bracing is not designed for each phase of the construction sequence.

The behavior of wood and wood systems in fire depends in large measure on the size of the members involved. The slow rate of fire penetration in thick wood members is one of the basic assets of wood which has been accepted and utilized for many years in heavy timber construction. The fire penetration rate of thinner wood members and panel products is considerably higher than that of thick wood members under severe fire conditions. For such systems, sprinkler protection and the possible use of fire retardants should be considered.

A brief look at wood and fire may serve to provide some basic understanding of the resulting behavior of wood structural systems in this environment. A comprehensive treatment of the subject is presented by Schaffer in *Wood Structures* [16]. A brief summary of this presentation will be given here. Other information concerning fire in relation to wood structures may be found in the *Timber Construction Manual* [42].

Wood surfaces are observed to char prior to ignition. The time to ignite wood with a small flame varies with temperature, from about one-half

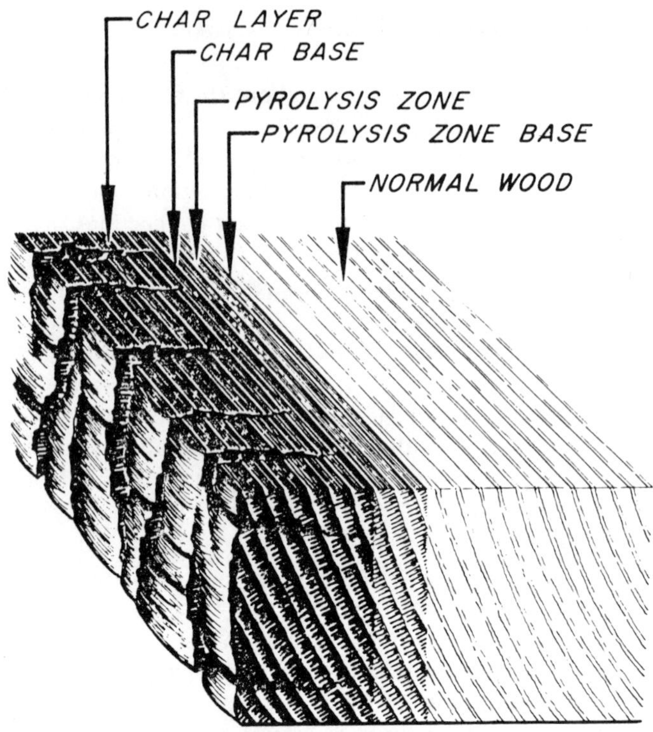

Figure 8 Degradation zones in wood exposed to fire on one surface (after Schaffer [16]).

minute at 800°F to 14 to 40 minutes at 350°F. With continued exposure to flame, the char layer increases in thickness. The char base is characterized by a temperature of about 550°F. The char provides an effective insulative effect and the rate of charring decreases as the char layer builds. Figure 8 shows the degradation zones in a solid wood member exposed to fire on one surface. It should be noted that the temperature of the wood even as near as $\frac{1}{2}$ inch to the char base is only about 200°F (for Douglas-fir), and thus the strength loss of the remaining cross section of wood member is not appreciable.

The particularly satisfactory performance of heavy timber construction with either solid sawn or glulam members is of special importance to structural engineers. Heavy timber construction is defined by Schaffer [16] as listed in Table 5. Most building codes recognize that the performance of heavy timber construction is similar to that of noncombustible construction with a 1-hour or higher fire endurance and generally permits its use in all fire districts for all types of occupancy.

A recent case demonstrating the performance of heavy-timber construction is the Filene Center for the Performing Arts, located at Wolf Trap Farm,

Table 5

HEAVY TIMBER CONSTRUCTION
(MINIMUM SIZES OF MEMBERS)

Member	*Inches, Nominal*
Columns	
Supporting floor loads	8 by 8
Supporting roof and ceiling loads only	6 by 8
Floor framing	
Beams and girders	6 wide by 10 deep
Arches and trusses	8 in any dimension
Roof framing, not supporting floor loads	
Arches springing from grade	6 by 8 lower half
Arches, trusses, other framing springing from top of walls, etc.	4 by 6
Floor (covered with 1-in. nominal flooring, $\frac{1}{2}$-in. plywood, or other approved surfacing)	
Splined or tongue-and-groove plank	3
Planks set on edge	4
Roof decks	
Splined or tongue-and-groove plank	2
Plank set on edge	3
Tongue-and-groove plywood	$1\frac{1}{8}$

Virginia. The roof framing plan on this project consisted of systems of unique glulam and steel queen-post and king-post trusses with spans ranging from 150 to 60 feet. Exposed 2×6 roof decking, 2-inch sheathing on wall surfaces, and steel H-columns, boxed-in-wood, completed the framing system. Just prior to completion of construction, a fire started at the base of the steel H-columns and spread quickly. Heat from the fire caused the H-columns to buckle, and steel plate girders were replaced with new steel trusses as a result of the fire. The condition of the major glulam members after the fire revealed a charring of the outside of the members. It was determined that these members were still satisfactory for their original purpose and after sand blasting were reinstalled.

Clearly, much additional research is needed to establish the fire performance of all types of wood structural systems. It has been shown, for example, that use of sprinkler systems may be more critical than for some other type of materials of construction.

Nonprismatic glulam beams—an example of structural mechanics applications in wood design

A myriad of types of nonprismatic glulam beams and arches are constructed. Double-tapered pitched and/or curved beams provide a sloping roof and maximum clearance within a building and are often chosen for aesthetic

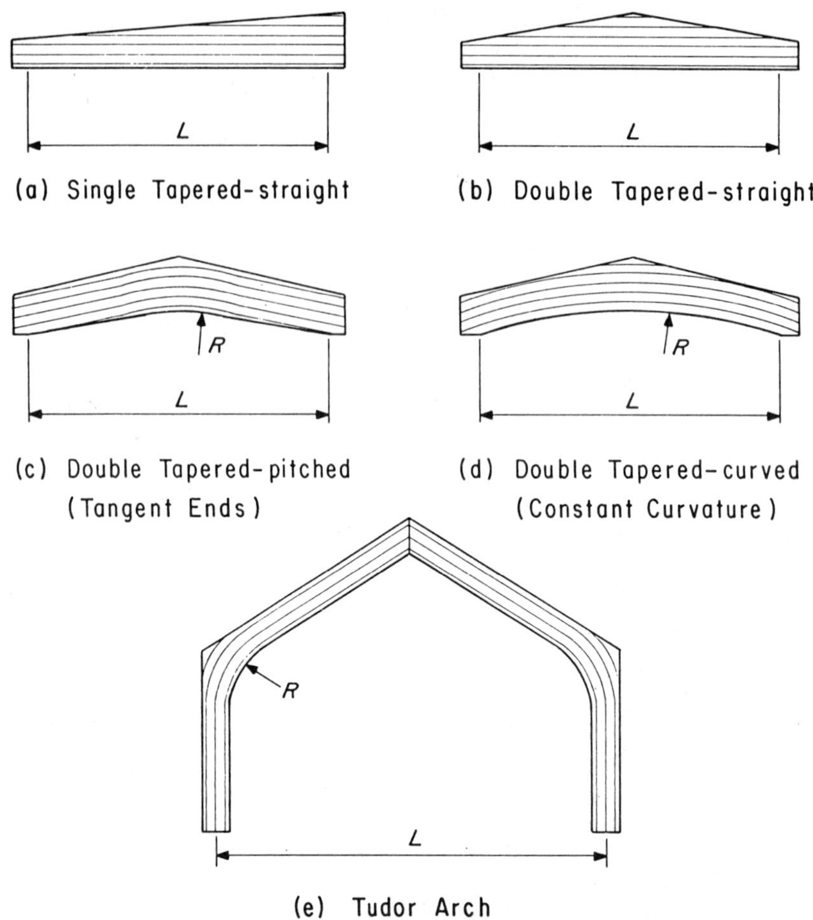

Figure 9   Nonprismatic glulam configurations.

considerations. Glulam arches are constructed in many shapes; however, probably the most widely used is the three-hinged tudor arch. Configurations of several types of nonprismatic glulam members are shown in Fig. 9.

The analysis of all these types of nonprismatic members present similar problems. Recently, a series of experimental and analytical studies has been conducted at Colorado State University in which the concepts of structural mechanics were applied in order to develops new design procedures for such members.

Interest in these problems was prompted by in-service, tension-perpendicular-to-grain separations which occurred in some pitched, tapered, and curved beams, as illustrated in Fig. 10. Thus tests and analysis of these types of members were conducted to recommend design procedures to avoid these field service problems.

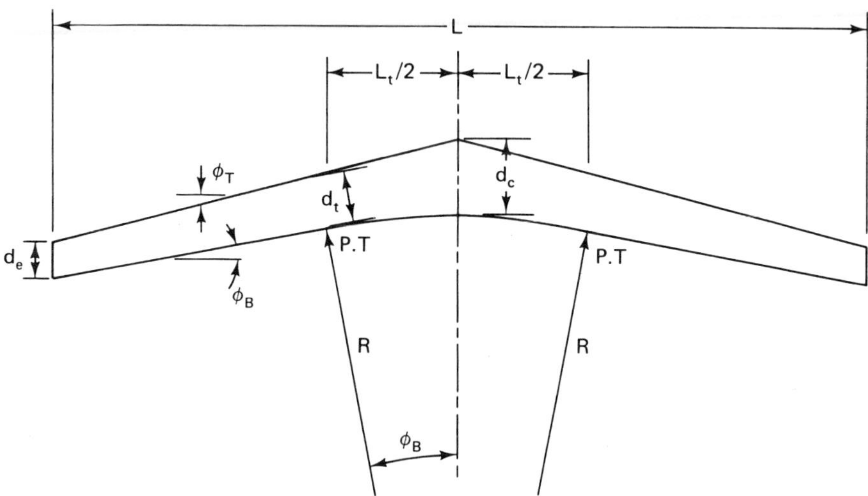

Figure 10 Double-tapered, pitched, and curved glulam beam.

The central portion of the member was analyzed using a special trapezoidal, orthotropic, plane stress finite element technique [18,19]. A special strain gaging method using strain transducers attached to the glulam test members was used to obtain experimental verification of predicted strain distribution [20]. A typical result of these measurements is shown in Fig. 11 for radial stress, the critical stress causing perpendicular-to-grain tensile failure. The result of the test program and the associated analyses confirmed that for Douglas-fir members, the failure mechanism was initiated only by radial stress and was inherently a brittle-fracture behavior.

As a result of the investigation on pitched, tapered, and curved glulam beams, a new design method has recently been approved by the American Institute of Timber Construction [21]. This procedure will replace a method that is presently in use, based on approximate formulas for stress distribution extrapolated from tapered beams. In addition, the concept that wood fails by distortion energy failure theory will be replaced by recognition of the brittle nature of tension failure perpendicular-to-grain. The new design procedure actually results in more efficient (less volume) wood members as opposed to members designed by the present method. Extension of these methods to single- and double-tapered straight soffit members and to tudor arches is currently under study.

This example illustrates several important considerations related to wood design. It is probably most important to recognize that concepts used in the design of ductile, isotropic materials such as steel cannot be directly translated to wood, an orthotropic material that exhibits brittle fracture behavior in at least some tensile modes. Much greater attention must be given to the inherent behavior of wood as a structural material to ensure proper, safe, and reliable designs.

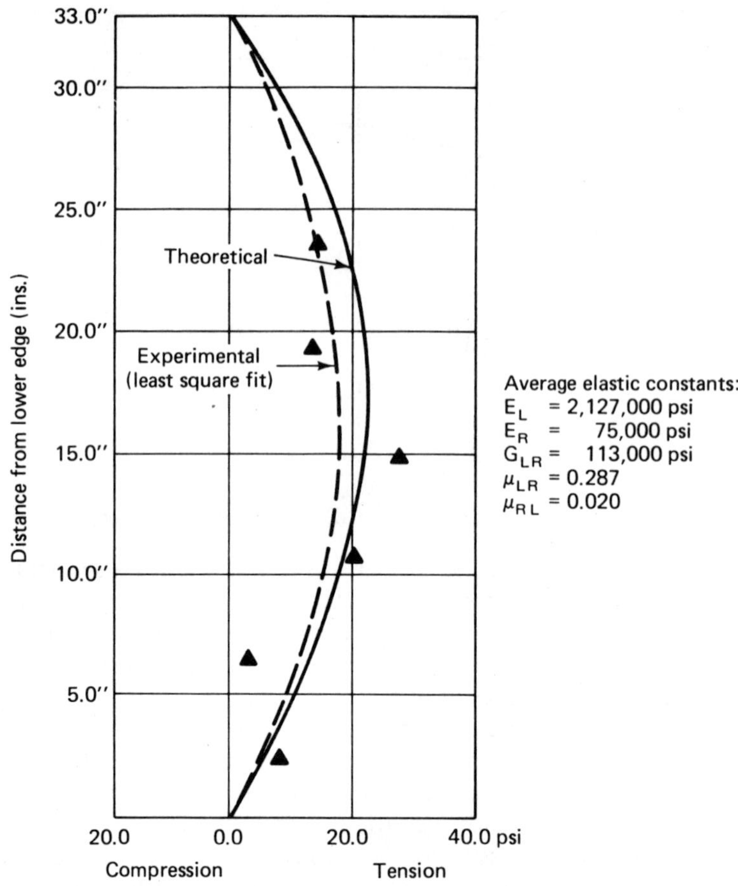

Figure 11   Radial stress distribution at the center line for a pitched, tapered, and curved glulam test beam (pitch 3 in 12).

Reliability and limit states design— new methods recognize the variability of wood

It has been established around the world that some form of combined Limit States Design (LSD) concept along with recognition of the probabilistic nature of material behavior and loading is a more rational design procedure than current methods. Wood design has long incorporated, albeit through a rather simplified approach, variability in strength behavior. A unified, consistent approach, which would also recognize variation in loading as well as the material resistance in the safety equation, is at this time under study [2,48]. Much information is missing on in-grade behavior of wood materials, on variations in loading history (especially for residential structures), and on resistance of systems as opposed to single members. Thus considerable research will be required to bring these more rational design methods to fruition.

As an example of some of the current research in this area, housing floor systems, wall systems, and trusses are being studied for their probabilistic performance as structural systems. These wood systems are subject to variable material strength and stiffness input for individual members, which in turn affect system performance. The common method of attack for these problems is the Monte Carlo approach, wherein a large number of structural systems (trusses, floors, or walls) are "constructed" in the computer. Member input for strength or stiffness is entered by random selection from known distributions of properties for individual members for the grade of wood presumed to be used in the structure. An example of the results obtained by such an analysis is given by Dawson and Goodman [7] and is shown here as Fig. 12. The influence of variability on performance (in this case maximum floor deflection) can be clearly seen. Increased variability of materials results in a greater probability of poor performance. It should be noted that the load was held constant in this analysis; thus the true safety or performance index was not obtained.

There are many implications of the use of reliability-based designs for wood structures. A means for seeking optimum economic efficiency through restricting joist MOE variability, by nondestructive testing, for example, to

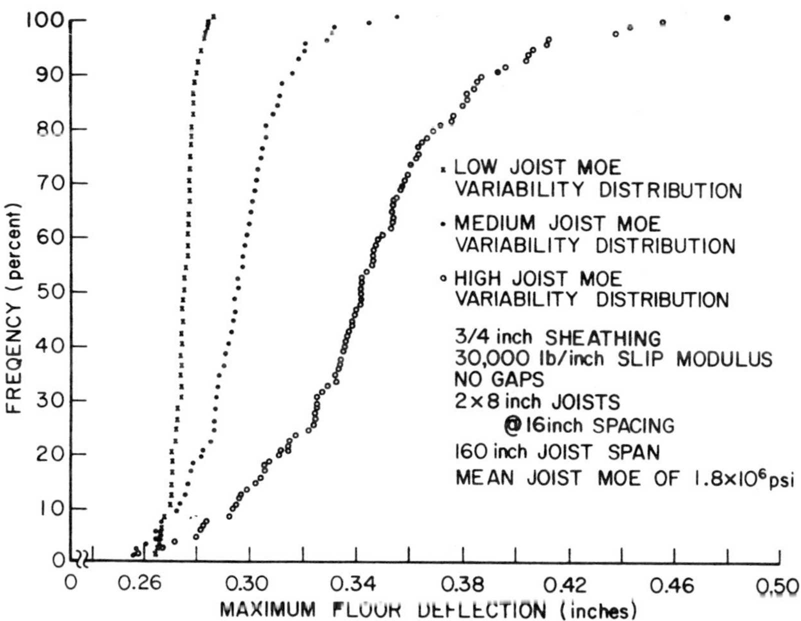

Figure 12 Wood joist floor system performance as affected by joist variability.

a value that yields the best floor deflection response to joist-cost relationship can now be formulated. Implications of changing grading methods for wood members can now be evaluated. An exciting new era in which engineers will have the tools with which to more nearly predict the true performance of wood structures has arrived with the advent to these new probabilistic methods. Hopefully, the research needed to bring these methods to full fruition will be organized and funded.

WOOD ENGINEERING EDUCATION

Introduction

As structural design with wood becomes more sophisticated and as understanding of the behavior of wood as an engineering material becomes more extensive, the need for engineers with training in wood design is becoming more evident. Currently, civil engineering programs in the United States generally do not offer a course in design of wood structures. This situation has reached a critical stage and, if continued, will impede the needed developments to ensure the efficient utilization of wood for structural purposes. The statistics cited previously in this paper attest to the importance of wood as a construction material and to its viability as a renewable source of energy. Thus, a three-part program is suggested here:

1. Wood design and wood materials information should be introduced into the present civil engineering curriculum.
2. A "Wood Engineering" discipline, perhaps at the graduate level should be created.
3. Continuing education programs for practicing engineers should be established.

Each of these concepts will be discussed briefly.

Introduction of wood design and wood materials information into present civil engineering curriculum

Recognizing that present civil engineering programs are extremely crowded, it is very difficult to add any new courses. Many schools do have some introduction to wood as a material within a basic materials laboratory course. The content of this introduction could no doubt be greatly improved. An attempt will soon be made to upgrade the information presented on wood as a material through the joint effort of universities and government laboratories by providing films and/or other aids.

The larger problem of introducing new wood structural design courses into the curriculum is more problematical. Most civil engineering departments are already finding it difficult to absorb the complexities of current steel and concrete codes in their structural design courses. In addition, many schools have reduced their emphasis on structures within the framework of the broad areas of civil engineering. Thus, it may be very difficult to introduce wood design at universities that do not now have an active program in this area. Lack of faculty competence is also a hindrance at many schools. These facts will place an increasing importance on the development of graduate-level programs at universities with special competence in the area of wood design and on continuing education programs for practicing engineers.

"Wood engineering" as a discipline

In recent years, a new discipline, new at least for the United States, has emerged which might be called "Wood Engineering." Individuals with graduate training which has included both a study of wood as a material and structural engineering have been entering professional fields to practice "Wood Engineering." These individuals have been in high demand by industry, government, and universities. Several of these specially trained engineers or wood scientists have become leaders in research and development areas within a very short time of service; others practice in structural engineering design with wood.

Unfortunately, few universities have the necessary combination of expertise and the desire by qualified professors in both structural engineering and wood science and technology areas to provide the required program. The need for individuals with "Wood Engineering" competence will continue to increase with further technological developments in the manufacture and utilization of wood. If civil engineering curriculums continue to ignore wood as a structural material, the country will depend even more in the future than now on these specially trained "Wood Engineers" for developing proper methods of utilizing wood—our most important renewable source of construction material.

Continuing education programs in wood engineering

The American Society of Civil Engineer's Committee on Wood has recently developed some new programs for reaching practicing engineers with information on structural design with wood. A publication of this committee, "Wood Structures—A Design Guide and Commentary" [16] has been well received and has helped to provide information on wood structures to many thousands of designers. Other efforts, particularly related to research needs, are planned by this committee.

In addition to these efforts, the Forest Products Research Society is active in efforts to publish and to organize meetings to disseminate information on wood. Other publications are available to practicing engineers through industry and trade associations. All such efforts are desirable, but may not suffice, particularly in view of the need for rapid transfer of the knowledge that is being developed by researchers at an increasing rate both in the United States and in many other countries. A program of technology transfer is badly needed to provide the newly developed information on wood to engineers in a form useful to them in their everyday practice.

SUMMARY

The intent of this paper has been to provide an overview concerning the role of the engineer in the design of structural systems with wood. The need for efficient utilization of every fiber of wood that can be produced under the constraints cited is clear. Several examples illustrating the application of currently available structural mechanics analysis procedures to problems in design with wood were presented. It is hoped that these examples will serve to point up the need for continued efforts by engineers, in cooperation with wood scientists, to develop additional new design methods using all available analytical tools. These additional design methods are needed to enhance wood as a construction material and to ensure the reliability of its structural applications.

The significant contributions of engineers to the full utilization of timber resources, together with possible increased supplies made available by proper management of forests, promise a new age of wood in the future. The abundance, renewability, low energy demand for production, and the negligible harvest impact of wood will ensure that it will continue to be a favored material of construction.

REFERENCES

[1] ANDERSEN, H. J., *An Introduction to Timber Engineering*, Pergamon Press, London, 1967.

[2] APLIN, E. N., AND F. J. KENNAN, "Limit States Design in Wood: A Canadian Perspective," *Forest Products Journal*, Vol. 27, No. 7, pp. 14–18, July 1977.

[3] BAXTER, J. W., "Timber as a Replaceable Resource," *Proceedings of Third Joint Conference of the American Society of Civil Engineers and the Institution of Civil Engineering*, World Resources—Engineering Solutions, Institution of Civil Engineers, London, 1976.

[4] BODIG, J., AND J. R. GOODMAN, "A New Apparatus for Compression Testing of Wood," *Wood and Fiber*, Vol. 1, No. 2, pp. 146–153, 1969.

[5] BODIG, J., AND J. R. GOODMAN, "Prediction of Elastic Parameters for Wood," *Wood Science*, Vol. 5, No. 4, pp. 249–264, April 1973.

[6] Committee on Renewable Resources for Industrial Materials, "Renewable Resources for Industrial Materials," National Research Council (CORRIM), National Academy of Sciences, Washington, D.C., 1976.

[7] DAWSON, P. R., AND J. R. GOODMAN, "Variability Simulations of Wood Joist Floor Systems," *Wood Science*, Vol. 8, No. 4, pp. 242–251, April 1976.

[8] GALLIGAN, W. L., AND D. V. SNODGRASS, "Machine Stress Rated Lumber: Challenge to Design," *Forest Products Journal*, Vol. 20, No. 9, pp. 63–39, 1970.

[9] GOODMAN, J. R., "Layered Wood Systems with Interlayer Slip," Ph.D. dissertation, University of California, Berkeley, 192 pp., 1967.

[10] GOODMAN, J. R., AND E. P. POPOV, "Layered Beam Systems with Interlayer Slip," *Journal of the Structural Division, ASCE*, Vol. 94, No. ST11, Proc. Paper 6214, pp. 2535–2547, 1968.

[11] GOODMAN, J. R., "Layered Wood Systems with Interlayer Slip," *Wood Science*, Vol. 1, No. 3, pp. 148–158, 1969.

[12] GOODMAN, J. R., AND J. BODIG, "Orthotropic Elastic Properties in Wood in Compression," *Journal of the Structural Division, ASCE*, Vol. 96, No. ST11, Proc. Paper 7682, pp. 2301–2319, 1970.

[13] GOODMAN, J. R., AND J. BODIG, "Orthotropic Strength of Wood in Compression," *Wood Science*, Vol. 4, No. 2, pp. 83–94, October 1971.

[14] GOODMAN, J. R., M. E. CRISWELL, M. D. VANDERBILT, AND J. BODIG, "Implications of Rational Analysis of Wood Joist Housing Floor Systems," *Third International Symposium of International Association for Housing Science*, Montreal, Canada, May 1974.

[15] GOODMAN, J. R., M. D. VANDERBILT, M. E. CRISWELL, AND J. BODIG, "Composite and Two-Way Action in Wood Joist Floor Systems," *Wood Science*, Vol. 7, No. 1, pp. 25–33, July 1974.

[16] GOODMAN, J. R., *Wood Structures—A Design Guide and Commentary*, Editor, Author of Chapter 2, Concepts of Solid Mechanics of Wood, Sections 2.1 and 2.2, Published by ASCE Committee on Wood, ASCE, 1975.

[17] GOODMAN, J. R., AND R. M. GUTKOWSKI, *Handbook of Composite Construction Engineering*, Chapter 10, Composite Construction in Wood and Timber. (edited by G. M. Sabnis), Van Nostrand Reinhold Co., New York, 1979.

[18] GOPU, V. K. A., AND J. R. GOODMAN, "Analysis of Double-Tapered, Pitched, and Curved Laminated Beam Section," *Wood Science*, Vol. 7, No. 1, pp. 52–60, July 1974.

[19] GOPU, V. K. A., "Behavior of Double-Tapered Curved Glulam Beams," Dissertation, Colorado State University, Fort Collins, Colorado, 1975.

[20] GOPU, V. K. A., AND J. R. GOODMAN, "Full-Scale Tests on Tapered and Curved Glulam Beams," *Journal of the Structural Division, ASCE*, Vol. 101, No. ST12, Proc. Paper 11794, pp. 2609–2626, December 1975.

[21] GOPU, V. K. A., AND J. R. GOODMAN, "Design of Double-Tapered and Curved Glulam Beams," *Journal of the Structural Division, ASCE*, Vol. 103, No. ST10, pp. 1903–2077, October 1977.

[22] GRANHOLM, H., "Om Sammansatta balkar och pelare med sarskild hansyn till spikade trakonstruktioner. (On Composite Beams and Columns with Particular Regard to Nailed Timber Structures)," Chalmers Tekniska Hogskolas Handlingar, No. 88, 1949.

[23] GURFINKEL, G., *Wood Engineering*, Southern Forest Products Association, New Orleans, Louisiana, 1973.

[24] GUTKOWSKI, R. M., AND J. R. GOODMAN, "Research on Glulam Bridge Systems," presented at the Forest Products Research Society Annual Meeting, July 1976 (publication pending).

[25] HENGHOLD, W. M., "Layered Beam Vibrations Including Slip," Ph.D. dissertation, Colorado State University, Fort Collins, Colorado, 1972.

[26] HODGES, R. D., JR., "America's Houses Grow on Trees," *National Journal*, October 30, 1976.

[27] HOYLE, R. J., *Wood Technology in the Design of Structures*, Mountain Press Publishing Co., Missoula, Montana, 1973.

[28] KO, M. R., "Layered Beams Systems with Interlayer Slip," M.S. thesis, Colorado State University, Fort Collins, Colorado, 1974.

[29] KUO, M. L., "Verification of a Mathematical Model for Layered T-Beams," M.S. thesis, Colorado State University, Fort Collins, Colorado, 1974.

[30] LIU, J. S., "Verification of Mathematical Model for Wood Joist Floor Systems," Ph.D. dissertation, Colorado State University, Fort Collins, Colorado, 1974.

[31] National Science Foundation, "Better Use of a Prime Resource," *Mosaic*, Vol. 5, No. 4, Fall 1974.

[32] NEWMARK, N. M., C. P. SIESS, AND I. M. VIEST, "Tests and Analysis of Composite Beams with Incomplete Interactions," *Proceedings, Society for Experimental Stress Analysis*, Vol. 9, No. 1, 1951.

[33] PLESHKOV, P. F., *Teoriia Rascheta Depeviannykh Sostavnnykh Sterzhnei* (Theoretical Studies of Composite Wood Structures), Moscow, 192 pp., 1952.

[34] POLENSEK, A., "Rational Design Procedure for Wood-Stud Walls Under Bending and Compression Loads," *Wood Science*, Vol. 9, No. 1, pp. 8–20, July 1976.

[35] RASSAM, H. Y., AND J. R. GOODMAN, "Buckling Behavior of Layered Wood Columns," *Wood Science*, Vol. 2, No. 4, pp. 238–246, 1970.

[36] RASSAM, H. Y., AND J. R. GOODMAN, "Design of Layered Wood Columns with Interlayer Slip," *Wood Science*, Vol. 3, No. 3, pp. 149–155, January 1971.

[37] Southern Pine Association, *How to Build Storm Resistant Structures*.

[38] STERN, E. G., "Anchorage of Light Frame Buildings in Storm-Resistant Construction," presented at Conference on Wind Loads on Structures, California Institute of Technology, December 1970.

[39] SUDDARTH, S. K., "A Computerized Wood Engineering System—Purdue Plane Structures Analyzer," USDA Forest Service Research Paper, FPL 168, 1972.

[40] THOMPSON, E. G., M. D. VANDERBILT, AND J. R. GOODMAN, "Finite Element Analysis of Layered Wood Systems," *Journal of the Structural Division, ASCE*, Vol. 101, No. ST12, December 1975.

[41] THOMPSON, E. G., M. D. VANDERBILT, AND J. R. GOODMAN, "FEAFLO: A Program for the Analysis of Layered Wood Systems," *Computers and Structures*, Vol. 7, pp. 237–248, 1977.

[42] *Timber Construction Manual*, American Institute of Timber Construction, Englewood, Colorado, 1974.

[43] TOUMI, R. L., "Lightweight Truss-Framed House for Safety and Energy Efficiency," *Agricultural Engineering*, May 1977.

[44] TREMBLAY, G. A., "Nonlinear Analysis of Layered T-Beam with Interlayer Slip," M.S. Thesis, Civil Engineering Department, Colorado State University, Fort Collins, Colorado, 1974.

[45] VANDERBILT, M. D., J. R. GOODMAN, M. E. CRISWELL, AND J. BODIG, "Development and Verification of a Mathematical Model of Wood Joist Floors Using Computer Analysis and Closed-Loop Structural Testing," Closed-Loop Magazine, MTS Corporation, 1973.

[46] VANDERBILT, M. D., J. R. GOODMAN, M. E. CRISWELL, AND J. BODIG, "Service and Overload Behavior of Wood Joist Floor Systems," *Journal of the Structural Division*, ASCE, Vol. 100, No. ST1, January 1974.

[47] VANDERBILT, M. D., J. R. GOODMAN, J. BODIG, AND M. E. CRISWELL, "A Rational Analysis and Design Procedure for Wood Joist Floor Systems," Final Report to the National Science Foundation Research Grant GK-30353, September 1974.

[48] ZAHN, J. J., "Reliability-Based Design Procedures for Wood Structures," *Forest Products Journal*, Vol. 27, No. 3, pp. 21–28, March 1977.

[49] ZORNIG, H. F., AND G. E. SHERWOOD, "Wood Structures Survive Hurricane Camille's Winds," USDA Forest Service, Rescarch Paper, FPL 123, October 1969.

BEHAVIOR OF STRUCTURES
IN FIRE ENVIRONMENTS

*B. Bresler**

INTRODUCTION

Technological progress during the twentieth century brought about many comforts, conveniences, and improvements in health and safety. It also brought changes in our built environment that pose new hazards and new problems. One of these problems, which has received considerable attention in recent years, is the increased fire hazard in buildings. Deaths, injuries, and property losses by fire in the United States are among the highest, per capita, of the world's major industrial nations [1].

The basic regulations for fire protection of modern buildings currently used by engineers and architects were developed at the turn of the century. While these regulations served their purpose reasonably well for about 50 years, they are no longer adequate, because of changes in buildings—their size, configuration, uses, contents, and methods and materials of construction. The new buildings are generally taller, with large open spaces on the inside and large window areas, providing more ventilation to the burning

---

*Professor Emeritus of Civil Engineering, University of California, Berkeley; Principal, Wiss, Janney, Elstner and Associates, Emeryville, California.

contents. Suspended ceilings, ducts and piping, electrical wiring and equipment, and interior finishes and furnishings often contain large amounts of combustible plastics. More complex ventilation and telecommunication systems require more penetrations in the walls and floors, which serve as paths for the spread of smoke and fire in the buildings. These changes in the built environment tend to increase fire hazards associated with the amount and nature of combustible contents, risk of ignition, fire intensity and duration, and human response to the critical fire event.

Deficiencies in current testing methods and fire endurance ratings include the following. First, *the rate of heating and the intensity and duration of fire exposure of building elements may differ substantially from the time–temperature curve used in a standard test.* Second, *most standard tests provide only for performance during the heating stage, neglecting the cooling stage.* Some wall tests call for hose stream cooling after removal of specimens from the furnace, but most slab and column tests are conducted only to record performance during the heating stage and do not determine residual performance after cool down. Third, *specimens are normally tested in isolation from structural systems.* A column tested for fire endurance is rated identically whether that column is exposed to fire uniformly on all sides or exposed to fire on one side only, or whether it is located at the ground floor or at the fiftieth floor of a high-rise building, neglecting the different restraints that will be developed on these elements during heating. Fourth, *the criteria for endurance ratings are often based on measurements of thermal transmission, without specific limits on various levels of damage, such as deflection, cracking, yielding, and residual load-bearing capacity after the fire.*

In recent years building officials, architects, and engineers have become increasingly aware of the obsolescence of current approaches to fire safety based on standard tests for fire endurance of isolated components and requirements for exits and details of construction without due regard to increases in levels of fire hazard. Improvements in fire protection measures, such as devices for automatic detection, alarm, and control of fire, as well as in equipment and operations of fire services, tend to offset some of the increases in fire hazards. Nevertheless, the current state of the art does not reflect the full potential of modern engineering capability.

While the emphasis in this paper is on the behavior of structures in fire environments, it is necessary to recognize that a holistic approach to fire protection engineering must comprise a number of interrelated concerns: ignitability, control of fire growth, toxic fire products, human response to fire, fire characteristics related to damage, and fire endurance. Reduction of fire hazards may be achieved in a variety of ways; however, all the factors listed above must be considered in the assessment of fire hazard (Fig. 1).

In a typical fire the initial phase represents a slow temperature rise during a growth period of varying duration. During this period, fire detec-

GOAL-ORIENTED CONCEPTUAL FRAMEWORK

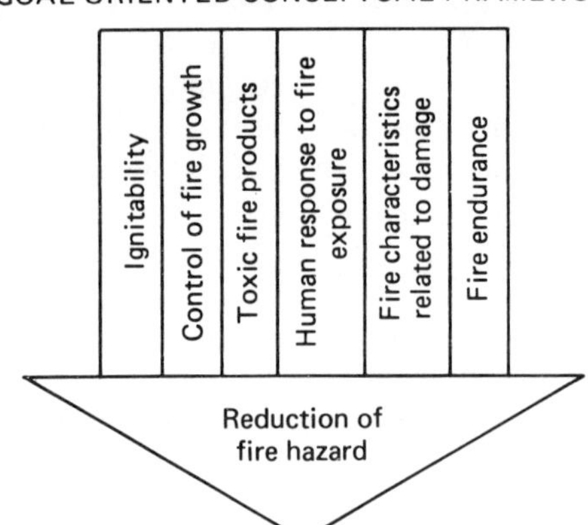

Figure 1   Reduction of fire hazard.

tion and suppression systems and evacuation of occupants may be effective. If a fire spreads rapidly and suppression systems are not effective, fire growth may peak in a very short time. Occupants would experience difficulty in escaping from a building, and by the time that the fire department had arrived the building might be fully involved.

Structural fire resistance is essential both during this fully involved period and after a fire has been controlled and a structure has begun to cool. *When other measures of containing the fire fail, structural integrity is the last line of defense.*

PREDICTION OF STRUCTURAL BEHAVIOR IN FIRE ENVIRONMENTS

In order to assess the integrity of structural and nonstructural elements during and after a fire, the performance of these elements as barriers to fire propagation as well as deformation and load-bearing capacity of these elements must be evaluated. During the past 10 years sufficient progress has been made in developing analytical methods for the prediction of structural response to fire so that a consistent process for the design of structures for all services as well as for extreme environmental conditions can be followed [2]. The basic approach for an analytical prediction of structural response to accidental fires is summarized below, while the details are published elsewhere [3–6].

To predict the fire response of structures, it has been assumed that the heat-flow problem is separable from structural analysis. This separation greatly simplifies the development of appropriate analytical models and related computer programs. In some cases however, changes in geometric characteristics associated with structural distress—such as spalling—may have to be accounted for in solving the heat-flow problems. Such interaction between heat flow and structural response can be handled by an iterative process.

Given a structure with specified geometry, material properties and fire exposure, thermal response can be calculated in terms of transient temperature distribution throughout the structure. Once the temperature distribution is known, structural response in terms of cracking, stresses, crushing, fracture and yielding can be calculated. The flow diagram illustrating this procedure is Fig. 2.

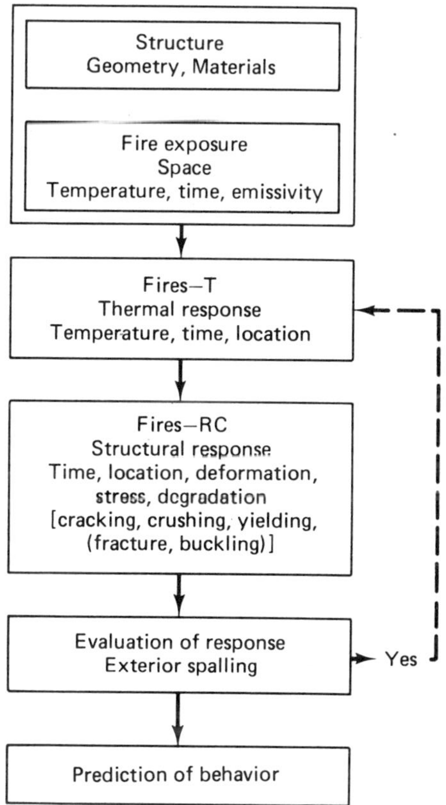

Figure 2 Procedure for thermal and structural response analysis.

Thermal response

Thermal response is defined as the transient temperature field throughout a structure. A finite element method coupled with time-step integration is used to solve a nonlinear heat-balance equation following formulations of Wilson [7] and Zienkiewicz [8]:

$$
\begin{array}{lll}
\text{rate at which heat} & \text{rate at which heat} & \text{rate at which} \\
\text{is stored in elements} + & \text{flows from elements} = & \text{external heat} \qquad (1)\\
\text{adjacent to a node} & \text{adjacent to a node} & \text{enters a node}
\end{array}
$$

In turn, this heat-balance equation can be expressed in the following matrix notation:

$$C[\rho(T), c_p(T)]\dot{T} + K[k(T)]T = Q[T, F(t)] \qquad (2)$$

in which C is the capacity matrix, K the conductivity matrix, Q the external heat flow vector, T the temperature vector, $\dot{T}$ the time rate temperature change vector, $\rho(T)$ the density as function of temperature, $c_p(T)$ the specific heat as a function of temperature, $k(T)$ the conductivity as a function of temperature, and $F(t)$ the external heat source as a function of time.

A fire zone (heat source) can be represented by a turbulent, well-mixed gas having single values of temperature and emissivity at any given time. This can be viewed as a model in which the effects of such factors as temperature variations within the fire, gas flow, fuel distribution, and enclosure wall characteristics are represented by a pseudo-fire with single values of temperature and emissivity. Variation in fire characteristics within a compartment can be handled by specifying different fire zones with appropriate boundary conditions. Exposure to nonfire conditions on the boundary, such as ambient atmospheric exposure, can be modeled as exposure to another pseudo-fire with appropriate values of temperature and emissivity. The boundary condition of an exposed surface, modeled by a heat flux expressed as a sum of convective and radiative terms, eq. (3), is generally nonlinear and may be expressed as follows:

$$Q_F = A_F[A_C(T_f - T_s)^N + V \cdot \sigma \cdot \epsilon \cdot a(\theta_f^4 - \theta_s^4)] \qquad (3)$$

where Q_F = heat flow caused by the exposure of the system to fire on the boundary

A_F = tributary area

A_c = convection coefficient

T_f = temperature of the pseudo-fire

T_s = surface temperature

N = convection power factor

V = radiation view factor

σ = Stefan–Boltzmann constant

a = absorption coefficient of surface

ϵ = effective emissivity associated with flame and enclosure

θ_f = absolute temperature of fire

θ_s = absolute temperature of surface

The nonlinearities associated with temperature-dependent material properties, such as C and K in eq. (2), and nonlinearities in the boundary condition, eq. (3), require an iterative solution within any given time step. Within a given iteration the problem is linearized about the current temperature distribution. For frames, a two-dimensional problem is solved assuming that no heat flows along frame member axes, while a one-dimensional solution often suffices for slabs. In corners and connections, a three–dimensional problem must be solved. When the presence of transverse reinforcing steel in concrete, such as ties and spirals, or discontinuities in structural steel sections, such as stiffeners or diaphragms, must be accounted for in heat-flow calculations, a three-dimensional problem must be solved. Member cross sections can have any shape and may be composed of several materials (concrete, steel, insulation), and it is assumed that there is no contact resistance to heat transmission at the interface between these materials.

A general computer program for solving the nonlinear heat-flow problem in a composite solid has been developed (FIRES-T3) which accommodates any combination of three-, two-, and one-dimensional isoparametric elements. The theoretical formulation, solution techniques, a user's manual, and sample problems with input and output, are contained in [3].

As noted above, the fire environment is modeled in FIRES-T3 by a pseudo-fire defined by a time–temperature curve and convective and radiative heat-flow terms in the fire boundary condition. The phenomenon associated with heat transfer in the turbulent environment of a fire is difficult to model exactly. The temperature dependence of the related parameters and the detailed temperature and flow fields in a fire compartment can be considered only when such information becomes available. The parameters critical to the pseudo-fire concept appear to be the emissivity of the flame and that of the surface of the structural element. Flame emissivity depends on the particulate content of the flame, which may vary greatly from the controlled fire of a test furnace to that of the uncontrolled environment of an actual fire. In addition, radiation sources and compartment enclosure surfaces may contribute to the overall fire boundary condition problem.

The radiation component of heat transfer at a boundary of an element exposed to fire depends on the spatial configuration of the enclosure, the geometry of the element under consideration, and the emissivities of the fire, the element surface, and the enclosure surface. The radiation boundary condition in eq. (3) is an attempt to represent the influence of the fire and element surface emissivities of the radiative heat transfer. For concrete structural elements surrounded by fire, the effective emissivity ϵ falls in the range 0.3 to 0.9 [3], and for concrete structures exposed to fire a value of $0.5 < \epsilon < 0.7$ yields good results.

Structural response

Structural response is defined in terms of transient deformations, forces, moments, and material degradation (cracking, crushing, yielding, or local instability) throughout the structure. These factors account for material behavior at varying temperature, including dimensional changes caused by temperature variation, changes in mechanical properties with changes in temperature, degradation of section through cracking and/or crushing, changes in rates of shrinkage and creep with change in temperature, and inelastic deformations associated with nonlinear stress–strain characteristics, yielding, unloading of portions of structures, and creep.

The basic nonlinear equation governing the behavior of a structural system is temperature and history dependent and may be written as follows:

$$F_i(U_i, T_i, H_{i-1}) = R_i \tag{4}$$

where F_i = internal forces at current time t_i
$\quad U_i$ = current deformed shape
$\quad T_i$ = current temperature distribution
$\quad H_{i-1}$ = prior response and thermal history of the structure
$\quad R_i$ = effects of external loading at current time t_i

Within a given time step, an iterative approach is used to find a deformed shape that results in equilibrium between forces associated with external loads and internal stress resultants. A finite element method using a nonlinear direct stiffness formulation coupled with time-step integration is used to predict structural response. The iteration method may be regarded as a sequence of corrections to obtain a convergent solution, so that

$$\delta U_i^j = [K_i^j]^{-1} \cdot \delta R_i^j \tag{5}$$

where δU_i^j = displacement correction for time step i and iteration j;
$\quad\quad = (U_i^{j+1} - U_i^j)$
$\quad K_i^j$ = tangent stiffness for time step i and iteration j;
$\quad\quad = \nabla F_i^j(U_i^j, T_i, H_{i-1})$
$\quad \delta R_i^j$ = out-of-balance forces for time step j and iteration j;
$\quad\quad = R_i - F_i(U_i^j, T_i, H_{i-1})$

Iteration continues until equilibrium is satisfied to within a specified tolerance; that is, until

$$\| \delta R_i \| < \text{permissible error} \tag{6}$$

at which point the solution for the time step i is considered complete $(U_i = U_i^j)$, and calculations for the next time step begin.

The overall system stiffness matrix K is assembled by incorporating the stiffness contribution of all the elements of the structural system. The assemblage of the stiffness matrix from the element stiffnesses requires that (1) the system be appropriately discretized, (2) modal displacements be transformed to strains in individual elements, and (3) these strains and prior strain history be related to the current stiffnesses, forces, and moments acting on discrete segments of the system.

The geometric discretizations for reinforced concrete frames [4] and slabs [5] are shown in Figs. 3 and 4. Considerable engineering judgment is needed to discretize a structure economically. The discretization must be fine enough to model nonlinearities (degradation) in structural behavior, such as cracking, crushing, yielding, local instability, and yet be as coarse as

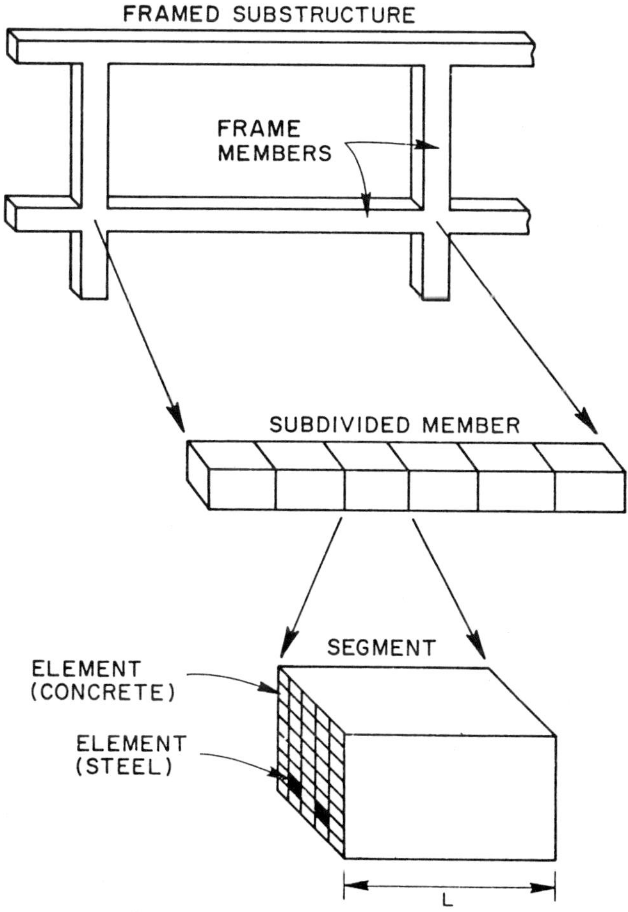

Figure 3   Geometric idealization of a frame.

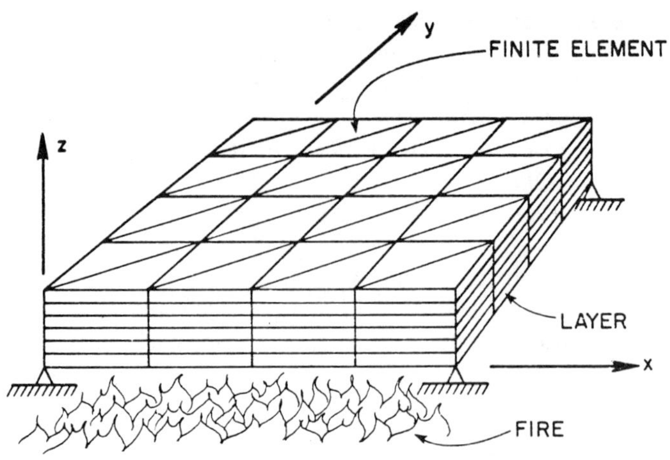

Figure 4 Geometric idealization of a slab.

is feasible, since solution cost increases rapidly with the number of segments and subslices. If a system has been discretized by too coarse a mesh, a convergent solution may not be obtained since a member will not have sufficient flexibility and degrees of freedom to deform easily into a shape and degradation pattern that satisfies equilibrium. It is sometimes necessary to redesign a discretization scheme where convergence is disproportionately slow.

The time discretization also requires considerable judgment. It must be able to accommodate changes in temperature (heating and cooling rates), changes in external loading, as well as changes associated with creep and other material characteristics. The basic assumption underlying the analytical procedures described here is that temperature, loading, and material properties remain constant during a given step. Therefore, the time steps should be smallest when these factors are expected to vary most. In most cases investigated to date, time steps of 0.05 hour have yielded apparently satisfactory results, although time steps in the range 0.01 to 0.1 hour have been used. The number of time steps should be as large as accuracy and rate of convergence permit.

When large structures are investigated for the effects of localized fires, structural partitioning may be advisable, so that the solution of the problem is restricted to response evaluation for a substructure with specified boundaries [9].

The number of iterations in each time step may be partly controlled by selecting a desired level of accuracy (i.e., the tolerance for out-of-balance forces). For frame problems a tolerance of 2 kips for forces and about 2 kip-ft for moments gives reasonable accuracy of solution.

STRUCTURAL FIRE DAMAGE

General remarks on fire damage

Two general categories of fire damage must be distinguished: cosmetic and functional. Cosmetic damage is easily, although not always inexpensively, repaired. Functional damage is more extensive; it may be reparable or it may require that the damaged system be partially or totally replaced. A further distinction must be made with respect to the fire damage of buildings: contents—furnishings, mechanical, electrical, architectural, and other service systems; secondary structural elements, such as partitions, stairways and enclosures, exterior walls, and other portions of the structure which, although severely damaged, will not lead to collapse of the building; and primary structural components. In this discussion of damage the focus is on the primary structural system. It should be noted, however, that in many cases the behavior of secondary structural elements exposed to fire may significantly influence the damage in the primary structural system.

The type of damage depends on the characteristics of the particular structural material and on the fire exposure. There are two principal mechanisms of damage: degradation of material during and subsequent to fire exposure, and changes in the internal forces and deformations during and subsequent to a fire.

In lieu of the present approach to fire safety of structures based on fire endurance, an alternative approach based on admissible damage levels and limit state criteria has been described by Bresler *et al.* [9] and is summarized below.

Damage due to fire may be expressed in terms of the physical condition of buildings and corresponding limit states:

Physical Damage	*Limit State*
1. Damage restricted to nonstructural elements; no significant fire exposure of structural elements.	1. The temperature in structural elements must not exceed a specified limit.
2. Minor damage to structural elements; no residual deformation and negligible degradation in stiffness and load-resisting properties after the fire.	2. No yielding of reinforcement; extent of cracking and crushing must not exceed specified limits.
3. Moderate but reparable damage to structural elements; negligible residual deformation.	3. Yielding of reinforcement may not exceed specified limit strain; extent of cracking and crushing must not exceed specified limits (higher than in 2 above).

4. Major structural damage; may or may not be reparable; no collapse within a prescribed time.

4. General yielding of reinforcement, reduction of bearing capacity, and/or excessive deformation, cracking, and crushing may occur only after specified endurance time.

Prediction of the first limit state requires that the temperature during the early stages of fire development as well as material properties under moderately elevated temperature be accurately known. For other limit states, early stages of fire development are less important, but material properties under a wide range of temperature must be specified.

For characteristic types of structures and specified fire environment models, these limit states of behavior can be related to specific temperatures and corresponding times. When different structural systems and configurations under specified time–temperature relations have been analyzed, times and corresponding temperatures associated with different limit states may be defined, providing the information necessary for rational fire-safe design. Only in special cases will full response analyses be required, particularly when unique structural systems or configurations are involved or when optimal fire protection with respect to structural design is desired.

CASE STUDY

A thirteen-story reinforced concrete frame was analyzed to illustrate the procedure for investigating thermal and structural response to fire and to suggest directions for future investigations. Figure 5 shows the plan and section of the building. Two fire levels were considered: one on the ground floor, the other on the twelfth story of the building. At each level, two fire-spread zones were considered: one with full involvement in at least a two-bay compartment, the other a local fire uniformly surrounding a column but affecting neither beams nor the floors above and below. Although the latter fire is totally unrealistic, it represents the condition of a standard fire test of a column. For each level and fire zone, two types of fire action—a long-duration, moderate-intensity fire (ASTM E-119), and a short-duration, high-intensity fire (SDHI)—were considered (Fig. 6).

In order to reduce computational effort, the multistory building was modeled by isolated substructures, as shown in Fig. 5. For the case of the ground-floor fire, the model included two stories above the fire level. The column loading caused by the upper nine stories was assumed to act on an infinitely stiff base, and variation in these loads during the fire process was thus neglected. It was furthermore assumed that in addition to permanent gravity loads, 60% of the design live load was acting during the fire process. All loads were assumed to be uniformly distributed over the floors of the

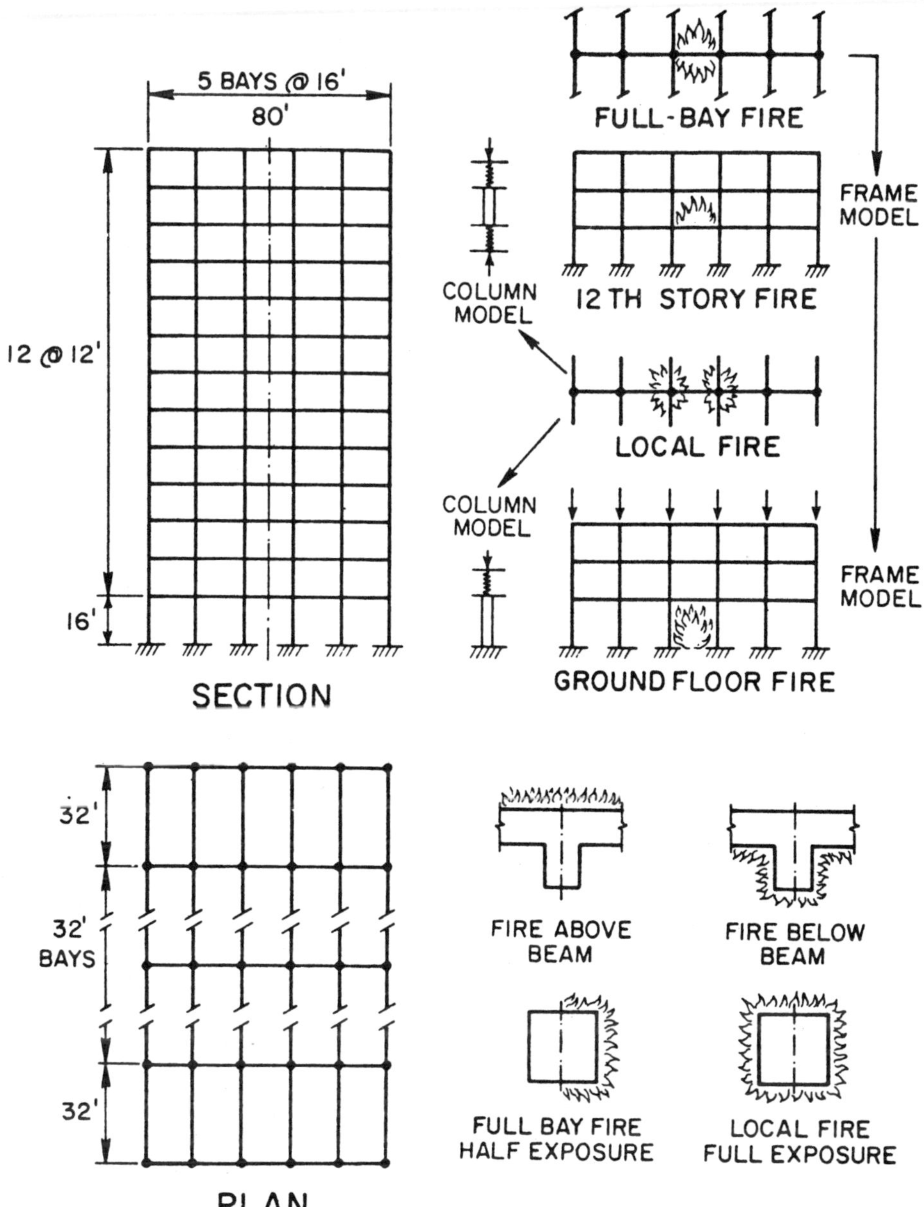

Figure 5 Building prototype, fire conditions, and structural models.

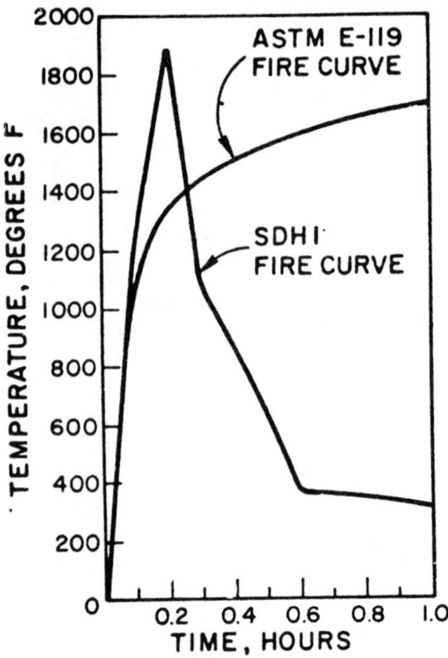

Figure 6   Time–temperature fire curves.

building, and the loads acting on the beams were calculated on the basis of tributary floor areas bounded by midbay and 45° lines. The rotational restraints caused by the upper nine stories on the three-story frame were neglected.

For the case of a fire on the twelfth story, the model included one story above and one below the fire level. The columns of the eleventh story were assumed to be fully restrained at the base, leading to a three-story frame model, shown in Fig. 5.

For the case of a local fire symmetrically enclosing the four sides of the center column, it was assumed that the column was fixed against rotation at both top and bottom and restrained axially by portions of the structure above and below which had constant stiffness during the fire. Consequently, the surrounding building was modeled by means of axial springs of appropriate stiffness as shown in Fig. 5. Details of cross-sectional characteristics as well as a description of thermal and mechanical properties and their variation with temperature are omitted here for the sake of brevity; the values are fully documented elsewhere [2–4].

Typical results of the thermal analysis from computer program FIRES-T3 are shown in Figs. 7 and 8. Isotherms for the ground-floor column are shown in Fig. 7 for the SDHI fire for 0.5-hour duration and for the ASTM fire for 1.0-hour duration. For the SDHI fire, the maximum concrete temperature barely exceeded 600°F, and the maximum reinforcing steel tem-

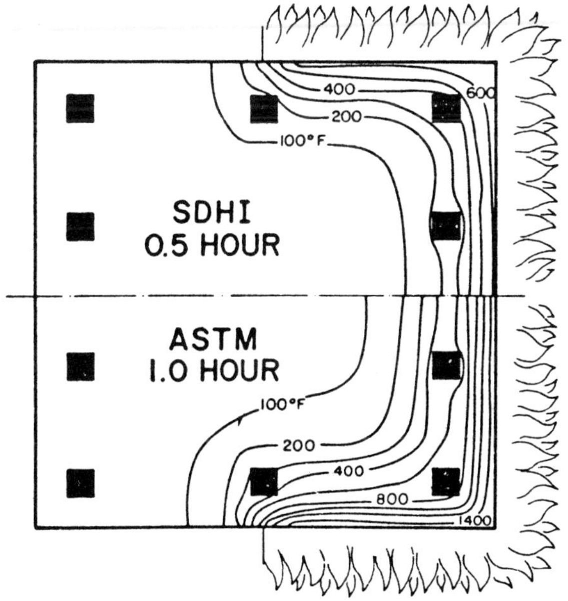

GROUND FLOOR FIRE

Figure 7   Typical column thermal response.

perature reached 400°F. For the ASTM 1.0-hour-duration fire, the maximum concrete temperature approached 1600°F, and the maximum reinforcing steel temperature reached 800°F in the corner bar only; remaining steel bars were between 300 and 600°F. It is important to note that the presence of reinforcing steel influenced the temperature field as reflected in the shape of the isotherms.

In Fig. 8, isotherms for two beam conditions are shown, one for the case of fire on the floor above the beam, and the other for the case of fire on the floor below the beam (i.e., for the twelfth-story case). Both cases are for the ASTM fire of 1.0 hour. In these cases, temperature distribution was again influenced by reinforcing steel as reflected in the shape of the isotherms.

Typical results from the structural analysis are shown in Figs. 9, 10, and 11 and in Tables 1 and 2. The frame deformations resulting from the ground-floor fire are shown in Fig. 9. The scale of deformations plotted is 100:1 with respect to the linear scale. Heavy dashed lines indicate the deformations due to normal service loads without fire. Column shortening is due to the load of the upper nine stories on this frame. Heavy solid lines show the deformed shape of the frame for the ASTM fire at 1 hour.

The thermal expansion of the floor above the fire pushed the girders in adjacent bays outward, causing large internal displacement of the top of the

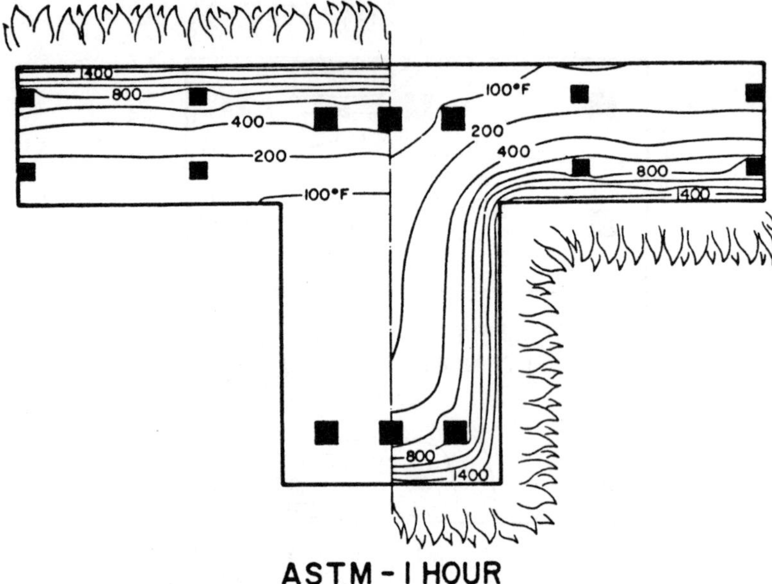

Figure 8   Typical beam thermal response.

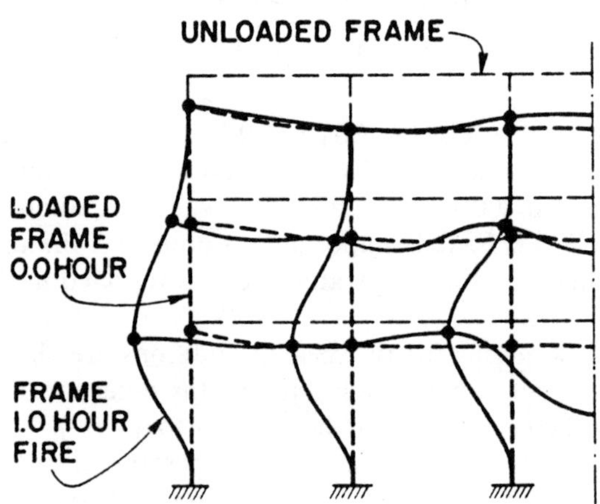

Figure 9   Fire-induced frame deformations—ground-floor ASTM fire.

column of the fire story (in this case the ground floor). The magnitude of this displacement was about 0.25 inch or a story drift of 1:750. The lateral displacement of the joints at this level was restrained by beam and column frame elements away from the fire zone. Thus, significant compression forces would be expected in the floor girders at a level just above the fire. The girders farther removed from the fire levels acted as ties restraining lateral

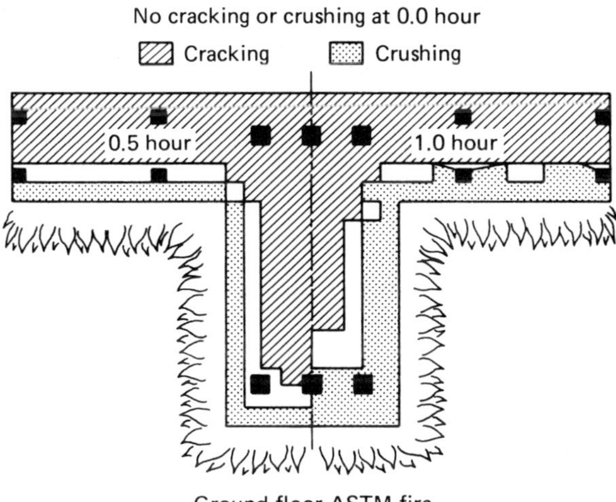

Figure 10 Typical column section degradation—cracking and crushing due to fire.

Figure 11 Typical beam section degradation—cracking and crushing due to fire.

Table 1

VARIATION IN JOINT FORCES AND MOMENTS—GROUND-FLOOR FIRE

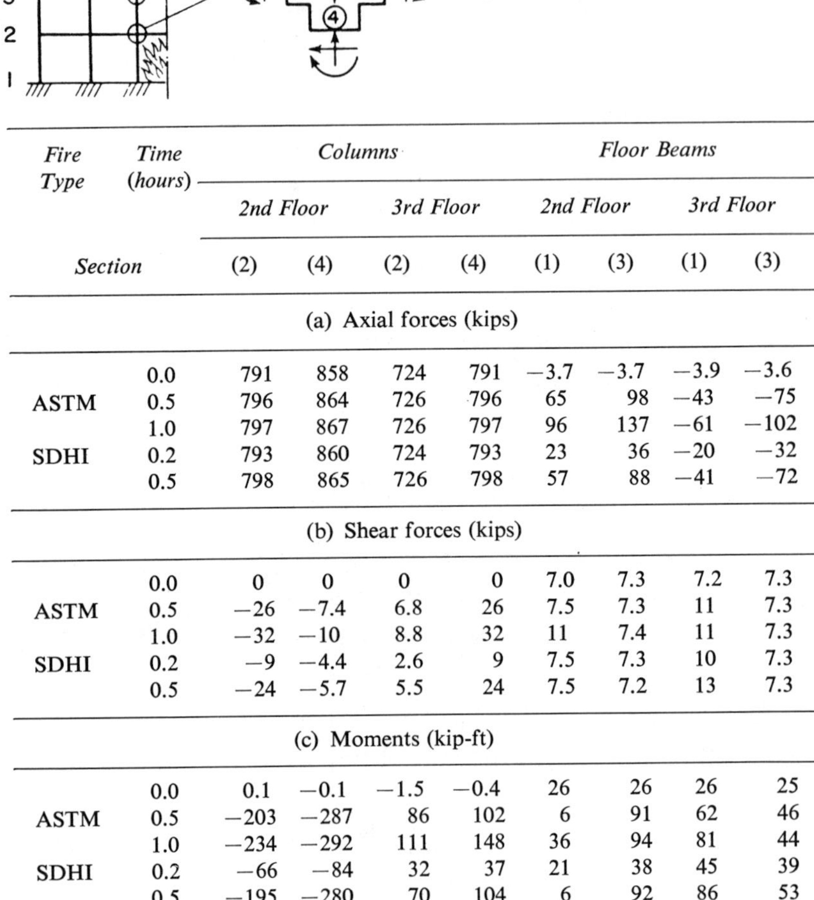

Fire Type	Time (hours)	Columns				Floor Beams			
		2nd Floor		3rd Floor		2nd Floor		3rd Floor	
Section		(2)	(4)	(2)	(4)	(1)	(3)	(1)	(3)
				(a) Axial forces (kips)					
	0.0	791	858	724	791	−3.7	−3.7	−3.9	−3.6
ASTM	0.5	796	864	726	796	65	98	−43	−75
	1.0	797	867	726	797	96	137	−61	−102
SDHI	0.2	793	860	724	793	23	36	−20	−32
	0.5	798	865	726	798	57	88	−41	−72
				(b) Shear forces (kips)					
	0.0	0	0	0	0	7.0	7.3	7.2	7.3
ASTM	0.5	−26	−7.4	6.8	26	7.5	7.3	11	7.3
	1.0	−32	−10	8.8	32	11	7.4	11	7.3
SDHI	0.2	−9	−4.4	2.6	9	7.5	7.3	10	7.3
	0.5	−24	−5.7	5.5	24	7.5	7.2	13	7.3
				(c) Moments (kip-ft)					
	0.0	0.1	−0.1	−1.5	−0.4	26	26	26	25
ASTM	0.5	−203	−287	86	102	6	91	62	46
	1.0	−234	−292	111	148	36	94	81	44
SDHI	0.2	−66	−84	32	37	21	38	45	39
	0.5	−195	−280	70	104	6	92	86	53

displacement at the fire level and significant tension forces would be expected in such girders. Large lateral displacements of columns at and adjacent to (above and below) the fire level were expected to generate significant shears and bending moments in these elements. Upward displacement of the top of the column of the fire story was relatively small, about 0.05 inch, but the vertical deflection at midspan of the floor beam exposed to fire was about 0.4 inch or (1/380) of the span.

Table 2

SELECTED RESPONSE HISTORY—TIME OF RESPONSE (HOURS)

| | Local | | 2-Bay | | | |
	ASTM	SDHI	ASTM	SDHI	ASTM	SDHI
Response / *Section*	Column		Column Top		Beam End	
1. Initial cracking	0.55	*a*	0.75	0.95	0.10	0.10
2. 30% Area cracked	0.60	*a*	*a*	*a*	0.10	0.10
3. Initial crushing	0.15	0.15	0.15	0.125	0.15	0.15
4. 30% Area crushed	0.80	*a*	*a*	*a*	1.0	*a*
5. Initial yielding	0.45	0.40	(0.5)[b]	(0.5)[b]	(0.5)[b]	*a*
6. 70% Axial stiffness (AE)	0.50	0.45	0.65	*a*	0.10	0.10
7. 70% Bending stiffness (EI)	0.20	0.15	0.40	0.35	0.10	0.10

[a]Response not attained within 1 hour.
[b]Estimated time of response.

The magnitude of all displacements as well as forces and moments in the frame elements are provided as output by the computer program FIRES-RC II. In Table 1, selected values for forces and moments on joints at the second and third floors (circled on the sketch) are given. Positive values of forces and moments are shown on the joint freebody.

The effect of fire on axial forces in columns was practically negligible. This may be due partly to the assumption made in modeling of the ground-floor fire by a three-story frame with constant loads above the fourth floor.

The effect of fire on axial forces in the floor beams is most important. Maximum compression in the second-floor beam exposed to fire reached 137 kips in the ASTM 1-hour fire, and 88 kips in the SDHI 0.5-hour fire. The adjacent floor beam, not exposed to fire, experienced slightly smaller compression forces, 96 and 57 kips, respectively, but even these axial forces are 20 to 30 times greater than the tensile axial forces in floor beams under normal service conditions.

Shear forces in floor beams were only slightly affected by the fire. However, in view of the potential degradation of concrete cross sections by cracking and crushing, the possibility of serious distress cannot be discounted.

Shearing forces in columns were greatly affected by the fire. Under normal service loads, shear forces in interior columns are practically zero, but under these fire conditions increased to about 25 to 30 kips, particularly in columns just above the fire level. Here again, because of significant degradation of concrete sections (cracking and crushing), particularly under long-duration fires, shear distress would be likely to occur.

Bending moments in floor beams increased significantly due to restraint of thermally induced curvatures. End moments increased by a factor of 2 to

4 for the SDHI 0.5-hour and ASTM 1-hour fires, respectively. Bending moments in columns increased by two or three orders of magnitude. Under normal service loads, column moments are practically negligible, but under fire conditions, these moments are significant and may cause flexural distress. In this building, however, the reserve flexural capacity of the columns was large, and distress due to bending moment for the types and durations of fire studied might not have developed.

The top section of the ground-floor column exposed to the ASTM fire after 0.5-hour and 1-hour durations is shown in Fig. 10. Owing to the deflected shape of the column (Fig. 9), tension stresses developed on the cool side. Although these stresses were not sufficiently high to cause cracking during the first 0.5 hour (the tension stresses were partly offset by the large compression load transmitted from the top 12 stories), tension stresses increased as the bending of the column increased with time and the cool outward face cracked. The cracking zone at 1 hour extended just beyond the inside faces of the steel reinforcement.

Large compression stresses due to thermal gradients developed in the concrete directly exposed to fire and it crushed early. In modeling the structural behavior of concrete in FIRES-RC II, it is assumed that a state of "crushing" corresponds to a large compression strain in concrete, ϵ_u, at which the concrete can no longer sustain any significant compression. In this study, a constant value of 0.004 was assumed for ϵ_u.

The end section of the floor beam just above the ground-floor ASTM fire for 0.5-hour and 1.0-hour durations is shown in Fig. 11. Characteristically, the concrete directly exposed to fire had already crushed at 0.5 hour, and the layer of concrete just inward of this crushed zone was subjected to some compression (unshaded area in Fig. 11). The concrete in the flange of the T-beam had cracked due to large restraining end moments opposing the end rotations which would result from thermal gradients in a free condition. After 1 hour, the crushing increased, and some of the previously cracked concrete was subjected to compression, but a large portion of the concrete remained cracked.

Using the data from the case study described above, several structural response limit states were explored. These limit states, as well as others, could be used as indicators of damage limits. The seven states studied here are:

1. Initial cracking in concrete,
2. Cracking of at least 30% of concrete area,
3. Initial crushing in concrete,
4. Crushing of at least 30% of concrete area,
5. Initial yielding of steel reinforcement,
6. Reduction of axial stiffness (AE) to 70% of its original value (unexposed to fire),
7. Reduction of bending stiffness (EI) to 70% of its original value.

The time of occurrence of these limit states during the two fires is shown in Table 2 for two elements in a ground-floor fire: (1) top of a column, and (2) end of floor beam section. The responses selected for Table 2 are entirely arbitrary. They have been used to illustrate the procedure by which analytical solutions can be used to develop appropriate limit-state criteria in terms of fire type and duration of acceptable performance within prescribed unambiguous limit states.

It is interesting to note that the response times corresponding to the local ASTM fire (standard column test) show little congruence with the response times of the column exposed to an ASTM compartment fire. For example, in a local fire simulating test conditions, 30% cracking and 30% crushing are calculated to have occurred at 0.6 and 0.8 hour, respectively; these states are not reached at all during the first hour of the ASTM fire in a two-bay compartment. The reduction in bending stiffness to 70% of original value occurred early in the local fire (0.2 hour), but in the compartment fire did not take place until much later (0.4 hour).

On the other hand, it is somewhat surprising that for the beam end section selected, the response times for the ASTM and SDHI fires are rather similar for concrete. An essential difference, however, is that steel yielding took place at about 0.5 hour of the ASTM fire but did not occur at all under SDHI fire exposure.

To develop fully such an approach to design, a specified number of fire categories and corresponding levels of tolerable damage must be identified. It has been suggested that four fire categories would provide the necessary criteria for a broad range of building types and uses [10]. These four fire categories and corresponding damage levels are shown in Table 3 and Fig. 12.

In order to determine probable damage levels, thermal and structural response to the critical fire environment expected in a particular building must be evaluated by calculating time variations in temperature distribution within the structural elements, as well as deformation and stresses in these elements, and initiation and extent of degradation (cracking, crushing,

Table 3

PROPOSED FIRE TYPES AND DAMAGE LEVELS

| | Fire | | Structural Response— |
Category	Intensity	Duration	Damage Level
1	Low	Short	Nonstructural damage only
2	Low	Long	Some structural damage; no collapse
3	High	Short	Some structural damage; no collapse
4	High	Long	Specified endurance (hours)

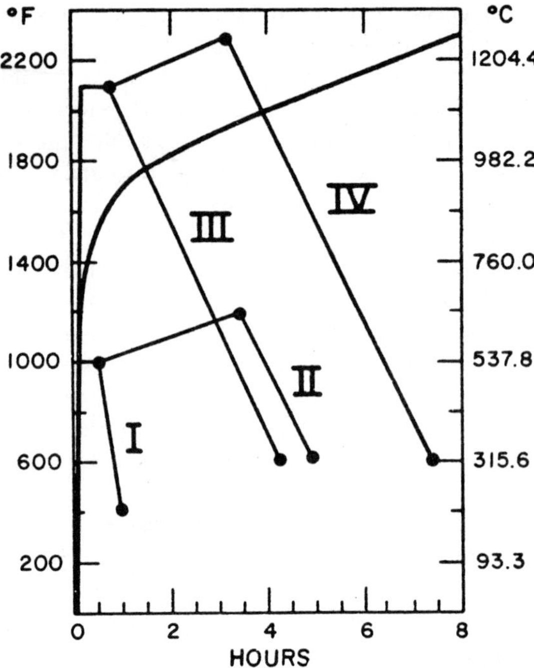

Figure 12 Proposed four categories of design fires.

yielding, or rupture) for different types of fire. Evaluation of structural response should also account for different types of fire, for the cooling as well as growth phase of a fire action, and for conditions of restraint by a building system. This is essential for economical and safe design, as such information is needed to select trade-offs between various means of fire protection versus additional structural integrity, and to assess in-place performance.

CONCLUDING REMARKS

The current practice of building design for fire safety ranges over a broad spectrum, with substantial disregard for fire safety at one extreme and a more-or-less empirical set of criteria for fire safety at the other. Fire design procedures based on calculation of critical fire environments and estimation of corresponding fire endurance have been proposed [11]. Alternatively, "runaway temperature" has been suggested as a design criterion [12].

Neither concept relates the fire endurance or runaway temperature to damage. Bresler *et al.* suggested using damageability limit-state criteria derived from analytical evaluations for appropriate fire categories. Four design fire categories have been suggested, characterized by two levels of intensity (low and high) and two duration periods (short and long). The acceptable damage levels for each design fire have been specified in general terms.

Selection of an appropriate design fire for a particular building would depend on combustible contents; potential ventilation of the fire; detection, alarm, and automatic extinguishment devices; and local fire control services. Selection of an acceptable damage level would depend on the number and safety requirements for occupants, evacuation routes, economic and social costs of damage, and cost of repair or replacement.

Having established a design fire category and an acceptable level of damage, evaluations of damage based on selected limit-state criteria could be used in design to provide a more reliable and consistent level of safety.

Limit-state criteria for fire safety and acceptable damage levels must be based on extensive investigation of damageability of differing structural systems in varying fire environments. Laboratory testing in a fire furnace greatly limits the scope of investigation due to cost, required time, and size and similitude of subassemblages. Furthermore, laboratory testing cannot readily provide details of behavior such as initial cracking, initial crushing, extent of cracking, extent of crushing, first yielding, influence of varying degrees of restraint, and influence of different fire environments. Use of analytical methods can, however, provide such information, leading to the development of design criteria based on structural response and limit states of damageability.

An essential part of the analysis described above is the evaluation of structural fire response. Since realistic fire environments virtually defy experimental modeling, an analytical approach promises to be more powerful provided that test data validate analytical simulations. Limited comparisons of analytical predictions to selected data obtained in the laboratory and observations of structural response in actual fires show good agreement [10,13]. Additional verification of the analytical methods by comparing predicted response to observed test data should lead to further improvements in reliability. In particular, further detailed investigation with respect to three-dimensional structural behavior and probabilistically based sensitivity analyses is of great importance.

Detailed information on constitutive laws and physical properties of materials under arbitrary temperature conditions, mathematical models of critical compartment fires, and corresponding boundary conditions, and computational algorithms are crucial preconditions for a reliable analytical approach.

Extensive tests have been carried out to determine material properties, and model constitutive laws of concrete and steel under a wide range of temperature have been proposed [3–5,14]. Available results allow analytical simulation of structural fire response at high temperatures, although other studies of material properties now under way will lead to further improvement in the reliability of analytical solutions.

The above suggestions regarding the overall approach to design for fire safety are based on limiting damage due to fire. The suggestions provide only a tentative framework; extensive analytical and experimental studies will be needed before acceptable practical recommendations can be specified.

REFERENCES

[1] BRESLER, B., "Fire Protection of Modern Buildings: Engineering Response to New Problems," Eleventh Annual *Henry M. Shaw Lecture*, Department of Civil Engineering, North Carolina State University, Raleigh, 1976.

[2] BRESLER, B., AND IDING, R., "Effects of Normal and Extreme Environment on Reinforced Concrete Structures," *Proceedings of the Douglas McHenry International Symposium*, American Concrete Institute, 1978.

[3] IDING, R., BRESLER, B., AND NIZAMUDDIN, Z., "FIRES-T3, A Computer Program for the Fire Response of Structures—Thermal (Three-Dimensional Version)," *Report No. UCB FRG 77-15*, Fire Research Group, Structural Engineering and Structural Mechanics, Department of Civil Engineering, University of California, Berkeley, 1977.

[4] IDING, R., BRESLER, B., AND NIZAMUDDIN, Z., "FIRES-RC II, A Computer Program for the Fire Response of Structures—Reinforced Concrete Frames (Revised Version)," *Report UCB FRG 77-8*, Fire Research Group, Structural Engineering and Structural Mechanics, Department of Civil Engineering, University of California, Berkeley, 1977.

[5] NIZAMUDDIN, Z., AND BRESLER, B., "Fire Response of Reinforced Concrete Slabs," *ASCE Preprint 2868*, Dallas Convention, April 1977.

[6] BECKER, J. M., AND BRESLER, B., "Reinforced Concrete Frames in Fire Environments," *Proceedings of American Society of Civil Engineers*, Vol. 103, No. ST1, January 1977, pp. 211–224.

[7] WILSON, E. L., AND FARHOOMAND, I., "Non-linear Heat Transfer Analysis of Axisymmetric Solids," *Report UC SESM 71-6*, Structural Engineering Laboratory, University of California, Berkeley, 1971.

[8] ZIENKIEWICZ, O. C., *The Finite Element Method in Engineering Science*, McGraw-Hill, London, 1971.

[9] BRESLER, B., THIELEN, G., NIZAMUDDIN, Z., AND IDING, R., "Limit State Behavior of Reinforced Concrete Frames in Fire Environments," *Proceedings of the Regional Conference on Tall Buildings*, Hong Kong, September 20–22, 1976.

[10] BRESLER, B., "Response of Reinforced Concrete Frames to Fire," *Preliminary Report*, 10th Congress of the IABSE, Tokyo, September 6–11, 1976.

[11] PETTERSON, O., MAGNUSSON, S. E., AND THOR, J., "Fire Engineering Design of Steel Structures," *Publication 50*, Swedish Institute of Steel Construction, Stockholm, 1976.

[12] SKINNER, D. H., "Runaway Temperature—A Design Criterion Based on the High Temperature Properties of Steel," *BHP Technical Bulletin*, Vol. 16, No. 2, November 1972.

[13] IDING, R., LEE, J., AND BRESLER, B., "Behavior of Reinforced Concrete Under Variable Elevated Temperatures," *Report UCB FRG 75-8*, Fire Research Group, Structural Engineering and Structural Mechanics, Department of Civil Engineering, University of California, Berkeley, 1975.

[14] NIZAMUDDIN, Z., *Thermal and Structural Analysis of Reinforced Concrete Slabs in Fire Environments*, Ph.D. Dissertation, Department of Civil Engineering, University of California, Berkeley, 1976.

BEHAVIOR OF STEEL STRUCTURES

Theodore V. Galambos, F. ASCE*

INTRODUCTION

The purpose of this paper is to present a critical appraisal of the state-of-the-art of understanding the behavior of steel structures and to define needed developments in research. The views expressed here are those of the author and thus they reflect his particular point of view and result from his particular experience. The field is so vast that it appears hardly possible to absorb all that is known about the structural behavior of steel, nor can one expect to express views on the state-of-the-art of current (1977) knowledge without some bias. However, such views are presented here in the hope that they may focus the attention of the readers on the problems yet to be solved. In the long run, society must conserve its resources and the structural engineering profession must direct its efforts toward better utilization of all construction materials. This can be achieved through a wide use of what is known and by intelligently channeling research and development in the direction of greatest need.

---

*The Harold D. Jolley Professor of Civil Engineering, Washington University, St. Louis, Missouri.

GENERAL COMMENTS
ON THE ROLE OF RESEARCH

In structural engineering there is no central overall agency or group that controls and directs research. There are no long-range plans or all-encompassing research directives. Research is performed because a need exists or because a certain set of ideas are interesting and intellectually stimulating.

The history of research in steel construction is rich in heritage and it has its origins in the beginning of the nineteenth century. During these two centuries, periods of intense development alternated with quiescent times. Almost half a century passed without much change since the recognition of the concepts of the tangent and reduced moduli by Engesser and Considere at the turn of the nineteenth century until Shanley's resolution of the column paradox in 1945. Once the theory was properly understood, further developments poured forth almost overnight.

The relationship of practice and research in steel construction is very complex. Research sometimes leads the way for new construction and design developments (e.g., plastic design, the "staggered truss," etc.), but often follows practice when a new scheme does not work, leading to difficulties or even failures (e.g., research on thin box girders after the failure of several bridges in the years 1969 to 1972, to mention only one notable example).

Structural steel construction and design practice, as well as research, has had some remarkable achievements of which the profession can be rightly proud. This has been true especially since the end of World War II in 1945, when such developments as cable-stayed bridges, high-strength bolted joints, composite construction, plastic design, and many other developments flourished and matured.

In structural steel construction, design, and research, we are encountering a very sophisticated and mature condition. Owing to the cost and importance of individual steel structures, much more thinking and money has been expended than, say, for masonry buildings which represent perhaps a much larger overall expenditure of construction dollars. For this reason, there is little that is not known about the behavior of steel structures and major new developments that will result in order-of-magnitude changes in understanding are probably not to be expected. The changes that can be expected will be in the areas of refinements, in optimization, in simplification, and in unification of the concepts.

Given the mature state of understanding, what, then, are the needs for further research? Some specific instances will be enumerated later in this paper, but there are some general areas that can be outlined here.

1. Discounting undiscovered and unexpected material defects and gross errors of omission and negligence, the major cause of catastrophic failures in steel structures has been the pushing of practice beyond theory.

That is, some effect, which in the normal previous experience was secondary, becomes of primary importance when construction practice is extended. There are many examples of this, but one example will suffice. During the late 1960s and early 1970s there were many near-failures, and four catastrophic bridge collapses (Fourth Danube Bridge, Vienna, Austria, November 1969; Milford Haven Road Bridge, Wales, United Kingdom, June 1970; West Gate Bridge, Melbourne, Australia, October 1970; Rhine River Bridge, Koblenz, German Federal Republic, November 1971), which caused the loss of some 70 lives. Each of these bridges failed during construction, and the specific causes of collapse could be traced to one or several specific shortcomings of the construction or the design and execution of the erection process. However, when all the excusable and inexcusable errors of calculation, fabrication, and erection were accounted for, it still remained that the design of the steel boxes was inadequate and that the reliability of the completed structures under extreme loads was deemed insufficient. The reason was that for the particular geometry of these thin-walled stiffened boxes, the linear buckling theory, which is satisfactory for many applications, was not adequate. A satisfactory design model must consider the influence of unavoidable initial plate imperfections. Another problem that could be cited is the effect of magnified forces in steel frames due to lateral deformations (the P–Δ forces).

The determination of such secondary effects is very expensive and time-consuming, even though research has provided adequate tools and realistic models. In view of the possible catastrophic consequences, the tendency is to try to compute these secondary effects in all cases. However, in most instances they prove to be, indeed, "secondary" and negligible. Thus one research need is to classify the types of secondary effects and to provide the designers with a knowledge of the parameter domains where they need not be considered, and, when they must be accounted for, to provide rigorous tools rather than grossly conservative simplifications. This process is under way for steel frames and for box girder bridges, but much more rigorous and extensive experimental and computational research is required before the problems are all covered.

2. The major effort of structural steel research, the exclusive emphasis of design specifications, and the primary concern of optimization schemes has concentrated on the strength of steel structures. Yet, the usual controlling design criterion is not a limit state of strength but some limit state of serviceability: deflection, drift, vibration. Steel can be made as strong as needed, but its stiffness is limited. Our understanding of the strength of steel structures is quite complete and very sophisticated; on the other hand, our concepts of serviceability limits is primitive. Research on a major scale should be performed on defining the functional, physiological, and psychological bases for serviceability limit states and on modeling design methodologies that

design and optimize for these limit states. In this area of research some major breakthroughs in economy could still be obtained. Some effort has gone into such research, but not nearly on the scale which the problems deserve nor on the scale extended for research on strength.

3. Properly designed, constructed, and maintained steel structures have a very long life. Some original nineteenth-century structures are still in service (e.g., the Brittania Tubes from the 1840s and the Eads Bridge from the 1870s), and there is no reason why some of our modern steel structures should not be around 300 years from now. Steel structures are very expensive, representing a major investment of society's resources, and they should be maintained through many generations. Time provides two enemies: fatigue and corrosion. While much sophisticated knowledge is available to deal with these problems, it is not fully used. We need reliable and sophisticated models for the concepts of damage, inspection, and repair. Much work needs to be done on such theories in order to make them practically operational and to employ these theories in optimization schemes.

Although it might be in the interest of society to retain some structures for centuries, the life span of many structures may be only 50 years. The functional usefulness is then exhausted, but the material is still good. Research should be conducted on developing optimal schemes of construction that would permit reuse without scrapping the steel. Such a scheme, together with a good theory of damage, would permit a reduction of energy use and a better utilization of financial resources.

4. Our knowledge of the behavior of steel structures under monotonic static loads is truly phenomenal, but real life offers few instances of such loads. Loads due to occupancy fluctuate, but they can be controlled by prudence and foresight through laws. Loads due to the environment also fluctuate, and this variation can be defined by observation through time. These fluctuations can be studied, and their effect on serviceability can be intelligently accounted for. However, once in a while the environment runs amok and then violent and uncontrollable forces tear and wrack the structure. There is a need to know the behavior of steel structures under two kinds of dynamic loads: the moderate kind, which are expected to occur annually or at least a few times during the life of the structure, and the violent kind, which is really not expected (i.e., it has a very low probability of occurrence) but for which the structure must be prepared. The behavior against relatively moderate dynamic loads due to wind or earthquake is well understood, and it is handled in design with reasonable success. However, the behavior against violent events is poorly understood. How does a steel structure resist a tornado or a large hurricane? How does it react to a large earthquake? How do these forces reach the structure and how are they distributed in time and space? How does the structure interact with its cladding, and how do the internal components (e.g., pipes) affect behavior? We need large-scale tests

in the field and reliable observations during the catastrophic events to generate realistic models before we can achieve a satisfactory understanding. Research in this area is under way, but a much greater effort is needed to arrive at workable results.

This enumeration of research needs could go on to include work to understand properly the effects of temperature and fire, to know the basic properties of the material, to be able to measure the forces in the structure, and so forth. Suffice it to say, the easy problems have been solved and the conceptually difficult ones must be thought out and formulated before our truly sophisticated methodology can be brought to bear on their solution.

SOME EXAMPLES
OF RESEARCH ACCOMPLISHMENTS

Research and construction developments on the behavior of steel structures has been performed on a major scale in all the industrialized countries of the world, especially since the end of World War II. There have been numerous major books and innumerable articles and conferences to disseminate this knowledge. In addition, there is an immeasurably rich fount of experience within various organizations concerned with steel construction. An enumeration of some of these topics of research follows:

1. High-strength bolted joints.
2. Welded joints and connections.
3. Plastic analysis and design.
4. Plate, member, and structure instability in the elastic and the inelastic range.
5. Fatigue of components, members, details.
6. Composite beams, columns, structures.
7. Computational schemes, linear, nonlinear.
8. Finite-element applications.
9. Member and joint behavior under repeated loads.
10. Curved girder bridges.
11. Steel arches.
12. Bridge dynamics under wind and vehicles.
13. Brittle fracture and fracture mechanics.
14. Thin-walled box girders.
15. Probability-based codes.
16. Optimization.
17. Decision tables for steel codes.
18. Building dynamics under wind and earthquake.

19. Behavior under earthquake loads.
20. Structural systems.
21. Energy absorption capacity.
22. Corrosion resistance.
23. Material properties.
24. Deflection stability.
25. Cable-stayed bridges.
26. Shell stability and analysis.
27. Reticulated roofs.
28. Tubular joints and structures, pipelines.
29. Tall buildings with unusual framing schemes.
30. Floor and frame vibrations and human response.
31. Welding technology.
32. Temperature and fire effects.
33. Transmission poles and towers.
34. Ship structures.
35. Underwater and offshore structures.
36. Field observations of bridges and buildings.
37. Full-scale tests of actual structures.

This list, by no means exhaustive, illustrates the wide range of interests and problems. The accompanying literature is vast and sophisticated, but a truly clear overview and a fully unified treatment has so far eluded the writers of textbooks. It would be impossible to discuss in detail all these areas. One area, the problem of unbraced beams, will be explored subsequently in some detail to illustrate several general aspects of the state-of-the-art of research.

First, however, one general observation: each of the research topics listed above has associated with it literature, specialists, committees, sponsors, and industry groups. Each was attacked to meet a need, but a given topic is rarely fully explored, and more work is therefore needed to gain more understanding or to bring consistency to research in overlapping areas.

LATERALLY UNSUPPORTED BEAMS

The problem area of laterally unsupported beams concerns the study of the behavior of unbraced beams or beam segments when failure is due to the inability of the structure to continue to support load because of the uncontrolled growth of lateral and torsional deformations. It is here selected to exemplify structural research: it is an old problem, it has a vast and continuous literature reaching back almost 100 years, it is practically important, and current (1977) research activity is in progress.

In the early decades of the nineteenth century, Navier presented engineers with a model of elastic in-plane bending which serves us well to this day for most practical applications. This model, however, did not include a means of predicting the limit state of lateral-torsional instability of slender iron and, later, steel beams when their compression flange was laterally unsupported. A suitable model for such instability was found in the Rankine–Gordon formula, which had been used by Tredgold since the 1830s for predicting the capacity of axially loaded struts and which was experimentally verified by Hodgekinson's experiments in the 1840s. It is based on a type of an approximate first-yield limit state for imperfect beam columns, the forerunner of many such approaches used since then. The Rankine–Gordon formula was adapted to beam instability by assuming the compression flange to be an axially loaded strut (Fig. 1). It was later used also for the shear panels of plate girders and the same type of formula, with different coefficients, was used in the Specifications of the American Institute of Steel Construction (AISC) of the 1920s and 1930s for columns, beams, and plate girders, and it has to this date (1977) not entirely disappeared from some codes [7].

The Rankine–Gordon column model was practically very successful and it dominated structural steel design for a century or more. Theoretically,

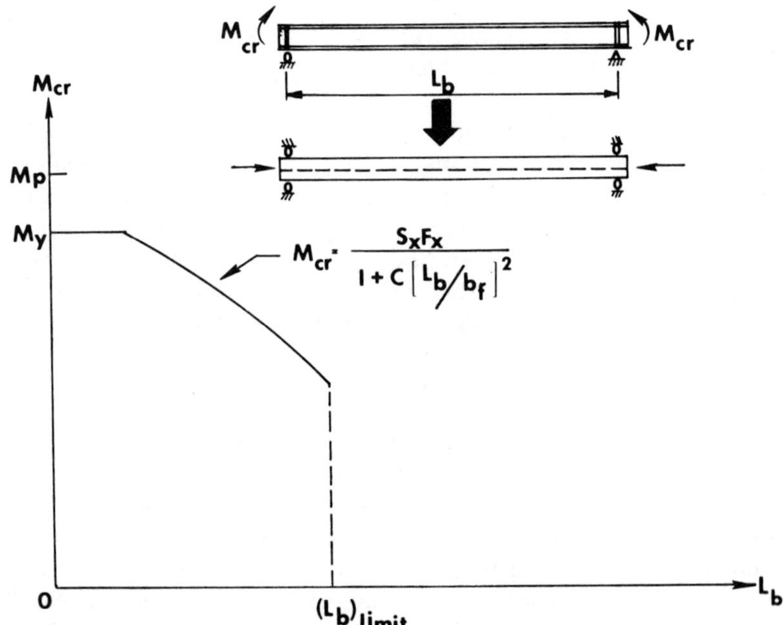

$$M_{cr} = \frac{S_x F_x}{1 + C\left[L_b / b_f\right]^2}$$

Figure 1   The Rankine–Gordon formula [7].

however, it was a very restrictive instrument. It could not handle a number of important situations: end restraint, moment gradient, and continuity. Even though a much more sophisticated and versatile model emerged at the end of the nineteenth century, it took the steel design profession some 50 more years to convert to this more rational theory in its daily design practice. This illustrates the danger in many other areas of structural steel research: a successful, simple, but theoretically rather superficial model can restrict development and use of better and more general models. Many other examples could be cited, especially in areas where design criteria are based almost entirely on experiment alone. The point is that the more complete and sophisticated the behavioral model, the more practical the outcome for the design profession in the long run. Experiments can be more readily interpreted, numerical studies can be made with confidence, and simple design approximations can be defined for practically dominant parameters.

Around the turn of the twentieth century, Prandtl, Michell, Reissner, and Timoshenko formulated the model of elastic lateral-torsional buckling (Fig. 2). This model conformed to the classical Eulerian bifurcation theory and was in the long run able to account for end-restraint, cross-section, material, loading, and in-plane continuity. The full development of this theory had to await the refinement of computational and mathematical techniques, but since the advent of the computer, a large share of all practical elastic lateral-torsional buckling problems has been solved. In fact, modern

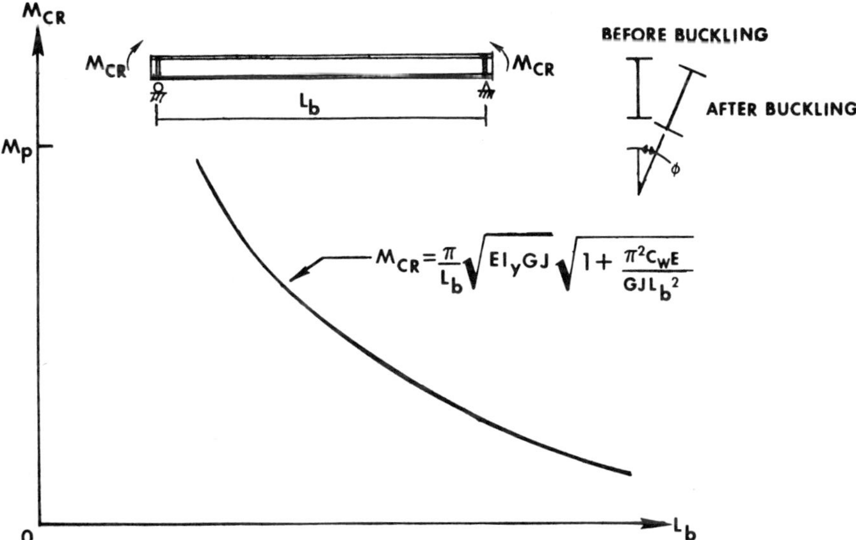

Figure 2　Elastic lateral-torsional buckling behavior of a simply supported beam [9].

finite element techniques no longer require that the complete original differential equations be considered [13]. However, such a situation could not
exist without the carefully thought out basic model with its complicated
equations.

Research and design literature of the years since 1945 is full of numerical results, tables, charts, computational aids, programs, and simplified
and tested design equations [10]. New problems continue to be solved and
published [16], and elastic lateral-torsional buckling problems serve as test
cases for computational schemes designed for more general problems. Experiments have provided excellent correlation with prediction (Fig. 3), with the
divergence between test and theory essentially reflecting the care with which
the experiments were performed [8].

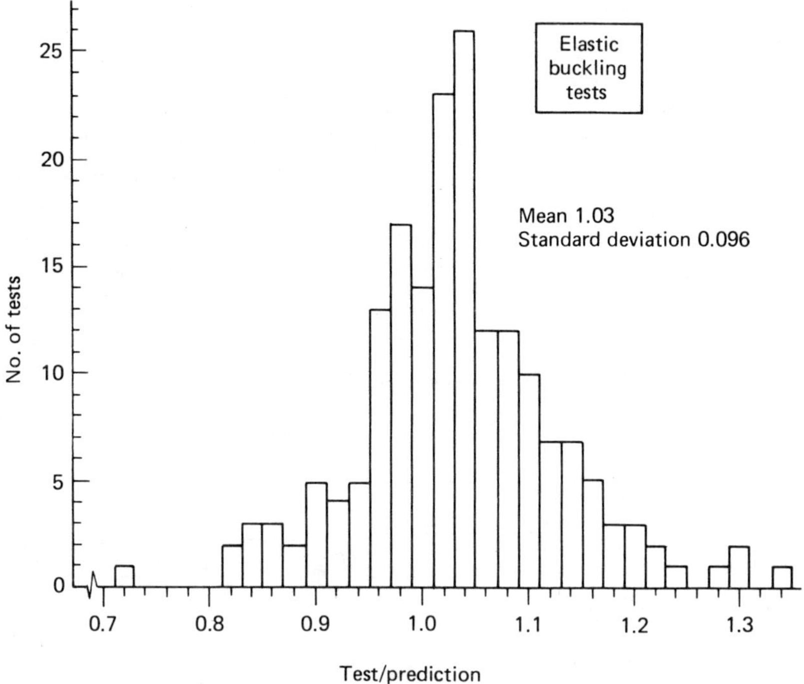

Figure 3   Elastic lateral-torsional buckling tests [8].

Prandtl's solution for the elastic critical moment of a simply supported
rectangular beam represented a very successful model for predicting beam
instability, leading not only to many practical results but also demonstrating
consistency with overlapping problems: beams proved to be a limiting case
of beam columns, lateral-torsional buckling was shown to be a limiting case

of general biaxial bending and torsion, and predictions from this theory and plate buckling theory coincided where they should (e.g., for angles, tees, cruciforms). This is characteristic of successful models, and it is absent from many other models in modern structural research, especially in areas of overlapping research efforts, such as in the case of steel-encased concrete columns (or is it steel columns with a concrete core?). Thus another need for modern research in steel structures is to achieve such consistency in the case of all types of overlapping problem areas.

The elastic lateral-torsional buckling model is by no means perfect: among other things it neglects cross-sectional distortion and the large-deflection postbuckling problem. These problems were subsequently examined and found to be, by and large, of little practical significance and so of no real detriment to the model [4,11]. However, interest in these problems has definitely revived in the 1970s, partly for theoretical reasons but also for practical considerations [3,15,17]. The former concern themselves with a better formulation of the total load response of structural systems, and the latter arise from a need to define energy absorption capacity and from the technology of light-gage cold-formed construction where interaction between overall and local deformation is of great importance. Much work needs to be done yet theoretically, computationally, and experimentally before a coherent picture will emerge. However, such work is important and should be vigorously pursued.

Except for the last-mentioned aspects of refined modeling, the problem of elastic beam instability is well advanced toward a complete solution. Elastic instability, however, is but one part of the total problem (Fig. 4). In construction, the highest danger of lateral-torsional buckling occurs in the erection phase, where slender beam elements are subject to awkwardly placed heavy loads. This dangerous phase is characterized by elastic buckling. The beams in the final structure are often fully braced and so are able to achieve their full plastic moment capacity (M_p in Fig. 4). There are, however, still many practical beams which fall in between; that is, they become unstable after a portion of their cross section has yielded and they cannot reach M_p. This problem is exacerbated by the presence of residual stresses which cause local yielding and a reduction in stiffness. We should then know where the end points of this inelastic region (L_{bp} and L_{br} in Fig. 4) are and what the form of this curve is (Fig. 5).

The history of the modeling of this beam curve to the left of point L_{br}, M_r in Fig. 4 is closely associated with research on the behavior of axially loaded columns and with research on plastic analysis and design. Research on inelastic axially loaded columns arose from the need for (1) reconciling the paradox introduced by Engesser's reduced modulus and tangent modulus models and (2) the development of rational design procedures

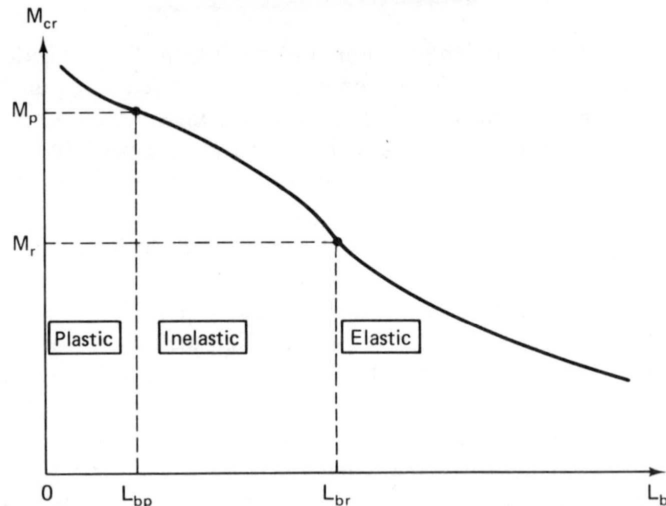

Figure 4 Elastic, inelastic, and plastic buckling of laterally unsupported beams [9].

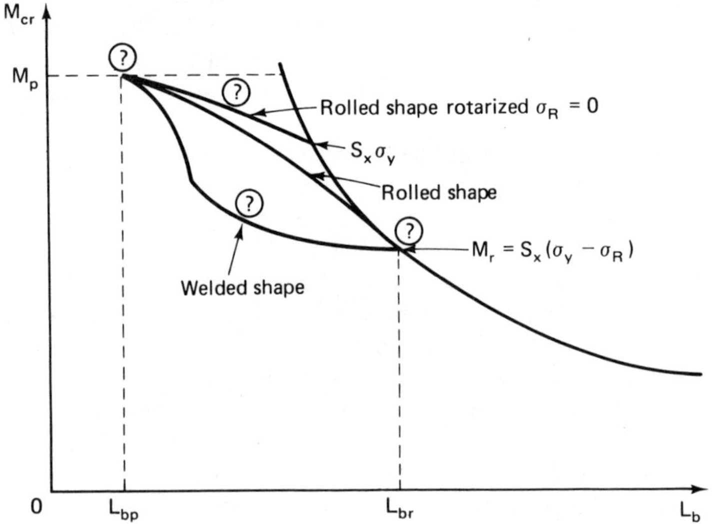

Figure 5 Effect of residual stress on inelastic lateral-torsional buckling [12].

[10]. With Shanley's resolution of the column paradox the way was opened for the formulation of a very good model of axially loaded column behavior, and a vast amount of developmental research, both experimental and computational, has placed this problem into the realm where we know what to do to fill in the gaps of practical needs and to react intelligently to new design requirements. Research on plastic analysis and design arose from the need

to find a rational method of defining the maximum capacity of steel structures [1]. Many valuable and interesting experiments were performed and simple intuitive methods and also very sophisticated computerized techniques evolved for both analysis and design, based on the models of the upper- and the lower-bound theorems. The problem of variable repeated loading was also successfully solved on the basis of the shakedown theorems, opening up the way to current research on the application of plastic theory to steel bridge design.

Research on plastic design and research on columns proceeded along two entirely different paths: for the former a simple overall model applicable to the whole structure was developed, while for the latter, methods to predict in detail what occurs in the fibers of a single member were established. The difficulty of the beam stability problem lies in the fact that the premises of these two extremes had to be bridged. This has not yet been successfully accomplished on a theoretical level, although for practical design purposes satisfactory resolution has been achieved.

The plastic analysis model simply assumes that member and local instability will not interfere with its proper functioning. This is insured by specifying limits of member slenderness and plate slenderness. The research on the definition of these limits concerned itself largely with the limits only, and little was done to relate what happens to the right or to the left of the point L_{bp}, M_p in Fig. 4. The problem is complex because of the interacting effects of strain hardening, cross-section distortion, discontinuous yielding, lateral and local buckling, and force redistribution in continuous beams. There were essentially three efforts to define the point L_{bp}, M_p in Fig. 4, that is, defining the geometric and material properties at the point where the plastic moment can be reached and sustained through an adequate rotation: (1) models that considered lateral buckling and flange and web buckling as independent effects, (2) models where the interaction between lateral and local deformation was considered, and (3) purely experimental means [1]. It is evident from the experiments that the predictions of the theoretical models are conservative. However, tests alone are not sufficient, and it is necessary that a better, more realistic and comprehensive model be found.

The model used most often to determine inelastic beam instability in the transition range (i.e., the curve between L_{br} and M_r in Fig. 4) is based on the tangent modulus concept of inelastic column theory [9]. The application of this theory is simple in concept, although the numerical work is tedious: for a given moment field, cross section, and residual stress pattern, the pre-buckling elastic core is established using in-plane theory and the ideal elastic plastic stress–strain law. Bifurcation–theory elastic lateral-torsional buckling methods are then used to investigate the stability of the member, which now has reduced lateral and torsional stiffnesses. By the definition of the tangent modulus, then, the critical moment corresponds to the beginning of lateral and torsional deformation for a perfect beam (Fig. 6). A great many solutions for beams were determined for various cross sections, loading cases, and residual stress patterns, and a variety of simple formulas were derived for the design of beams. A few solutions also were established for imperfect beams (Fig. 6), and some results for the reduced modulus were also obtained.

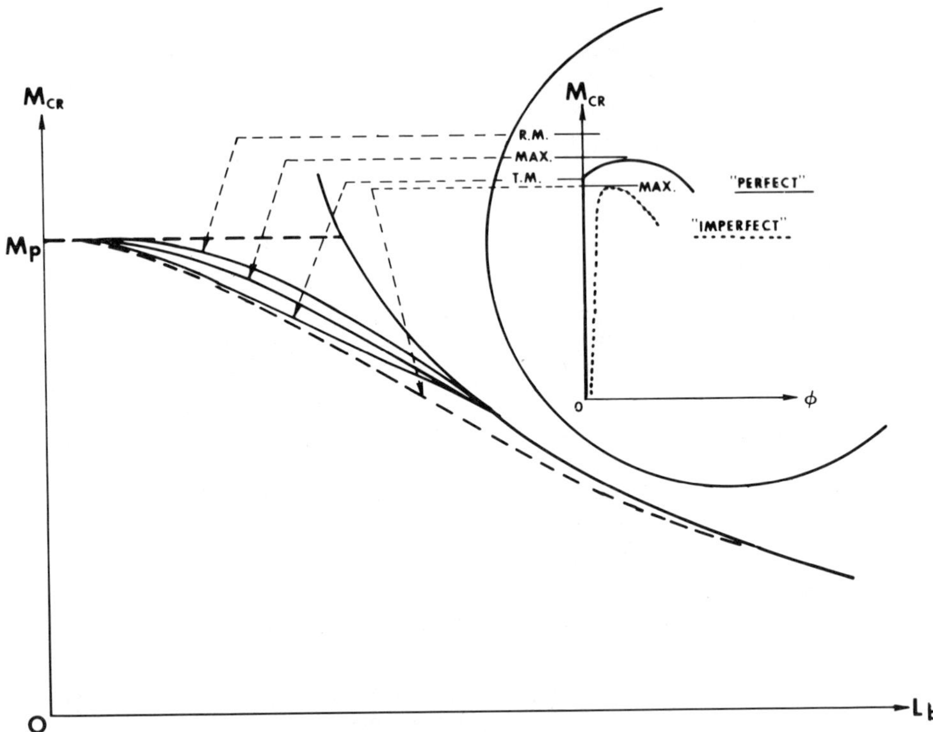

Figure 6   Models of inelastic beam behavior.

There are, however, some problems with the inelastic model based on the tangent modulus concept. While many parametric studies have indicated that from a practical design point of view the faults of the model are of no great significance, the shortcomings of the theory are nevertheless serious and the model should be improved. The main problem is that the solution based on the buckling of the elastic core ignores the differences in the change between flexural and torsional stiffness. Various assumptions with regard to the plastic shear modulus seem to have little effect on the final result for the cases studied, but this does not assure us that it will be so for other cases. What is needed is a model that can account for the simultaneous bending and torsion of a partially yielded beam. Such a model would need to be inclusive enough to incorporate as special cases pure torsion and pure flexure, to account for strain hardening and discontinuous yielding as well as local deformations. If such a model were available, the various problem areas, which now seem isolated, could be combined: torsion, lateral-torsional buckling, and biaxial flexure.

Whether or not such a comprehensive model becomes available, the present simpler model should be further exploited to clarify some practical points. What is the role of residual stress versus initial imperfection—need they be combined, or is one or the other predominant? The schematic curves in Fig. 5 show the effect of the type of residual stress, based on theoretical calculations [12]. According to these predictions welded girders with unfavorable residual stress patterns are relatively weaker than rolled beams (Fig. 5). This prediction appears to be borne out by experiment [6]. All available tests for rolled and welded beams, when averaged, tend to confirm this trend. However, further careful research needs to be performed, as follows: (1) measurements and statistical studies on residual stresses and initial curvatures in rolled and welded beam-type sections to establish reliable means of estimating these properties (most of our knowledge about them comes from column research and it is possible that they are different for beams); (2) careful tests of beams and corresponding analytical predictions based on the actual measured properties; (3) parameter studies to establish realistic prediction curves. In short, what is suggested here is a retracing of the type of research that was performed on axially loaded columns [10] so that an equally well-established theoretical, analytical, and experimental base also exists for beams. Plans for this work are in progress through a Task Group of the Structural Stability Research Council.

In addition to the two major needs for research presented above (i.e., a better model and data for more reliable design criteria) there is a need for a thorough review and expansion of the sporadic work that has been done on problems where torsion and flexure of open section members interact. This is important where torsion, due to skew loading, direct torsional force, or in horizontally curved members, combines with planar or biaxial flexure [14]. The test points shown in Figs. 7 and 8 show that torsion radically reduces flexural capacity.

While the previous comments relate largely to individual beam problems, there has been considerable investigation of problems of continuous beams and frames. Many numerical studies have yet to be performed to answer some unresolved questions. Test points on the nondimensional plot of Fig. 9 indicate that continuous beams have a considerable reserve of strength due to inelastic planar force redistribution [2,5]. Analytical studies to interpret this phenomenon still need to be performed [16].

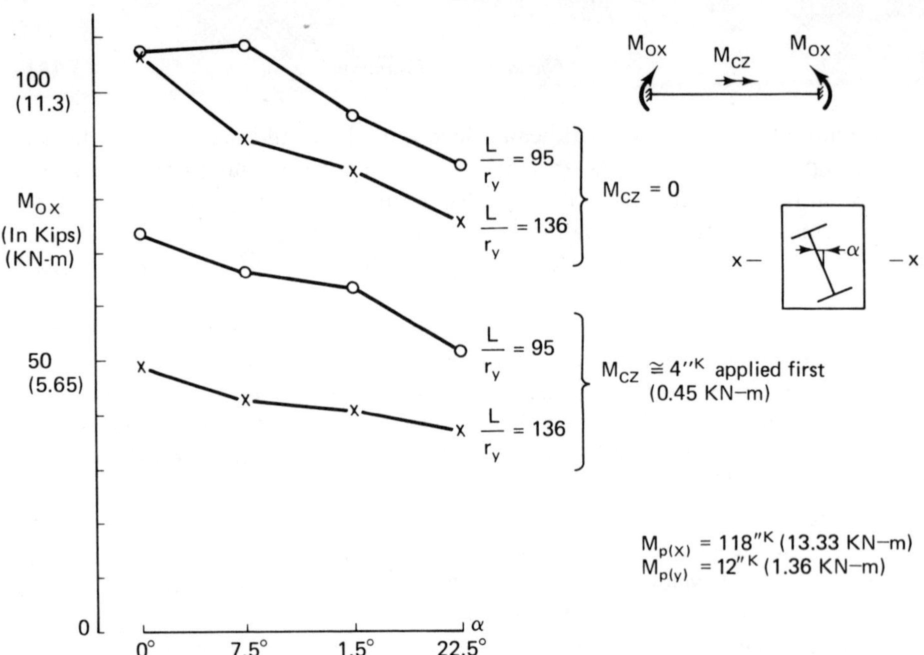

Figure 7 Test results on skew-loaded beams, with and without torsional moment [14].

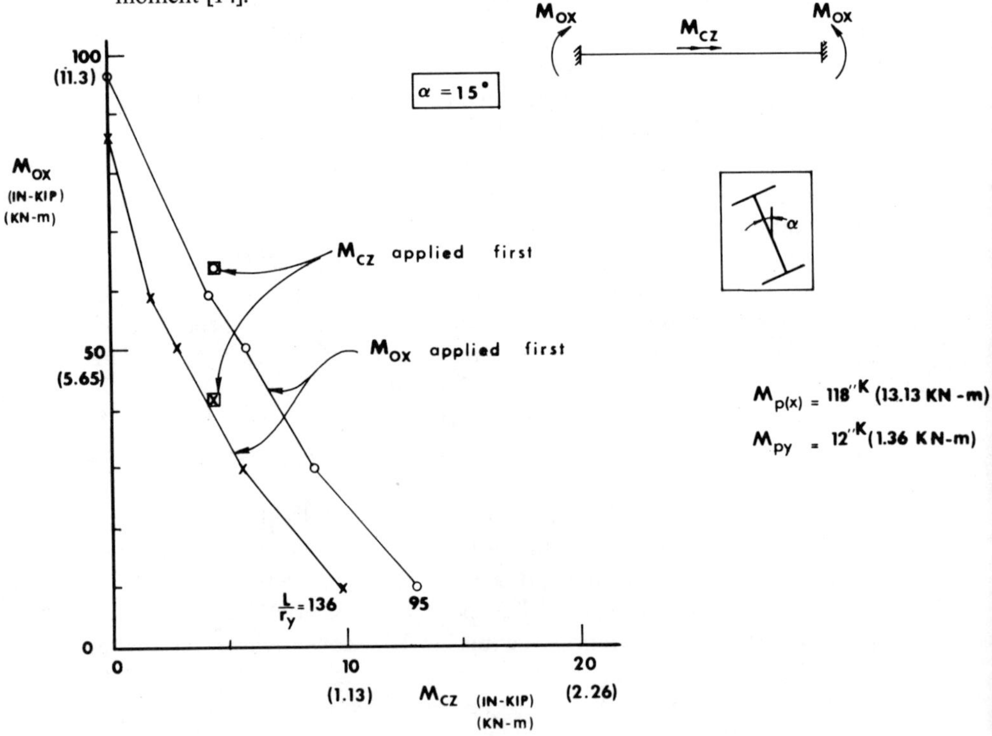

Figure 8 Effect of torsional moment on flexural capacity [14].

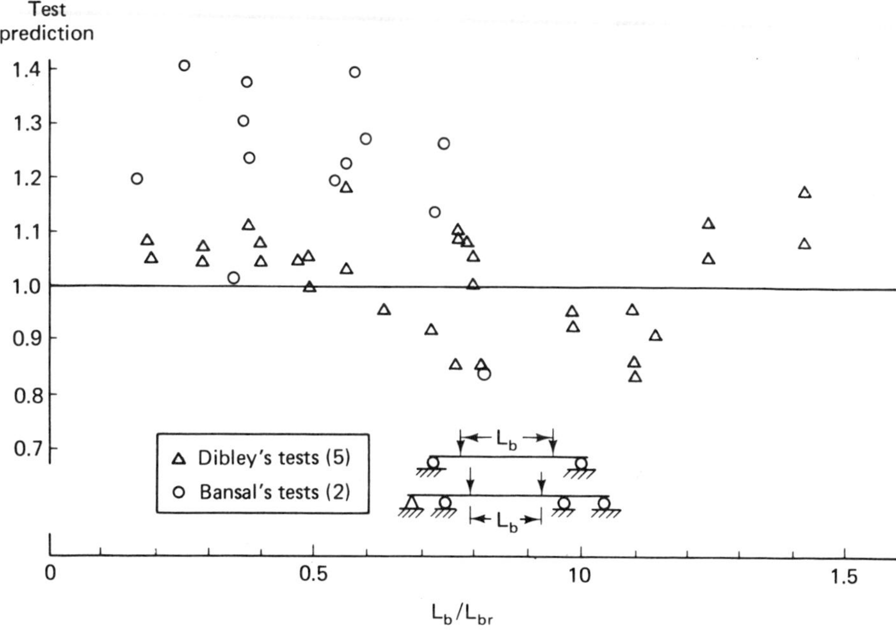

Figure 9 Continuous beam tests.

EVALUATION OF RESEARCH NEEDS
IN STEEL STRUCTURES

The brief review of the state-of-the-art of research on laterally unsupported beams illustrates the situation in many other areas of research on steel structures. While present knowledge is quite extensive, more needs to be done to refine and unify models of behavior; this refined modeling can lead to improvements that eventually will enhance the reliability and economy of construction. More specifically, the needs can be classified into the following categories:

1. Development of a hierarchy of engineering models that are consistent with reality, that coincide at points of overlap, and that can accommodate interaction problems (e.g., plate and member buckling, imperfection and residual stress, torsion and flexure).

2. Testing of the hierarchic model through comparison with available experiments or by conducting new, sophisticated "landmark" tests (in fact, modeling and testing would be iterative).

3. Classification of the various problem areas in steel construction into the applicable models for their solution, and performing the necessary numerical studies and physical experiments to complete a consistent data base

for use in design: charts, tables, simplified formulas, statistical data for probability-based design, information for use in optimization.

Much of the work outlined above has been done, but much still needs to be done in collecting, unifying, and completing. The difficult job ahead is to make a complete and unified picture from the many pieces of the puzzle. I believe that this is the major task of future research in understanding the behavior of steel structures.

ACKNOWLEDGMENT

This paper is a printed version of a talk given in the Symposium on Structural Engineering and Structural Mechanics honoring Professor Egor Paul Popov, to whom I owe much in my development as a structural engineer.

NOMENCLATURE

b_f	:	flange width
C	:	constant in Rankine–Gordon equation
C_w	:	warping constant
E	:	modulus of elasticity
F_y	:	yield stress
G	:	shear modulus
I_y	:	moment of inertia about y-axis
J	:	St. Venant torsion constant
L	:	length
L_b	:	unbraced length
L_{bp}	:	unbraced length when $M = M_p$
L_{br}	:	unbraced length when $M = M_r$
M_{cr}	:	critical moment
M_{cz}	:	torsional moment
M_{ox}	:	end moment about x-axis
M_p	:	plastic moment
$M_{p(x)}$	:	plastic moment about x-axis
M_r	:	moment at first yield in the presence of residual stress
M_y	:	yield moment
$M_{y(x)}$	:	yield moment about x-axis
R.M.	:	reduced modulus
r_y	:	radius of gyration about y-axis
S_x	:	elastic section modulus about x-axis
T.M.	:	tangent modulus
α	:	angle of inclination of skew-loaded beam
ϕ	:	angle of twist

REFERENCES

[1] ASCE-WRC, "Plastic Design in Steel" A Guide and a Commentary, *ASCE Manuals and Reports on Engineering Practice*, No. 41, 1971.

[2] BANSAL, J. P., "The Lateral Instability of Continuous Steel Beams," Department of Civil Engineering, University of Texas at Austin, Aug. 1971.

[3] BAZANT, Z. P., AND NIMEIRI, M., "Large Deflection Spatial Buckling of Thin-Walled Beams and Frames," *Journal of the Engineering Mechanics Division, ASCE*, Vol. 99, No. EM6, Dec. 1973.

[4] BLEICH, F., "The Buckling Strength of Metal Structures," McGraw-Hill Book Company, New York, 1952.

[5] DIBLEY, J. E., "Lateral-Torsional Buckling of I-Sections in Grade 55 Steel," *Proceedings ICE*, Vol. 43, Aug. 1969.

[6] FUKUMOTO, Y., AND KUBO, M., "An Experimental Review of Lateral Buckling of Beams and Girders," Proceedings of the International Colloquium on Stability of Metal Structures, May 17–19, 1977, Washington, D.C.

[7] GALAMBOS, T. V., "History of Steel Beam Design," *Engineering Journal of the American Institute of Steel Construction*, Vol. 14, No. 4, 1977.

[8] YURA, J. A., GALAMBOS, T. V., AND RAVINDRA, M. K., "The Bending Resistance of Steel Beams," *Journal of the Structural Division, ASCE*, Vol. 104, No. ST9, Sept. 1978.

[9] GALAMBOS, T. V., "Structural Members and Frames," Prentice-Hall, Englewood Cliffs, N.J., 1968.

[10] JOHNSTON, B. G., editor, "Guide to Stability Design Criteria for Metal Structures," 3rd ed., Structural Stability Research Council, John Wiley & Sons, New York, 1976.

[11] MASUR, E. F., "Strength of Very Slender Beams," *Journal of the Engineering Mechanics Division, ASCE*, Vol. 83, No. EM4, Oct. 1957.

[12] NETHERCOT, D. A., "Buckling of Welded Beams and Girders," *IABSE Publications*, Vol. 34-I, 1974.

[13] POWELL, G., AND KLINGNER, R., "Elastic Lateral Buckling of Steel Beams," *Journal of the Structural Division, ASCE*, Vol. 96, No. ST9, Sept. 1970.

[14] RAZZAQ, Z., "Theoretical and Experimental Study of Biaxially Loaded Thin-Walled Beams of Open Section with and Without Torsion," doctoral dissertation, Washington University, St. Louis, Mo., May 1974.

[15] WOOLCOCK, S. T., AND TRAHAIR, N. S., "Post-Buckling of Redundant Rectangular Beams," *Journal of the Engineering Mechanics Division, ASCE*, Vol. 101, No. EM4, Aug. 1975.

[16] YOSHIDA, H., NETHERCOT, D. A., AND TRAHAIR, N. S., "Analysis of Lateral Buckling of Continuous Beams," *IABSE Proceedings*, p-3/77.

[17] ZAMOST, G., AND JOHNSON, E. R., JR., "Post Lateral Buckling of Beams," *Journal of the Engineering Mechanics Division, ASCE*, Vol. 97, No. EM4, Aug. 1971.

STRENGTH AND DEFORMATION CAPACITIES OF BUILDINGS UNDER EXTREME ENVIRONMENTS

*Vitelmo V. Bertero**

INTRODUCTION

Statement of problem

The possible occurrence of a severe event, such as an earthquake, wind, flood, or fire, poses special problems when designing new buildings or evaluating the adequacy of existing buildings. There is the understandable desire to construct a building that can withstand the worst event without damage, but the combination of the expense that this would entail and the low probability of such an event occurring suggests less rigid performance criteria. The question is whether it is necessary to prevent collapse and subsequent loss of life or whether expensive damage should be limited as well. Answers to these questions are needed to achieve optimum design. A solution is offered by the philosophy of comprehensive design.

Sawyer has discussed a comprehensive design procedure in which the resistance of a structure at various failure stages is correlated to the probability that possible disturbances can reach the required intensity to induce such failure stages so that total cost (including the first cost and the expected

---

*Professor of Civil Engineering, University of California, Berkeley.

losses from all the limit stages) is minimized [1]. Under increasing loads, structures generally fail at successively more severe failure stages under increasingly less probability that the load will reach the required intensity levels. To illustrate this point, the relationship shown in Fig. 1 gives possible structure failure stages versus a monotonically increasing pseudostatic load for a typical statically indeterminate reinforced concrete (R/C) building [1].

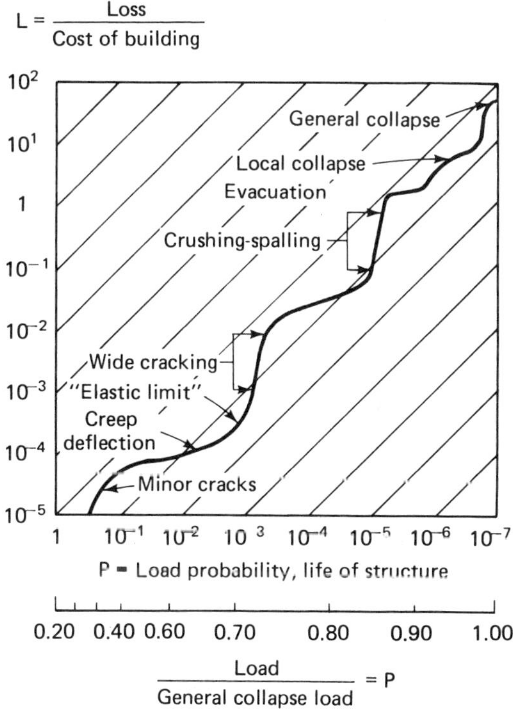

Figure 1 Losses versus load probabilities during the service life of a R/C structure [1].

Owing to the variability of loss for a given load (or the variability of load for a given loss), this relationship represents the mean values of the random variables involved. The full redistribution, as shown in Fig. 2, can sometimes involve large variances [2].

In comprehensive design, identification of potential failure modes requires the prediction of the mechanical behavior of a building structure at each significant level of critical combinations of all possible excitations to which the building may be subjected. For severe environmental conditions, current design methods require that the strength and deformation capacity of a building be established.

The major threat to life and loss of property under natural hazards

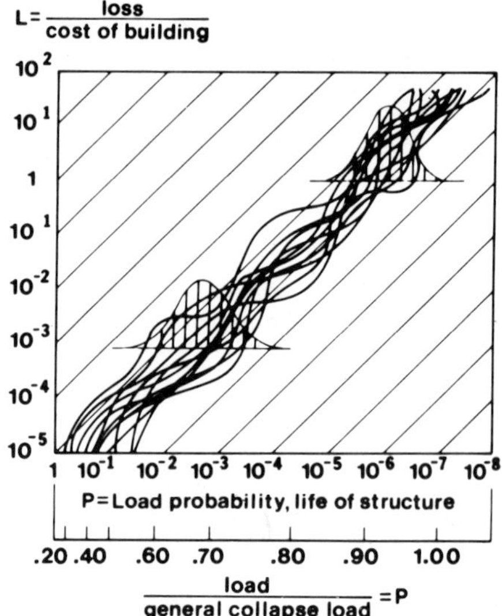

Figure 2   Distribution of losses versus load probabilities during the service life of a R/C structure [2].

stems from existing buildings. To resolve the problem of hazard abatement, existing buildings must be continuously assessed for danger under extreme environmental conditions, necessitating comprehensive analyses of the buildings to predict their strength and deformation capacities.

This subject is particularly appropriate for this symposium because of the very significant contributions that Professor Popov has made in this area. His early work on the inelastic behavior of steel structures [3] and later work on the hysteretic behavior of structural steel material, structural elements, and subassemblages of structures have contributed significantly to our understanding of the complete hysteretic behavior of different types of civil engineering structures. Professor Popov has been greatly responsible for advances in the seismic-resistant design of steel and R/C structures, as well as designs for other severe environments, such as the Alaska Pipeline system [4].

Objectives and scope

The main objectives of this paper are to discuss the importance of predicting the strength and deformation capacities of buildings, those to be designed (new) as well as those existing; to review the state-of-the-art and practice of predicting these capacities; and, finally, to formulate educational, research, and development needs for improving such predictions.

The importance of predicting the strength and deformation capacities of new buildings is demonstrated by comparing currently accepted philosophies for design against extreme environments with design methods presently used in practice. The same need arises for existing buildings in order to resolve the problem of hazard abatement.

The state-of-the-art in predicting strength and deformation capacities of buildings is briefly discussed by reviewing available methods and indicating unresolved problems. The state-of-the-practice is then assessed by comparing it to our present state of knowledge. Based on these findings, recommendations for educational, research, and development needs in this field are offered.

Coverage of this important subject is admittedly biased. The presentation is influenced strongly by the research and design problems on which the author has recently been working, which are in the area of severe seismic excitations. Emphasis is given to discussing concepts rather than to specific predictions of the strength and deformation of a particular building.

IMPORTANCE OF PREDICTING STRENGTH AND DEFORMATION CAPACITIES

New buildings

The importance of accurately predicting the strength and deformation capacities of buildings is evident from reviewing widely accepted design methods based on the philosophy of limit states.

DEFINITION OF LIMIT STATES. Structures must be designed to sustain safely all possible loads and deformations that are liable to occur during construction and use, and to have adequate durability during service life. A structure, or any of its parts, is unfit when it reaches a "limit state" at which it ceases to fulfill the function or satisfy the conditions for which it was designed [5]. To define the different limit states, the various events that might lead to some cost or "disutility" to the occupant, owner, or designer must be identified. The different limit states are presently grouped as either service or ultimate limit states. References [5–7] discuss events normally considered in limit-state design philosophy and applications of this philosophy to practical design methods.

The format used in formulating the limit-state design philosophy encourages the use of probabilistic methods when sufficient statistical information is available [6,7]. Owing to uncertainties in defining the design (critical) excitations due to earthquakes (or other extreme environments) and the structural parameters controlling the mechanical behavior of a building, a probabilistically formulated limit-state design philosophy is well suited to developing seismic-resistant and other design methods for extreme environments. In the comprehensive design procedure, which is one such method, potential modes of failure are identified according to the mechanical behavior of a structure at each significant level of critical combinations of all possible excitations to which the structure may be subjected.

Structural design is commonly based on idealized mechanical behavior under simplified excitations because it is not usually possible to consider actual behavior and the true history of disturbances. Sources, treatment, and effects of the different types of excitations are summarized in Fig. 3 [8]. Structures are usually subjected to unpredictable fluctuations in the magnitude, direction, and/or position of each of the individual excitations that may act on them during their service life; and the extreme values between which each of these excitations will oscillate are the only characteristics that may be estimated accurately. These types of actions are classified in Fig. 3 as generalized or variable-repeated excitations.

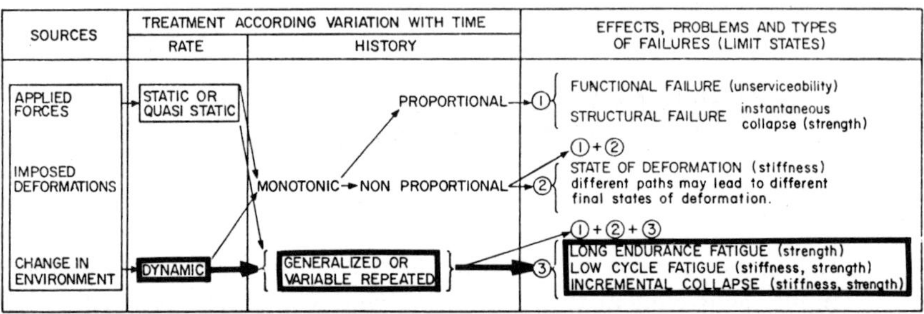

Figure 3   Sources, treatment, and effects of excitations on structures [8].

The types of failures associated with variable-repeated excitations are classified as long-endurance fatigue, low-cycle fatigue, and incremental collapse. Long-endurance fatigue is only critical for special structures. Review of results from experimental data shows that the real danger in low-cycle fatigue, which is associated with repeated reversed inelastic actions (alternating excitations), is not material fracture, as is usually reported in the literature, but deterioration of stiffness, particularly in reinforced concrete [8]. Incremental collapse is related to the progressive development of excessive deflections which occur under the cyclic applications of different combinations of peak actions. Because deterioration of stiffness can lead to an undesirable increase in deformations, the effects of alternating excitations cannot be treated independently, as is usually done, from those caused by excitation patterns leading to incremental deformations [8].

Failure prediction is clearly essential in designing against extreme environments and requires knowledge of the strength at different levels of structural deformation. The discussion above points out the difficulty in predicting strength and deformation capacities and the need for a probabilistic approach, or, at least, for considering the range of probable mechanical behavior and possible excitations. Some pertinent problems encountered in seismic-resistant design follow.

GENERAL GOALS AND CURRENT PRACTICE OF SEISMIC DESIGN. The general philosophy of earthquake-resistant design for nonessential facilities is well established and proposes to prevent: (1) nonstructural damage in frequent, minor earthquake ground shaking; (2) structural damage and to minimize nonstructural damage during occasional, moderate earthquake shakings; and (3) collapse or serious damage in rare, major earthquake ground shakings. This philosophy agrees completely with the concept of comprehensive design, but current design methodologies fall short of realizing its objectives. Practical application of the comprehensive design approach to seismic-resistant design is complex owing to the difficulty of assessing the relationship between loss and seismic excitation. According to comprehensive design, the ideal design is that which results in minimal total cost, including possible losses, at all limit states. This ideal is still to be realized because no practical design method has been developed to satisfy simultaneously all requirements imposed by the different limit states. In practice, the most critical limit state is used as the basis for proportioning members in the preliminary design, and all other main limit states are then checked through a comprehensive analysis. The advantages of developing a design method based on two failure stages have been discussed by Sawyer [1], and a design method based on the two behavioral criteria of collapse and loss of serviceability and on four optimizing criteria has been developed [9]. This method seems feasible and practical for the seismic-resistant design of ductile moment-resisting frames [10,11].

Because current building design practice in regions of high seismic risk focuses on collapse of the main structure as the controlling limit state, prediction of strength and deformation capacities is essential to the design procedure. The resulting design must then be checked for serviceability requirements under normal loading conditions. Before discussing in more detail the failure modes and safety requirements against collapse involved in current seismic design criteria, it should be pointed out that although most buildings damaged in recent severe earthquakes were far from reaching the collapse limit state, structural, and particularly nonstructural, damages were great enough to constitute failure, which indicates a need for a new group of limit states based on damage [12].

Ultimate failure of a building can be induced by different mechanisms acting independently or together. Certain limit states appear to be critical under pseudostatic loads, but negligible under dynamic ones. For example, under a sustained pseudostatic overload, the limit state caused by transformation of the structure into a mechanism leads to instability of the whole structure, which is not usually the case under dynamic loading. Present seismic-resistant design methods are based on the assumption that large displacement (ductility) develops after the structure is transformed into a mechanism. The distinction between pseudostatic and dynamic effects also applies to the case of ultimate limit states caused by deformational instability.

FAILURES UNDER GENERALIZED DYNAMIC EXCITATIONS. Structural collapse can occur from low-cycle fatigue or incremental deformation under excitation intensities lower than those required to induce instantaneous collapse if these generalized excitations are considered as monotonically increasing. As pointed out in [8,12,13], cumulative damage resulting from a severe, long-duration ground motion, a short main shock followed by a succession of aftershocks, or a combination of the main shock and a consequent event or environmental exposure such as fire can lead to either one of the above two phenomena; it is therefore important that these types of failure under generalized excitations be explored further.

Yamada and Kawamura discuss an ultimate seismic-resistant philosophy of reinforced concrete based on low-cycle fatigue [14]. This type of failure is very sensitive to the detailing and the quality control of materials and workmanship used in construction. If proper attention is given to these aspects and errors in design and/or construction are eliminated, the use of present seismic design provisions, with some recently suggested possible improvements [15], would result in structural designs in which low-cycle fatigue is not a controlling parameter. By detailing critical regions of structural members according to recent seismic code provisions [16] and latest recommendations for their improvement [17–19], the energy absorption and dissipation capacities that these members could develop under cyclic reversals of deformation would be so large that they could withstand the effects of the strongest credible seismic motions. Even under the most severe ground motions recorded, the number of possible reversals between maximum opposite peak deformations is not usually large enough to be of serious concern [15]. It should also be noted that the overall P–Δ effect is canceled under full reversals of symmetrically yielding and strain-hardening or strain-softening structures (Fig. 4).

Studies carried out at Berkeley show that low-cycle fatigue must be considered in designing members that are used as structural dampers to dissipate energy [20]. One typical example involves coupling girders in coupled wall systems. Since these elements act as structural safety fuses between two different structural-resistant systems, their failure would not cause brittle collapse of the complete system but only a change in its dynamic characteristics.

Figure 5(a) shows a schematic illustration of incremental ("crawling") collapse. Recent studies indicate that this type of failure must be considered in designing structures at sites near the sources of seismic ground motions with severe, long-acceleration pulses [21]. For example, a multistory steel

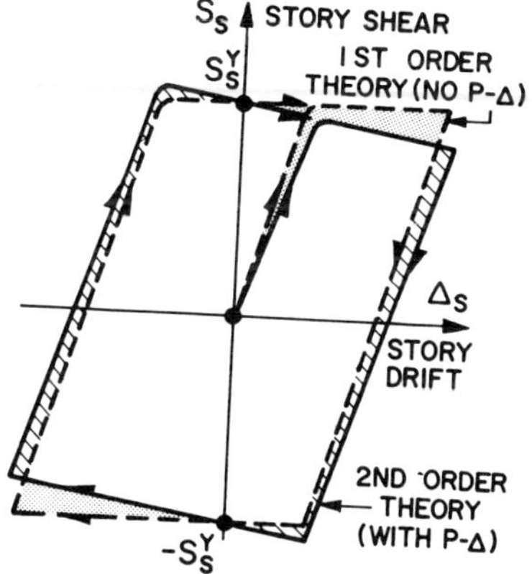

Figure 4 Effect of P–Δ on hysteretic behavior involving full deformation reversals (low-cycle fatigue) [12].

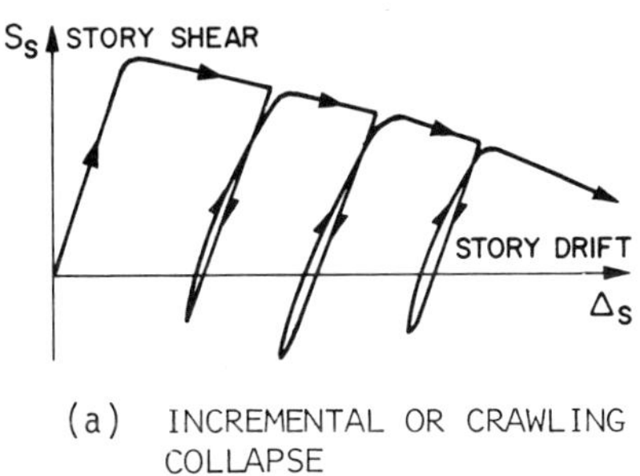

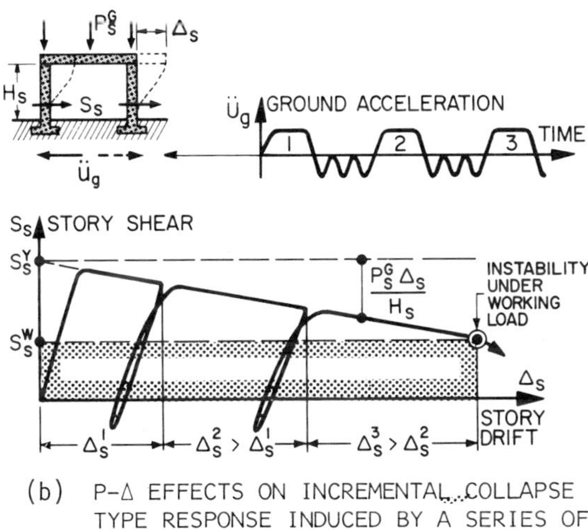

Figure 5 Incremental collapse: (a) incremental or crawling collapse, (b) P–Δ effects on incremental collapse-type response induced by a series of severe acceleration pulses [12].

frame, optimally designed using a nonlinear method and subjected to seismic ground motions derived from those recorded during the 1971 San Fernando earthquake, will collapse from incremental deformations, as illustrated by the second-floor displacement time-history responses shown in Fig. 6. Incremental collapse is aggravated by the high probability that several aftershocks of the same intensity and dynamic characteristics of the main shock will occur. As Newmark and Rosenblueth point out, it is not unusual for a structure that can withstand a major shock to collapse from an aftershock [13].

Although the P–Δ effect is negligible in failure from low-cycle fatigue, it is paramount in incremental collapse. If a structure is subjected to a series

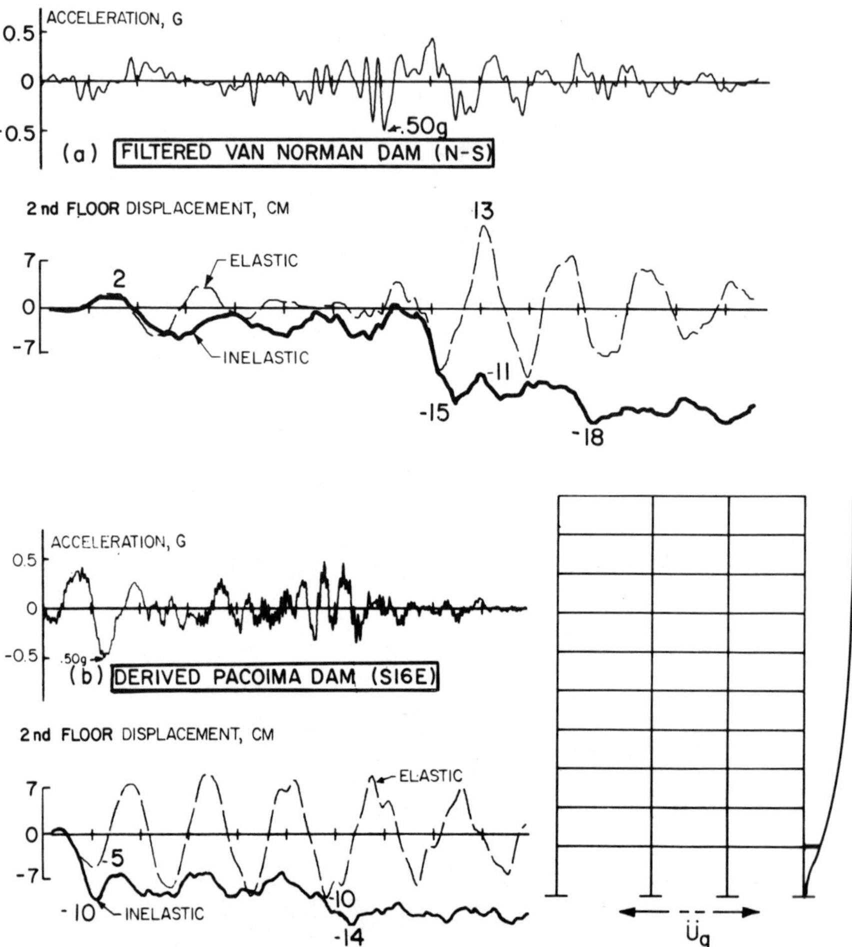

Figure 6 Second-floor displacement time-history responses of a multistory steel frame subjected to the following ground motions: (a) filtered Van Norman Dam (N–S), (b) derived Pacoima Dam (S-16°-E) [10].

of similar acceleration pulses as it is deflected away from its original vertical equilibrium position, the increment in sidesway deflection under each of the successive pulses will increase because the P–Δ effect reduces the structure's available net yielding resistance against lateral inertial forces [Fig. 5(b)]. Accumulation of these increasing incremental deflections can lead to instability under a working-load combination, such as gravity forces plus wind or minor earthquake. From Fig. 5(b) it is clear that structural instability under working loads may be delayed or prevented by reducing the maximum tolerable story drift and/or increasing the yielding strength against lateral forces. As illustrated by Fig. 7(a), if yielding strength or maximum story drift is not modified, the only advantage in augmenting initial stiffness is a small increase in energy absorption and dissipation capacity. This figure also indicates that if initial stiffness is increased without a reduction in tolerable story drift, ductility demands will increase considerably and greater structural damage will occur. A reduction in the acceptable story-displacement ductility generally lessens the danger of instability because it implies an increase in the structure's required strength, which, in turn, usually implies an increase in the required initial stiffness. The end result is a story drift at yielding equal to, or less than, that of a structure with a lower yielding strength and a considerably smaller story drift at the ultimate condition.

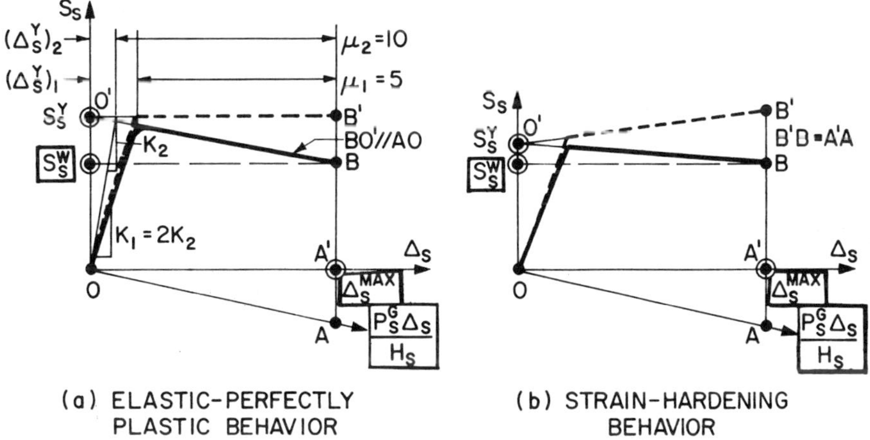

Figure 7 Determination of required yielding strength to avoid instability due to P–Δ effect [12].

The behavior depicted in Fig. 5 suggests an approximate design method, illustrated in Fig. 7, for delaying or preventing deformational instability under working-load levels. The method is based on the assumption that maximum tolerable story drift, Δ_s^{max}, and the story shear due to lateral working loads, S_s^W, are known. These values define point B in Fig. 7. Total axial force acting on a story during severe lateral seismic shaking is also assumed to be known, since it depends only on the gravity forces acting above that story, P_s^G. Thus the reduction in the net lateral resistance of the structure due to the P–Δ effect for any given value of Δ_s can be computed as $P_s^G \Delta_s / H_s$, line OA in Fig. 7.

Two different examples of possible inelastic behavior are considered in Fig. 7. If the mechanism deformation is perfectly plastic, it is sufficient to draw a line, BO', parallel to OA through point B [Fig. 7(a)]; but if it develops with some strain hardening, it is first necessary to locate point B'. If $B'O'$ is drawn with a slope equal to the expected rate of strain hardening, intersection O' gives the required mechanism yielding strength, S_s^Y, as shown in Fig. 7(b). Comparison of Fig. 7(a) and (b) illustrates the advantage of having a structural system whose mechanism deforms with some strain hardening.

Experimental results have shown that requirements for preventing instability of structural members depends on the desired level of ductility [22]. The larger the tolerable ductility, the more stringent the requirements should be. Under loading reversals, when ductility values exceed a certain limit, there is a sudden drop in resistance against instability, particularly in reinforced concrete buildings.

It is clear from the discussion above that there are no unique values for the strength and deformation capacities of a structure. Depending on the problem at hand, the strength and deformation will vary according to the type of failure (limit state) controlling the design.

Existing buildings

As pointed out in the introduction, the major threat to life and loss of property in the United State and elsewhere from natural hazards such as earthquakes, hurricanes, and tornadoes largely stems from the inadequate performance of buildings under these severe events. Continued updating of building codes and standards resulting from research and past experience will improve future buildings but not affect existing ones.

The response of a building to a severe environmental event reflects not only the performance level required by codes, standards, and construction practices prevalent at the time of design and construction, but also its maintenance and durability. Buildings deteriorate, which affects the margin of safety, so they must be continually assessed for danger under extreme environmental conditions. Following such evaluation, appropriate rehabilitative or abatement procedures can be taken. Thus for any building it is of paramount importance to predict the variation of its strength and deformation capacities with time (aging).

Since current programs for evaluating existing buildings accord with the requirements of current building codes, they do not provide adequate standards for life safety, protection of property, and maintenance of vital functions, nor do they provide an estimate of the amount of building damage to be expected. Researchers and professionals recognize that "there presently are no nationally recognized effective, systematic and economical procedures for assessing the hazards of existing buildings for future loads" [23]. The seriousness of this problem is also reflected in a report, "Meeting the Challenge," filed by the Joint Committee on Seismic Safety to the California Legislature [24]:

> Past performance in seismic safety has clearly been inadequate. Estimates vary widely, but California has perhaps more than 100,000 unreinforced brick or hollow-tile buildings, most of them built before 1933. These buildings are prime earthquake hazards, most of them have not been specifically designed and built to withstand seismic forces. Further, recent earthquakes show that many other types of buildings fall substantially short of providing adequate seismic safety. Some modern buildings are hazardous, as demonstrated by the damage the Los Angeles County's brand new Olive View Hospital sustained in the San Fernando earthquake of 1971. Furthermore, we have continued to build on or across faults, and to erect high-rise structures under conditions that leave authorities fearing for the fate of occupants, should an earthquake be followed by fire. Thus, not only has past performance fallen short of meeting the need for earthquake safety, but present performance is still clearly unsatisfactory in many respects.

Ferver reported that in *A Study of Earthquake Losses in the Los Angeles, California Area*, prepared in 1973 for the Federal Disaster Assistance Administration, Department of Housing and Urban Development, the potential effects of a magnitude 7.5 earthquake on the relatively close Newport-Inglewood Fault and a magnitude 8.3 event on the more distant San Andreas Fault were studied [24]. It was estimated that 1,889,827 housing units would be damaged by the former, with a repair cost of $2,527,000,000 (1967 cost basis), while the latter would damage 1,526,561 housing units, with a cost of $1,492,000,000. These two earthquakes would cause a loss of major general hospital beds amounting to 13,305 and 7,463; deaths, depending on the time of day at which the earthquake struck, of 4,090 to 20,728 and 2,790 to 12,385. Major government buildings would remain operational, although sustaining heavy damage, as would fire and police stations, railroad and highway bridges and tunnels, and various utility systems.

The "Los Angeles County Earthquake Commission Report," March 1972, prepared after the San Fernando earthquake, contains an excellent discussion of the problem of removal or repair of hazardous buildings [25]. In the volume prepared by Task Force A, it is stated that:

> There exist approximately 28,000 hazardous old buildings within Los Angeles County; 20,000 to 24,000 of which are located in the City of Los Angeles; 750 in the unincorporated area; and the remainder in other cities.
>
> These are occupied by an estimated 35,000 businesses and 50,000 families with somewhere between 30,000 to 200,000 persons affected county-wide (2,500 to 7,500 in the unincorportated area). Involved are all of the imagi-

> nable relocation and housing problems which would be attendant to the significant decrease in available housing units resulting from the elimination of these buildings.
>
> Possibly $8.5 billion would be required countywide to bring these buildings up to modern earthquake standards ($80 million in the unincorporated area).
>
> Perhaps upwards of $40 million in annual property tax revenues are currently being generated by these properties, with an unknown tax generation potential if they are repaired, demolished, or rebuilt.

As pointed out by Ferver, although these figures refer to Los Angeles with nearby faults and a large population and building development, midwestern and eastern earthquakes, while perhaps of lower magnitudes, have very high peak intensities and much larger felt areas than western earthquakes [24]. This fact, along with an almost complete lack of concern for earthquake-resistant design of buildings in midwestern and eastern states, invites major disasters like the 1811–1812 New Madrid or 1886 Charleston earthquakes.

The discussion and quotations above indicate the enormity of the problem; the question now is how to improve these seismically hazardous buildings. There are many ways of mitigating the problem, including demolition, retrofitting (strengthening, stiffening and/or toughening) and rehabilitation, reduction of building life, converting usage, and reducing occupancy. Whatever technique is to be used, it is essential that it be able to predict the behavior of an existing building, particularly its strength and deformation capacities and how these capacities will vary with time. The problems encountered in predicting strength and deformation capacities for new buildings also apply to existing buildings. The only difference is that estimates may be more reliable for the latter case since geometric and material characteristics, as well as details of elements and connections, are more readily obtained after construction [26].

STATE-OF-THE-ART
IN PREDICTING THE STRENGTH
AND DEFORMATION CAPACITIES OF BUILDINGS

Nature of problem

Strength and deformation capacities depend upon the interaction of a "building" and "extreme environments." Before discussing this interaction, it is necessary to define these terms on physical grounds.

BUILDINGS. Buildings are, unfortunately, usually represented by simple and, in most cases, unrealistic idealizations of their bare structure, that is, the skeleton that will be built using the main structural material that has been chosen. It must be recognized that a building rests on soil, which is usually deformable, and that while this deformability may be justifiably neglected under service conditions, it must be considered under extreme environments. Thus the whole soil–building system, rather than the building alone, must be considered.

Furthermore, in predicting the mechanical behavior of buildings it is necessary to distinguish clearly between components that constitute the resisting structure and elements that can be considered nonstructural components. Once this is done, the mechanical characteristics of the structural components must be defined clearly.

Within the building itself, the distinction between what are commonly called, or defined as, structural and nonstructural components is not as clear and as simple as is commonly assumed. Usually the partitions, claddings, and so on, are considered to be nonstructural components; but, because of the way they are constructed, they interact with the structure when the latter is deformed. Interaction is so important in most cases that it can completely change the strength and deformation capacity of the bare structure.

In reality, a building is a complex three-dimensional arrangement of structural components founded on soils that may interact with both components and equipment installed within it. When there is interaction, the components should be called "unintentional structural components" rather than "nonstructural elements."

EXTREME ENVIRONMENTS. Predicting the strength and deformation capacities necessitates consideration of all possible severe excitations that might occur during a building's service life, not just the one thought to be critical. Under a severe excitation, strength, and particularly deformation, capacity will depend on the damage (residual state of stress and strain) accumulated from previous medium or severe excitations caused by such natural hazards as earthquakes, windstorms, floods, and other severe events such as blasts and/or fires. Predictions of a building's response to any environmental hazard requires that the static and/or dynamic characteristics of the resulting excitations be determined (i.e., the intensity and its variation with time while acting on a building). In cases involving wind, the intensity and its variation can be defined with acceptable accuracy, given the site and shape of the building; the problem is more complicated in other cases since the actual characteristics of the excitations depend on the building's response.

INTERACTION OF BUILDINGS AND EXTREME ENVIRONMENTS. Even if the mechanical characteristics of a given building (soil–building system) are known, its strength and deformation capacities will necessitate specifying first the environments, and then the pattern of effects on the building. In most cases it is difficult to define this pattern because of the uncertainties involved in estimating the environment (excitation) and its attendant interaction with the building. To illustrate this point, consider the specific case of a severe earthquake ground motion.

The general problems involved in predicting the seismic response of a building are symbolically defined and schematically illustrated in Fig. 8.

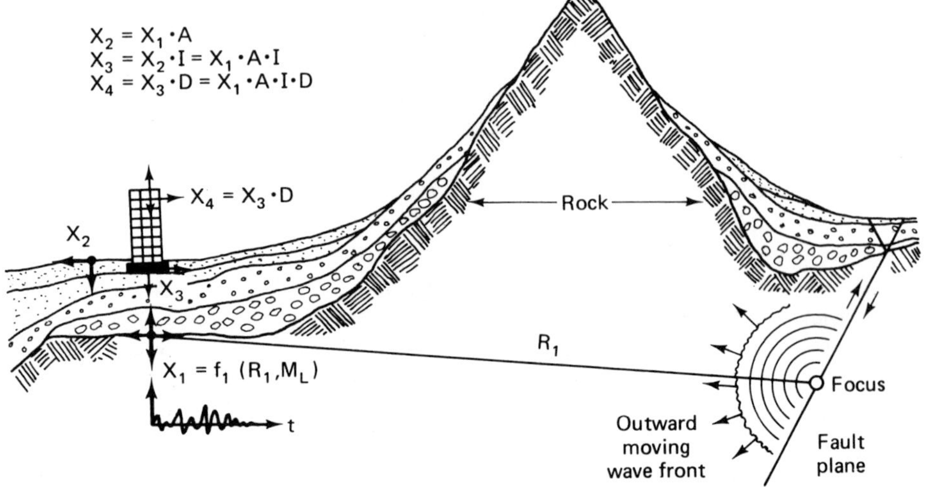

Figure 8   Illustration of problems and factors involved in predicting seismic response of a building [21].

The first problem is to estimate accurately the ground motion at the foundation of the building, X_3. For an earthquake of specified magnitude, M_L, and focal distance, R_1, it is analytically feasible to estimate the base rock motion at the given site of the building, X_1, if the fault type is known [$X_1 = f(R_1, M_L)$]. Prediction of X_3, however, must account for the effects of the soil layers underlying and/or surrounding a building. These effects can be classified in two groups: one is related to the influence of the dynamic characteristics of the different soil layers during the transmission of X_1 to the free ground surface, indicated in Fig. 8 by an attenuation or amplification factor, A, [$X_2 = A \cdot X_1$]; the other is related to the interaction between structure and soil foundation, symbolically represented by a factor I. There are presently large uncertainties regarding the realistic values of A and I, and major errors could be introduced by trying to quantify these two factors using available analytical techniques. Even if X_1 could be predicted with engineering accuracy, attempts to quantify the influence of soil conditions on X_1 to attain X_2 and X_3 would result in a wide range of predicted values. Soil behavior can be very sensitive to the intensity of the seismic waves, as well as to the rate of straining they could induce. Thus the analyst or designer should not rely exclusively on results obtained from a single deterministic analysis. Bounds on the possible variations in A and I should also be considered.

The second problem is to predict the deformation, X_4, from shaking at the foundation, X_3. As indicated in the figure, term X_4 can be obtained by multiplying X_3 by a dynamic operator, D. Although this is a simple expression, the uncertainties involved in realistically estimating X_3 and D give rise to serious difficulties in obtaining an accurate numerical evaluation of X_4.

Even if it were possible to predict the mechanical characteristics of a building, there remain many uncertainties in establishing the six components of the critical X_3, and attempts to predict the response (strength and deformation) of the building should consider the complete range, or at least the bounds of the dynamic characteristics of the possible excitations, X_3. Owing to these uncertainties, the ideal solution would be a conservative one which would identify the critical excitation X_3 for the given building. Although it is easy to define this critical X_3 as that which drives a structure to its maximum response, its specific quantification is more complicated. Quantification is feasible for elastic response [27] but is complicated for cases involving nonlinear response.

The precise evaluation of X_4 at any point in a structure requires that its three translational and three rotational components be established. Furthermore, the prediction of X_4 for a particular building to a specific ground motion depends on the combined effect and dynamic characteristics of all excitations acting on the building. Usually the main excitations on a structure during a severe earthquake are due to: (1) gravity forces, $G(t)$, with the associated effects of volumetric changes produced by creep of the structural material, especially in concrete; (2) changes in environmental conditions, $E(t)$, such as stresses produced by variations in temperature; and (3) at least the three translational components of the foundation shaking, $X_3(t)$.

As shown in eq. (1a), the dynamic characteristics of the whole system, which can change continuously as the structure is deformed into its inelastic range, can be summarized by denoting them as the instantaneous values of: (1) mass, $M(t)$; (2) damping coefficient, $\xi(t)$; and (3) resistance function, $(R$ versus $X_4)(t)$. They can also be represented as illustrated in eq. (1b), by the instantaneous values of: (1) fundamental period, $T(t)$; (2) damping coefficient, $\xi(t)$; (3) yielding strength, $R_y(t)$; and (4) energy absorption and dissipation capacity as denoted by instantaneous available ductility, $\mu(t)$, which is a function of $X_4(t)$.

$$X_4(t) = F\{[G(t), \Delta E(t), X_3(t)], \quad [M(t), \xi(t), (R \text{ vs. } X_4)(t)]\} \quad \text{(1a)}$$

$$X_4(t) = F\{\underbrace{[G(t), \Delta E(t), X_3(t)]}_{\substack{\text{dynamic charac-}\\\text{teristics of excita-}\\\text{tions}}}, \quad \underbrace{[T(t), \xi(t), R_y(t), \mu(t)]}_{\substack{\text{dynamic characteristics}\\\text{of whole soil–building}\\\text{system}}}\} \quad \text{(1b)}$$

Analysis of the parameters in eqs. (1a) and (1b) indicates the magnitude of the problems involved in predicting response to earthquake ground motions. The first problem is that to predict X_4, X_4 must be known. Another problem is that all such parameters are functions of time, although the gravity forces and environmental conditions tend to remain nearly constant for the duration of an earthquake. It should be noted that the value of $\Delta E(t)$ represents more than just the stresses induced by environmental changes that occur during the critical earthquake ground motion, X_3; it also accounts for stress and strain existing at the time of earthquake due to (1) previous thermal changes or shrinkage, which cause residual stress or distress, and deterioration from aging and corrosion; (2) degradation in strength and stiffness caused by previous exposure to high winds, fires, or earthquakes; (3) other disturbances or movements of the foundation; and (4) changes in strength and stiffness caused by alteration, repair, or strengthening. Since any one of these conditions can significantly affect structural response, factors that must be considered in determining the strength and deformation capacities include the variations in loading and environmental histories during the service life of the building and their effects on the structure's condition at the time of the occurrence of the extreme environment. To this end it should be noted that $\Delta E(t)$ also affects $(R$ versus $X_4)(t)$.

Prediction of strength capacity

As discussed above, prediction of the strength and deformation capacities requires (1) that the dynamic characteristics of the probable critical excitation, X_3, be defined; (2) that reliable data be available on the mechanical (dynamic) characteristics of the building–soil system to model the building's structural and nonstructural components, as well as its foundation, and to formulate its resistance function, that is, its load-deformation function (mathematical modeling), R versus X_4; and (3) that interaction between building and critical excitation be established.

It should be emphasized that a building's strength and deformation capacity cannot be predicted solely on the basis of the building's mechanical characteristics; the pattern (shape function) of forces from which strength and deformation are to be determined must also be defined. Establishing the probable critical excitation is one of the more difficult problems and involves perhaps the largest uncertainties. For the sake of simplicity, however, it will be assumed that the dynamic characteristics of the critical excitation, that is, its time variation and intensity, are known, which reduces the problem to that of finding the strength and deformation capacities when the building is subjected to the assumed critical excitation. It is also assumed that the intensity of the critical excitation is sufficient to induce the building's collapse; it would otherwise be necessary to assume that the time variation (shape) of this excitation were constant and to increase the intensity to that value necessary to induce the collapse (deformation capacity) of the building.

Owing to the complexity of any real building's three-dimensional inelastic behavior when subjected to a severe earthquake ground motion, numerical prediction should be carried out by computer programs. Unfortunately, no presently available program is capable of accurately predicting that behavior. Computer programs capable of carrying out sensitivity analyses of the different parameters affecting the three-dimensional behavior of buildings must be developed. It is well known that even symmetric buildings, as far as geometry, structural properties, and mass distribution are concerned, can be subjected to considerable torsional effects from inelastic deformations that can concentrate at a construction weakness. Even if such a computer program were developed, however, it is doubtful whether it could be used for reliable predictions of the three-dimensional hysteretic behavior of real buildings owing to a lack of data on the hysteretic behavior of structures under three-dimensional deformations. There is an urgent need for integrated experimental and analytical studies of such behavior.

Inelastic analysis of pseudo-three-dimensional building models can be performed with the program DRAIN-TABS [28]. A few general-purpose programs can be used to perform three-dimensional, direct integration time-history analyses of nonlinear models. These include NONSAP [29], ANSR-I [30], ADINA [31], and NASTRAN [32], which are generally inconvenient to use [33,34]. Special programs such as DRAIN-TABS are more convenient, but nonetheless require specialized skills and are costly for large practical applications. As a result, there have been few three-dimensional analyses of real buildings. For the most part, these programs neglect the effect of nonstructural elements. This is not surprising because few, if any, realistic idealizations (mathematical models) of the nonlinear behavior of common nonstructural elements have been formulated.

Thus, most predictions of strength and deformation capacity of buildings are now based on two-dimensional inelastic analyses that are carried out on models of the planar frame and/or shear-wall building components. These models can be used to idealize only regular buildings with negligible or no torsional eccentricity. All frames and shear walls in one direction are treated as planar systems tied together at each floor level by rigid links representing the floor diaphragm (Fig. 9). These idealizations thus assume that floor diaphragms are essentially rigid [34,35]. Semi-rigid floor diaphragms cannot be adequately represented by planar frame or shear-wall models. Only floor diaphragms that are very flexible in relation to the frame or shear wall can be modeled by uncoupling each plane of frames or shear walls and performing separate solutions. The main disadvantage of this model is that it neglects the effect of torsion; even in a building with no eccentricity between mass and resistance in the linear-elastic range, torsion could develop due to

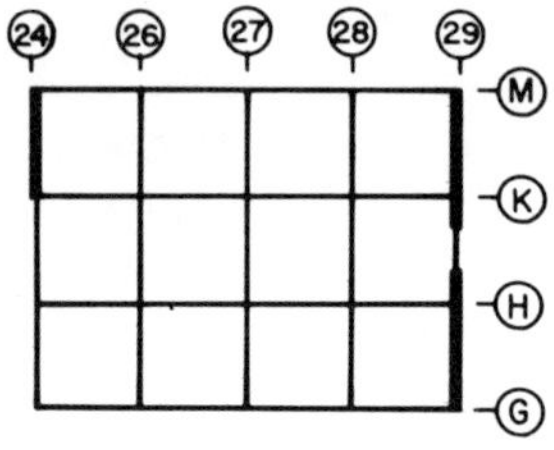

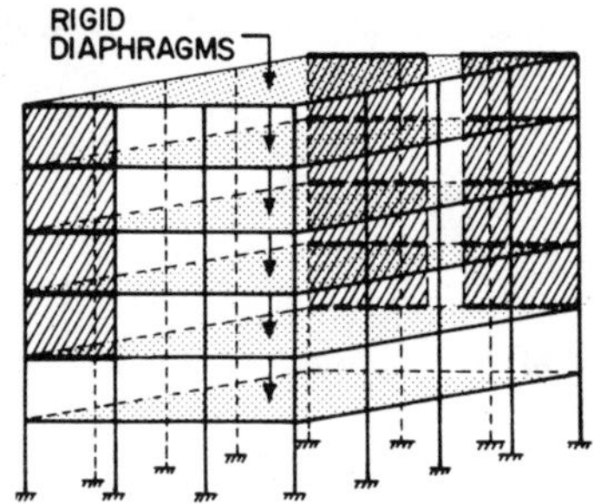

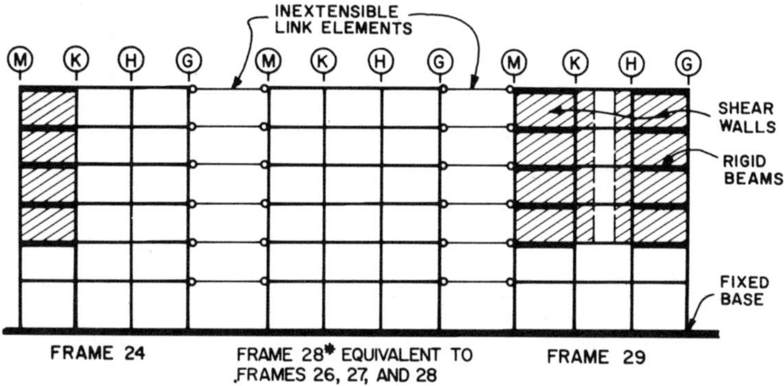

(c) ELEVATION OF COMPUTER MODEL

Figure 9 Schematic representation of modeling used in nonlinear analysis of seismic response of real building [35]: (a) plan view of real building, (b) isometric of real building, (c) elevation of computer model.

the early inelastic deformation of one or more weakly constructed region. There is an urgent need to study the effect of inelastic torsion on the overall strength and deformation capacities of buildings.

In spite of this disadvantage, the two-dimensional frame and shear-wall model can and has been used to clarify some problems in predicting the strength and deformation capacities of buildings. One such problem that will be discussed is the sensitivity of strength capacity on loading conditions.

IMPORTANCE OF LOADING CONDITIONS ON STRENGTH. Results obtained in the analyses of a two-dimensional frame model of the bare building structure shown in Fig. 10, which was designed with a computer-aided procedure [11], will be used to illustrate the importance of loading conditions. Figure 11

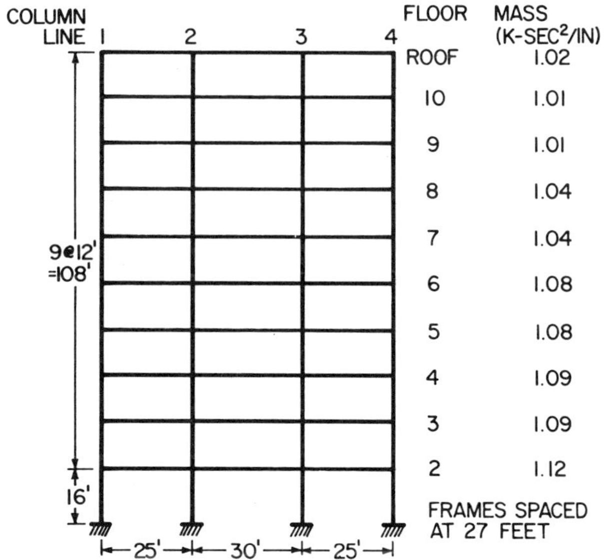

(a)   FRAME ELEVATION AND DESIGN DATA

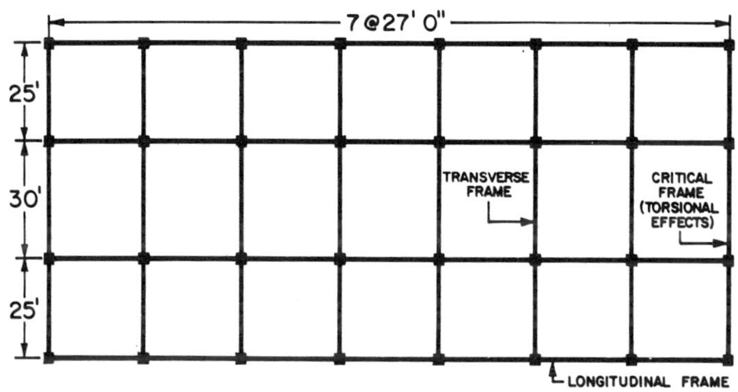

(b)   TYPICAL FLOOR PLAN

Figure 10   Plan and elevation of a building designed using a computer-aided procedure [11]: (a) frame elevation and design data, (b) typical floor plan.

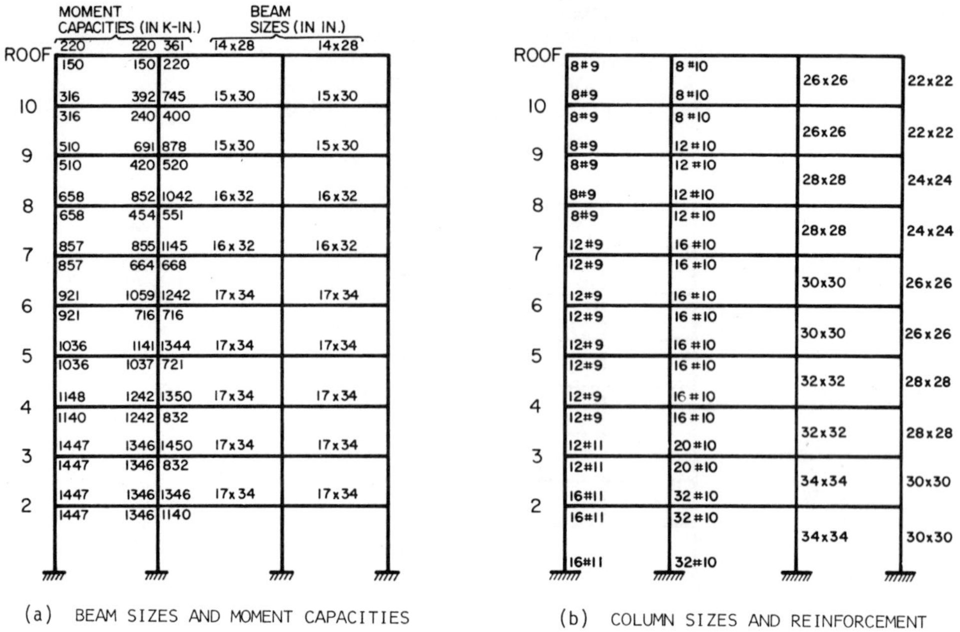

(a) BEAM SIZES AND MOMENT CAPACITIES

(b) COLUMN SIZES AND REINFORCEMENT

Figure 11 Optimum preliminary design of a frame of building of Figure 10 [11]: (a) beam sizes and moment capacities, (b) column sizes and reinforcement.

illustrates the member sizes and reinforcement and the beam moment capacities resulting from application of this computer-aided procedure; strength and deformation capacity of the so-designed structure were predicted using different procedures.

Static Analysis. The nonlinear static behavior of the frame was investigated using a modified version of ULARC [36]. The frame model of Fig. 11 was loaded with design gravity loads and then subjected to the effects of a monotonically increasing seismic base shear distributed through the height of the frame according to two lateral force patterns, one obtained from spectral modal analysis (spectral pattern) using the design spectra shown in Fig. 12, the other assuming a uniform distribution of forces through the height, which was selected to illustrate the sensitivity of the overall strength of the frame to the distribution of forces through its height. Results of these nonlinear static analyses are summarized in Figs. 13 to 16.

The variation of roof and second-floor lateral displacement, obtained with increasing value of base shear and assuming a spectral lateral force pattern, $\mathbf{\Phi}^s$, is shown in Fig. 13. Two analyses were carried out to investigate the P–Δ effect. As illustrated in Fig. 13, if the P–Δ effect is neglected, strength capacity may be overestimated by about 8%. On the other hand, roof dis-

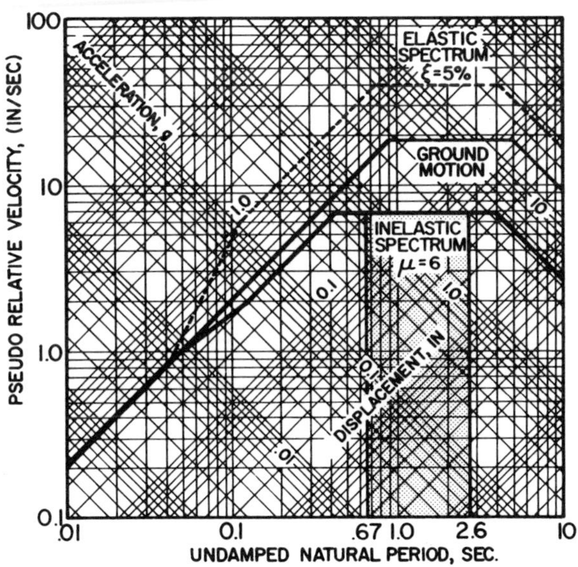

Figure 12 Spectra used in design of frame of Fig. 11 and in obtaining ϕ^s of Fig. 13 [11].

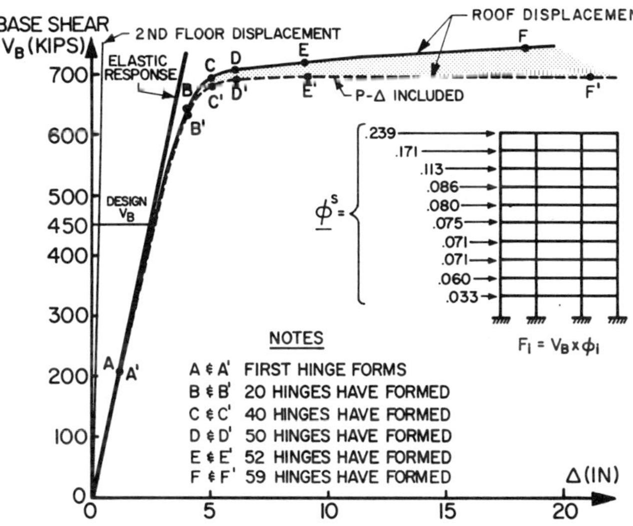

Figure 13 Lateral load-displacement relationship for roof and second floor assuming spectral lateral force pattern, ϕ^s [11].

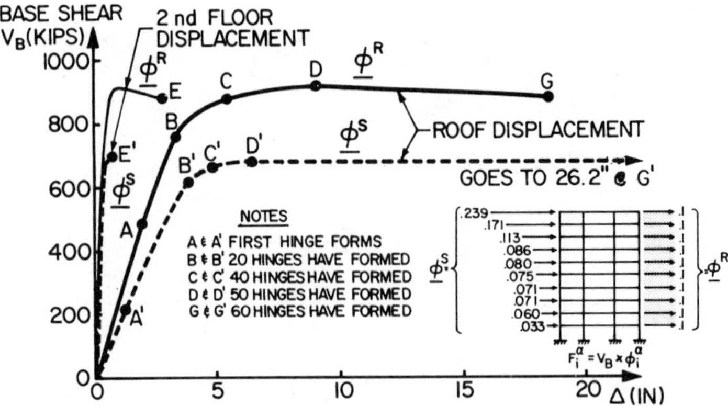

Figure 14 Lateral load-displacement relationships—effect of different loading patterns.

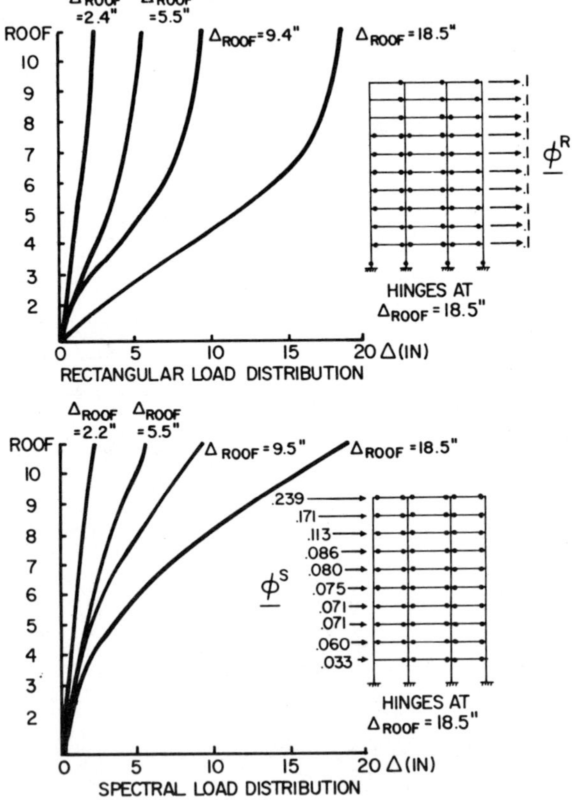

Figure 15 Plastic hinges and displacement patterns for different static lateral loading patterns.

210

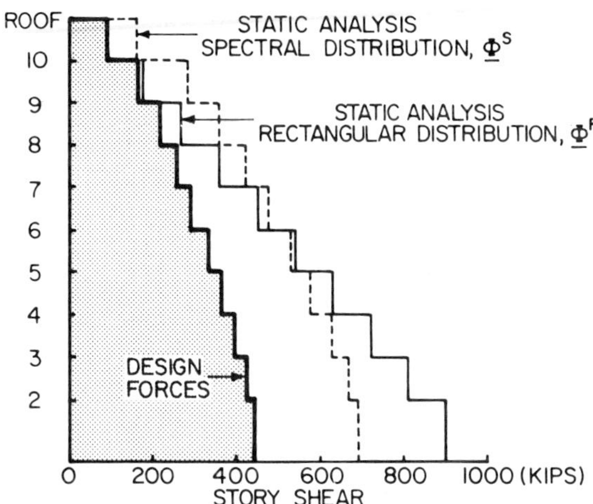

Figure 16 Effect of different lateral loading patterns on the strength (base shear) and distribution of story shears—static analysis.

placement at which maximum resistance is reached may be overestimated by a considerably larger percentage.

The results presented in Fig. 14 show that the strength or resistance capacity of the frame was sensitive to the lateral force pattern. Resistance capacity under the uniform (rectangular) load pattern, Φ^R, was about 20% higher than that under a spectral pattern, Φ^S. The displacement at which maximum base shear was obtained was considerably larger in the latter case. These discrepancies can be explained by examining Fig. 15, in which the order and pattern of plastic hinge formation is shown to have been sensitive to the lateral force pattern. The displacement pattern through the height of the frame was even more affected.

It is interesting to note that whether or not P–Δ effects were included, strength estimates were significantly higher than design forces for which the building had been designed. Even under Φ^S and when P–Δ effects were considered, the base shear strength of the frame was more than 55% greater than the design base shear. Two factors contributed to this overstrength: (1) final beam strengths were significantly larger than required owing to code capacity reduction factors, column slenderness effect, and the use of practical formwork sizes and limited sizes of available reinforcing bars; and (2) determination of the design forces (story shears) was based on the assumption that the entire frame was transformed into a mechanism by simultaneous formation of the necessary plastic hinges. Hinge formation was not simultaneous, however; the first plastic hinge formed at state A–A', which corresponded to a base shear of approximately one-half the design value (Figs. 13 and 16). Subsequent hinge formation was gradual, as depicted in Fig. 13.

The results above show that the strength capacity developed during a severe earthquake cannot be accurately determined from an analysis based just on code-specified lateral force patterns; the critical force pattern must be determined. Furthermore, the predicted strength is that of the frame model, not that of the actual building frame. This actual resistance differs from the model's due to the different material mechanical characteristics (which are usually larger than the minimum specified), different sizes of actual members and reinforcement, different foundation conditions and soil behavior, joint flexibility, reinforcement slipping at joints, and failure of one or more critical regions before reaching strength capacity.

Dynamic Analysis. It is well known that different ground motions produce different patterns of inertial forces through the height of the building. Since static analysis indicates that the strength of the frame was sensitive to the lateral force pattern, the strength capacity of a frame or any structure should depend on the type of ground motion to which the building is subjected. Thus dynamic analysis to determine the strength capacity of a given frame or structure must address the following questions. (1) For a specified time variation, where frequency content is defined but not the intensity of the ground motion, how can the strength of the frame and the corresponding intensity of the ground motion be determined? (2) What is the critical ground motion or, more specifically, what are the dynamic characteristics of the ground motion that lead to minimum strength? (3) What are the best parameters for determining the required intensity of that ground motion—is it the peak acceleration, as commonly assumed? To answer these questions, the following procedure can be used.

Conduct a series of nonlinear dynamic analyses of the response of a given frame to a specified ground motion under progressively increasing intensities. The minimum intensity for which no significant increase in the strength (maximum story shears) occurs is that which the structure can resist. The computed base shear and envelope of story shears under this minimum intensity represent the strength of the structure. Since strength depends on strain hardening of the mechanical (mathematical) model adopted, high values of strain hardening and analyses based on infinite ductility models will result in unrealistically high values of strength. Only nonlinear models that account for progressive collapse of inelastic regions as their deformation exceeds their available capacity can give reliable estimates of strength. These analyses should be carried out for a large enough number of different types of ground motion. Such analyses have been carried out for the designed frame of Fig. 11; some results are shown in Figs. 17 and 18, which indicate the following:

1. Strength of the frame model, including strain hardening, depended on the time variation of the ground motion. (Compare strengths under El Centro and Pacoima motions for the same intensity.)

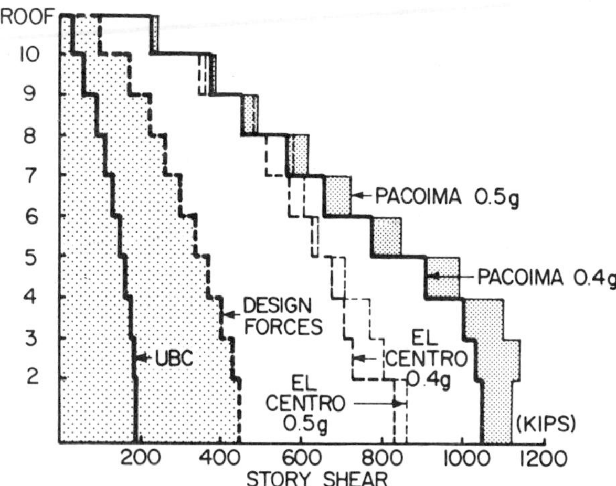

Figure 17 Effects of different ground motions and their intensity on strength (base shear) and on the envelope of maximum story shears [11].

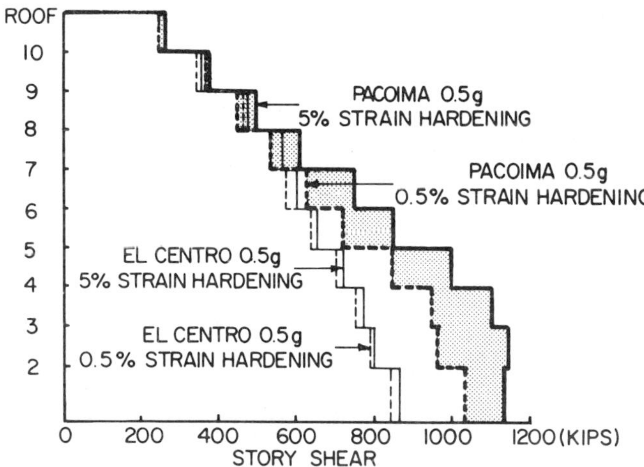

Figure 18 Effects of different earthquake ground motions and strain hardening on the strength (base shear) and on the envelope of maximum story shears.

2. Under El Centro-type ground motions, the increase in the strength was negligible once intensity reached a certain value (i.e., 0.4g). On the other hand, a model with 5% strain hardening subjected to a severe, pulse-like ground motion such as the Pacoima showed nearly a 10% increase in strength when intensity was increased from 0.4g to 0.5g (Fig. 17).

3. While an increase in strain hardening had little effect on the strength of the frame model when subjected to ground motions such as those of the El Centro, it led to an increase in strength under ground motions such as those of the Pacoima (Fig. 18).

Peak acceleration is generally the only parameter used to describe the severity of a seismic ground motion. Recent studies show that the use of this single parameter is generally inadequate, however [21]. Although it appears that peak incremental velocity is a better parameter, generally the severity of a seismic motion cannot be specified by just one parameter.

COMPARISON OF STRENGTH UNDER EQUIVALENT STATIC LATERAL LOADS AND TIME VARIATION GROUND MOTIONS.   In Fig. 19 the story shears obtained under static lateral forces are compared with those obtained under time histories of seismic ground motions. If overall strength is measured only by base shear, strength capacity predicted from dynamic analysis of the response to time-variation ground motions will be significantly higher. Maximum base shear of a frame model with 0.5% strain hardening was about 50% higher under the 0.5g derived Pacoima Dam Record than that developed under the static lateral force pattern obtained from modal spectrum analysis. Two interrelated factors account for this discrepancy: (1) migration of plastic hinges, and (2) time variation of inertial forces through the height of the frame.

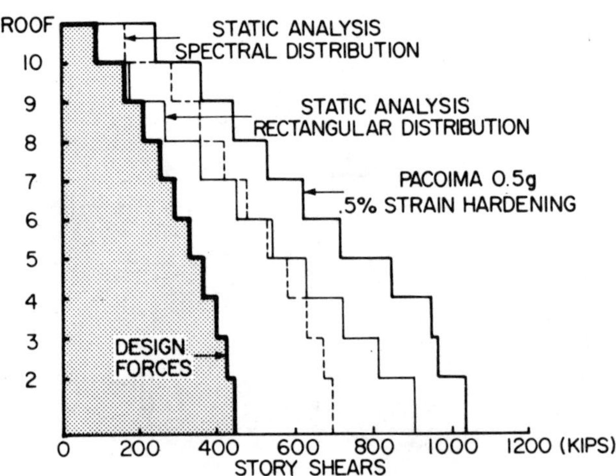

Figure 19 Comparison of strengths (base shears) and distribution of story shears obtained using different types of analyses (static versus dynamic).

Migration of Plastic Hinges During Dynamic Response. Plastic hinges did not form simultaneously and some canceled out (unloaded) as new hinges developed (Fig. 20). Formation of a complete or partial collapse mechanism under dynamic ground motions thus requires larger base shear than that under increasing static lateral forces in which unloading of plastic hinges is rare. Development of a collapse mechanism in tall buildings during dynamic response is possible only under ground motions containing severe, long-duration acceleration pulses or when enough hinges form in the columns

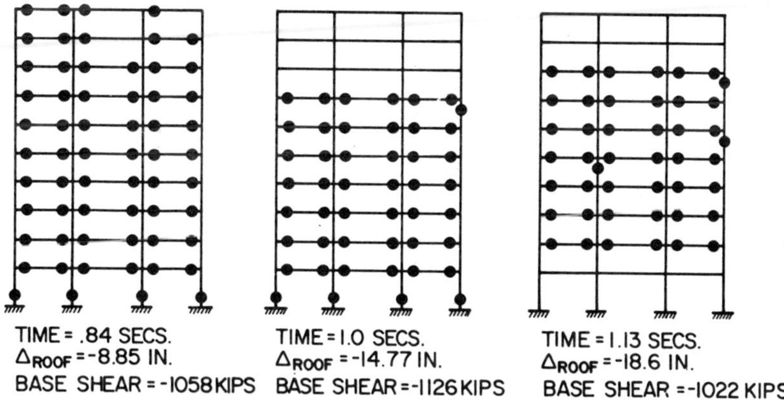

Figure 20 Plastic hinge patterns close to the time when maximum base shear is attained under 0.5g derived Pacoima ground motion for a model with 5% strain hardening [11].

of one story to constitute a partial sway mechanism. However, a design that permits formation of plastic hinges in columns leading to a partial sway mechanism is a poor seismic-resistant design.

Time Variation of Inertial Forces through Height of Frame. Owing to plastic hinge migration, the pattern of loading during dynamic response varied continuously through the height of the frame, and the base shear could reach considerably higher values than those predicted on the basis of a constant lateral force pattern. Note that the dynamic story shears plotted in Figs. 17, 18, and 19 are maximum values, attained at different times for each story during the response, and, as such, do not correspond to the same distribution of inertial forces through the height of the frame.

The discussion above shows that the strength capacity of a structure subjected to dynamic excitations usually cannot be predicted using statically equivalent forces, and strength cannot be defined solely on the basis of base shear. At the very least, it is necessary to define the maximum base shear and simultaneous overturning moment, or the maximum overturning moment and simultaneous base shear. Since deformation also affects strength through P–Δ and strain-hardening effects, the prediction of strength capacity is necessarily interrelated with that of deformation.

ANALYTICAL PREDICTIONS VERSUS EXPERIMENTAL RESULTS. Prediction of strength and deformation capacities requires consideration of nonstructural elements that are present when an extreme environment occurs. The need to include such nonstructural elements is illustrated by several examples.

1. Partitions decrease the slenderness of members, causing the well-known problem of short columns and short beams (Figs. 21 and 22)
2. Partitions or walls infilling frames significantly change the dynamic characteristics of frames.

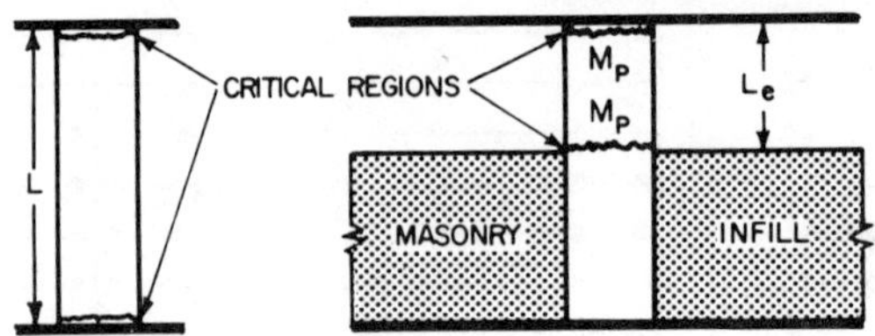

Figure 21   Reduced column effective length due to infills.

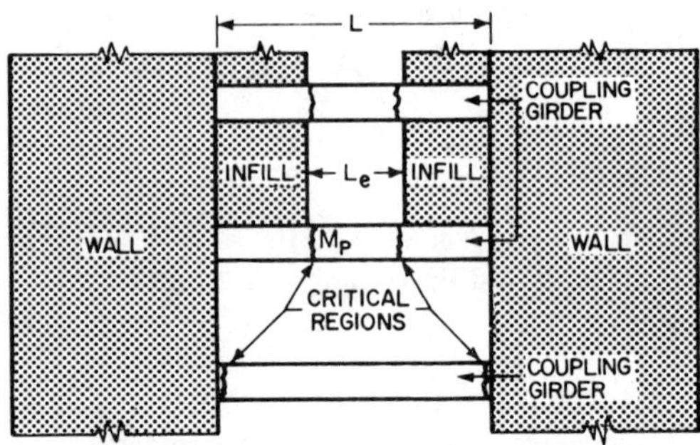

Figure 22   Reduced effective length of coupling girder due to infills.

To determine the effect of infilling partitions and walls, experimental and analytical work was conducted at Berkeley [37] on one-third-scale subassemblage models of a prototype eleven-story reinforced concrete frame building (Fig. 23). A prototype subassemblage of the lower three stories and of one-and-a-half bays of the end frame was studied (Fig. 24). The fundamental period of the building, assuming bare frame, was estimated as 1.3 s. Preliminary sizing of members was carried out using UBC-specified base shear. Final design, using $f_y = 414$ MPa (60 ksi) and $f'_c = 27.6$ MPa (4 ksi), was based on reduced inelastic design spectra obtained from ground motion spectra with a maximum ground acceleration, $\ddot{u}_g = 0.50g$; maximum ground velocity, $\dot{u}_{g_{max}} = 610$ mm/s (24 in./s); and maximum ground displacement $u_{g_{max}} = 457$ mm (18 in.); and assuming a damping ratio $\xi = 5\%$ and a ductility displacement of $\mu = 5$. To emphasize the importance of infills, usually considered to be nonstructural elements, the results obtained for a bare frame and an infilled frame will be compared. Also, the behavior of R/C walls and frame-wall systems is discussed.

Bare Frame. Figure 25 shows the load-deformation curve from the test carried out on the bare frame. Strength capacity was 50 kN (11.2 kips) and initial tangential stiffness was a very low 50 kN/cm (29 kips/in.). Based on rigid plastic first-order theory, the predicted strength was 70 kN (15.7 kips), an overestimate of 40% due to lateral flexibility and attendant P–Δ effect.

A more accurate method for predicting strength is based on the intersection of the diagram representing the response of a rigid plastic mechanical model using second-order theory with linear-elastic response predicted using

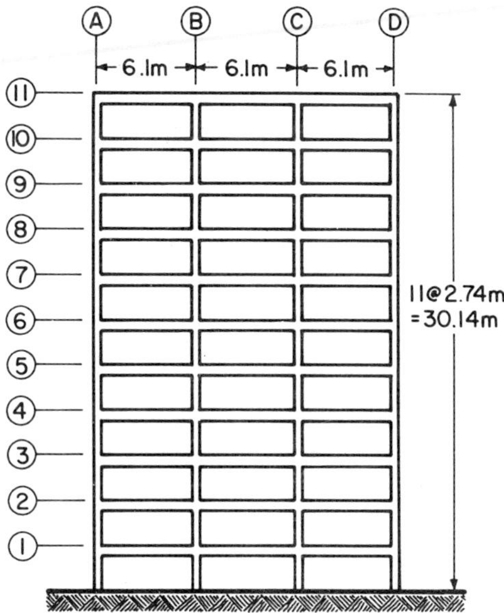

(a)   ELEVATION OF PROTOTYPE FRAME

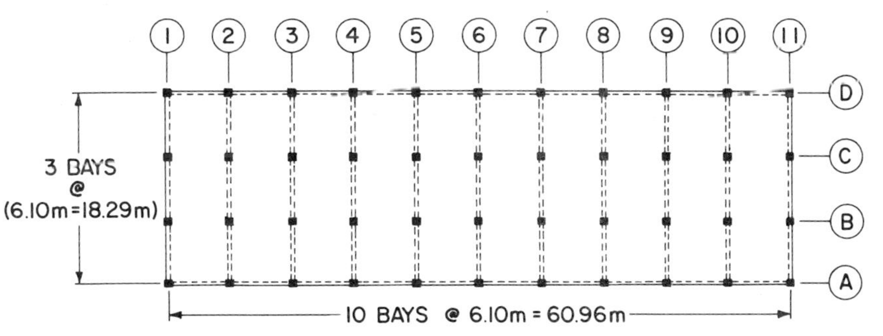

(b)   PLAN OF BUILDING

Figure 23   Plan and elevation of R/C frame building [37]: (a) elevation of prototype frame, (b) plan of building.

an effective moment of inertia of the members. Figure 26 shows that this intersection resulted in 60 kN (13.5 kips), which is 20% higher than the experimentally obtained value of 50 kN (11.2 kips). Thus for flexible structures, in which plastic hinges do not form simultaneously, prediction on the basis of simplified idealizations (rigid plastic, first-order theory) can lead to overestimates of strength.

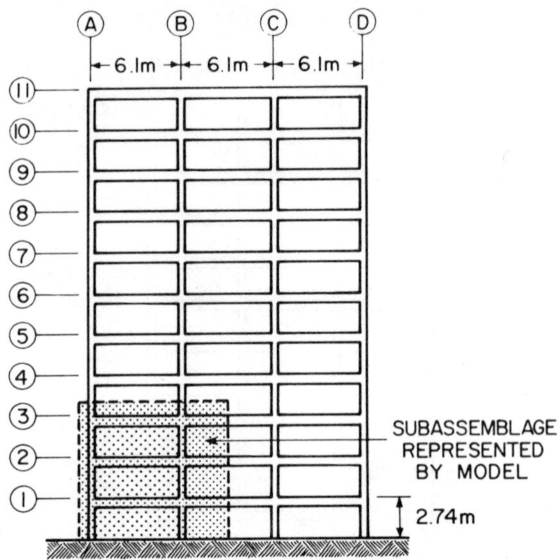

Figure 24 Prototype frame and subassemblage represented by specimens tested [37].

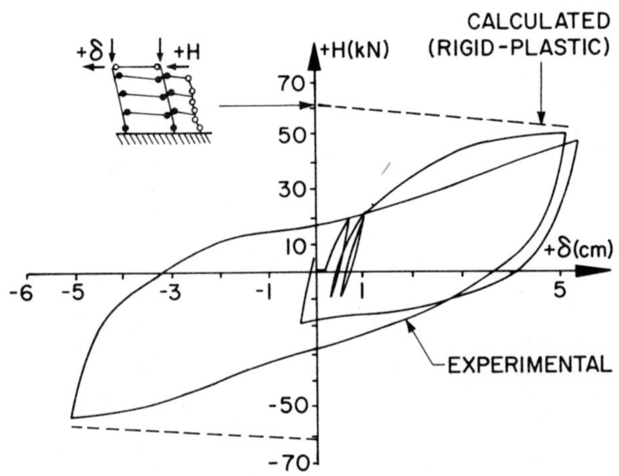

Figure 25 Lateral load-deflection relationship—bare frame [37].

It should be noted that the analytical curves shown in Fig. 26 were predicted on the basis of actual mechanical characteristics of the materials at the time of testing. Such predictions can be made for existing buildings if samples of the material can be extracted and tested. In designing new buildings, prediction must be based on the specified properties of the material. The diagrams corresponding to the latter are plotted in Fig. 27. Estimated strength, based on the intersection of linear-elastic and rigid plastic, second-

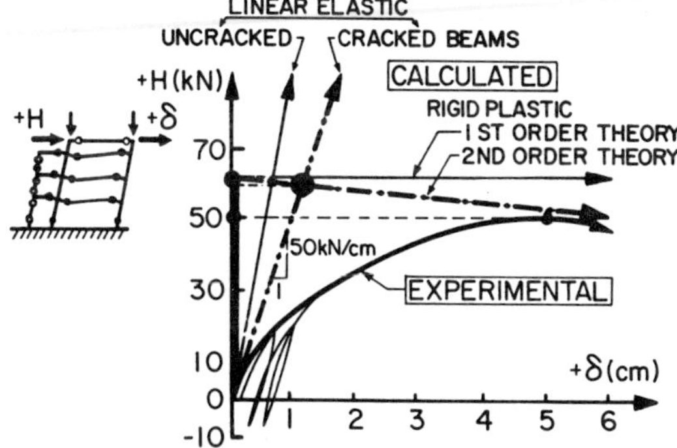

Figure 26 Comparison of experimental and analytical response of bare frame.

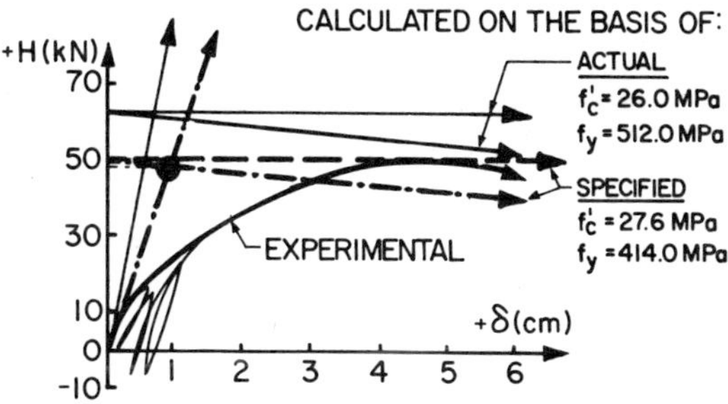

Figure 27 Comparison of experimental and analytical response of bare frame.

order theory lines is about 48 kN (10.8 kips), which is very close to the experimental value. This is fortuitous because these estimated values are based on different mechanical characteristics, which, strictly speaking, nullify the comparison.

For new buildings the only means of predicting strength is by using probable bounds of the mechanical characteristics of the material to be used in construction. The results presented in Figs. 26 and 27 indicate that accurately predicting a building's strength requires that the complete resistance–deformation relationship (function) be determined. For any real structure, this requires the use of computer programs. From the comparison and discussion above it is clear that prediction of the strength capacity of ductile moment-resistant frames cannot be separated from prediction of its deformation.

Agreement between analytically predicted and experimental results required modeling the beams using either three or four parallel beam elements each having essentially a linear, elastic perfectly plastic moment-curvature relationship (Fig. 28). The columns were modeled using only single

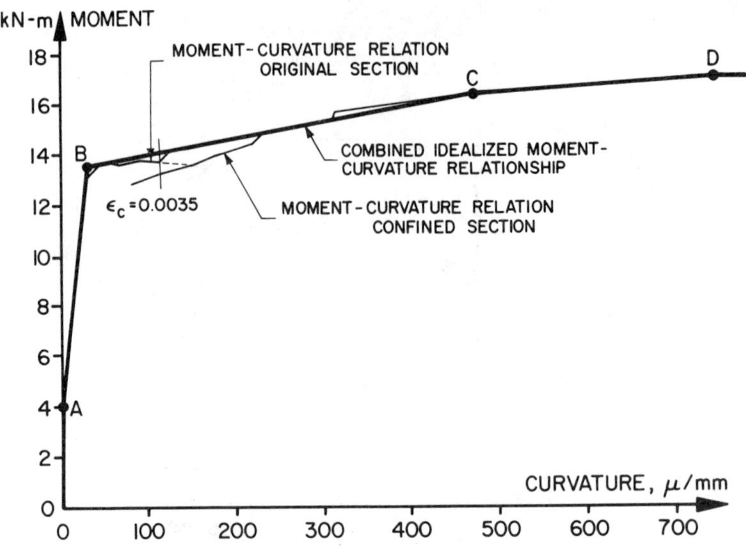

Figure 28   Idealized moment curvature for beam of specimen tested [37].

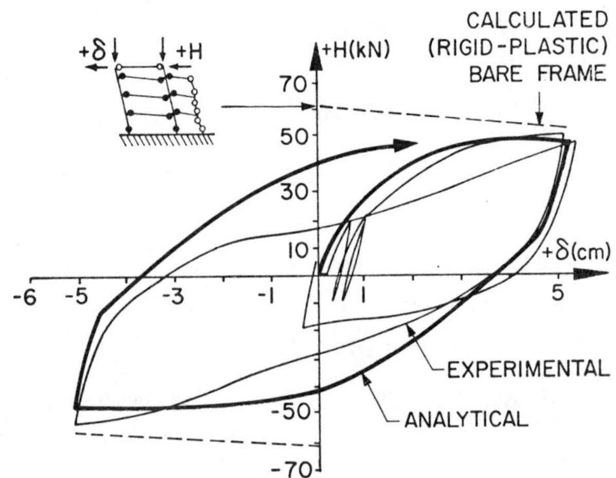

Figure 29   Comparison of experimental and analytical results [37].

bilinear elements. The ANSR-I program [30] was used. Figure 29 compares analytical results with experimental curves for top lateral deflection as a function of lateral load. Agreement was excellent except at the reloading portion.

Infilled Frame. The effects of two types of infill are illustrated in Fig. 30. The curves shown in this figure were obtained by drawing the envelope of results reported in [37]. Comparison of the curves corresponding to the infilled frame with those of the bare frame indicates that the effects of infilling walls

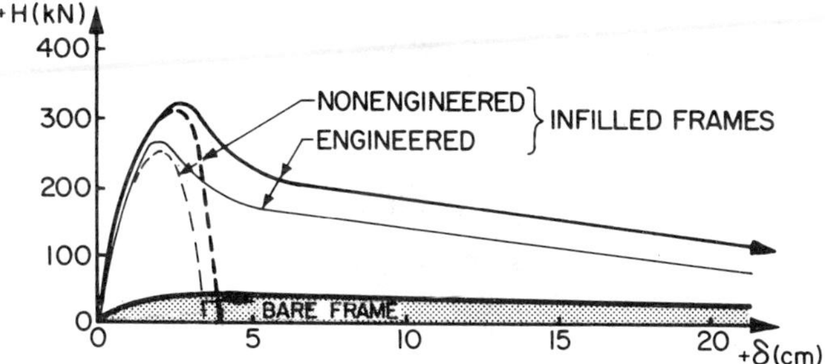

Figure 30   Comparison of behavior of infilled frames and bare frame.

must be considered in estimating the strength and stiffness of the infilled frame. Infilled frames under service loads behave as monolithic deep beams with an initial lateral stiffness of at least 250 kN/cm (143 kips/in.) or an increase of at least 500% with respect to the bare frame. After the addition of clay infill, the strength was increased from 50 kN (11 kips) to about 320 kN (72 kips), an increase of more than 500%.

The results above were obtained from properly reinforced infilled and special R/C (engineered-infilled) frame members. If the infills had been unreinforced and the member improperly reinforced, particularly against shear, the initial stiffness and maximum strength of the infilled frame under monotonic loading would have been close to values obtained with the rein-forced infills. This is because in the latter case failure started at crushing or shearing of the infills and, as a result, the contribution of reinforcement was negligible. However, post maximum strength behavior, particularly under seismic excitations, would be quite different. There, strength would drop abruptly after reaching maximum strength because of the sudden and brittle crushing or shearing of the unreinforced infill; all the shear forces developed at this point due to the inertial forces would then be resisted by the R/C columns or beams. If these elements had not been designed to resist the total shear (non-engineered infilled frames), they would have failed in a more-or-less brittle manner, as illustrated in Fig. 31.

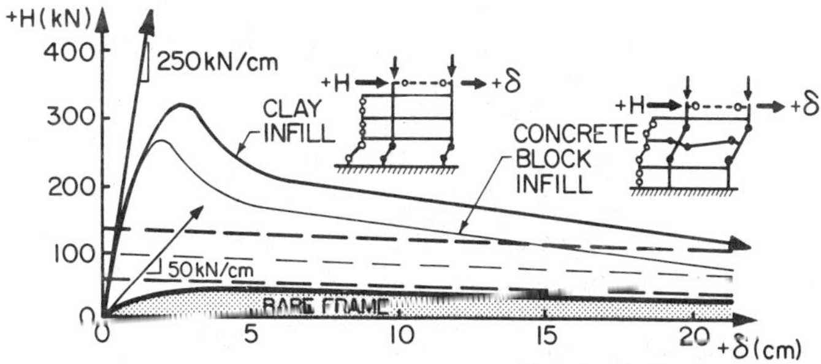

Figure 31   Comparison of behavior of engineered and nonengineered infilled frames.

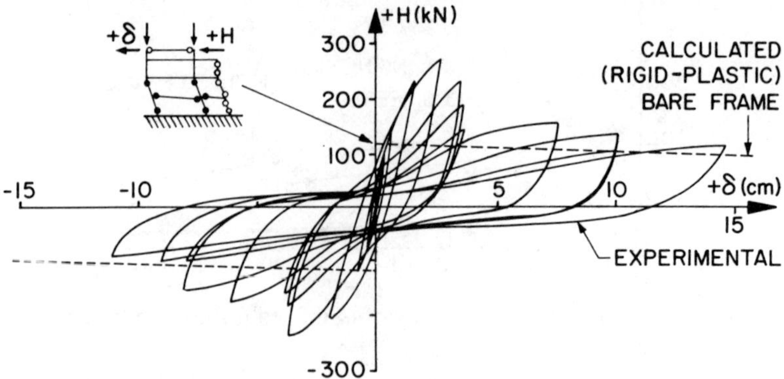

Figure 32   Lateral load-deflection relationship—infilled bare frame [37].

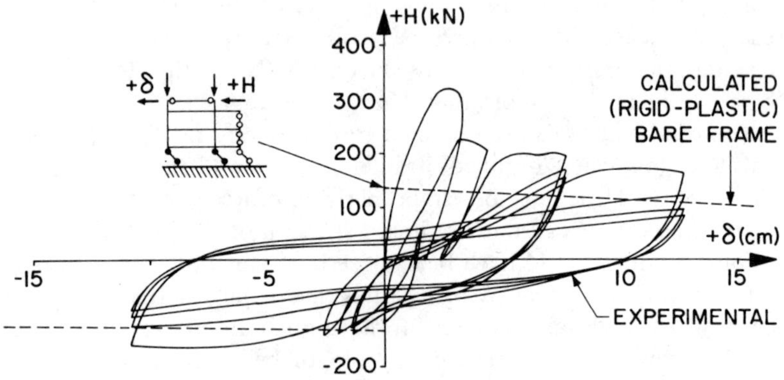

Figure 33   Hysteretic behavior of a frame infilled with clay bricks [37].

Figures 32 and 33 illustrate the lateral load-deformation curves obtained in two of the specimens tested under seismic loading conditions by Klingner and Bertero [37]. Maximum strength was unaffected by cyclic loading. Under service loads the infilled frames behaved as monolithic deep beams. Increased loads caused infill panels to separate from the frame except at the two diagonally opposite compression corners, which led to the development of equivalent, diagonal compression struts in the panels and, therefore, to the braced behavior of the subassemblage. Further increases in load caused crushing of some of the struts, which indicated that maximum strength had been obtained. Crushing of an equivalent strut (usually the one located in the weakest panel of the subassemblage) marked the start of serious panel degradation. Thereafter, specimens behaved as frames braced by gradually degrading struts in one or more panels.

The relative amount of damage in each panel determined the locations of the hinge regions that then developed in the frame members near the beam-column connections. The number of hinge regions increased enough to form a collapse mechanism and the strength of the infilled frame subassemblages gradually decreased to the second-order rigid plastic collapse load corresponding to that of the bare frame mechanism. The presence, behavior, and failure of the infill panels did not significantly reduce the rotational ductility of the frame members.

Relatively simple macroscopic mathematical models based on the equivalent strut concept were developed to predict the observed behavior [37]. These mathematical models were incorporated into ANSR-I [30]. Figure 34 shows the analytical results together with the experimental behavior

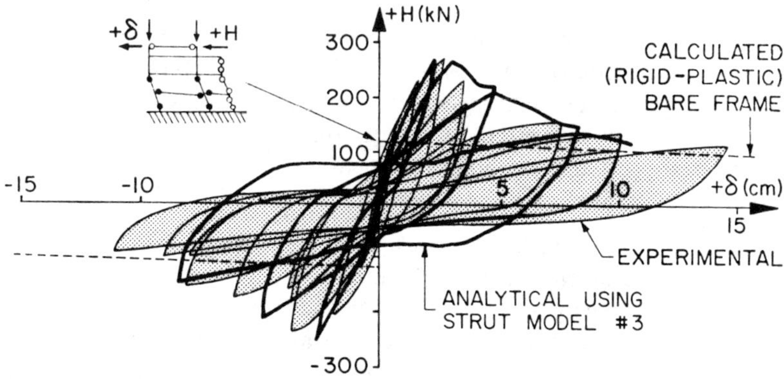

Figure 34 Comparison of experimental and analytically predicted hysteretic behavior for an infilled frame specimen [37].

of one of the specimens tested. Comparisons of the two curves show that the strut model developed gives an excellent representation of the experimentally observed stiffness, strength, and degradation characteristics of the entire infilled frame subassemblage over a wide range of load and displacement reversals. It is doubtful, however, that the same agreement could be obtained if this strut model were applied to other types of infilled frames. More refinements are needed or a more general model should be developed that is capable of predicting, with sufficient engineering accuracy, the behavior of frames with partial infill or infills with openings.

R/C Structural Walls. Proper reinforcing of masonry infills is expensive and in many cases it is more economical to replace these infilled frames by R/C walls, because they are easier to construct. The considerable stiffness and strength inherent in R/C walls can be used advantageously in seismic-resistant construction if they are properly located in the building and are adequately designed and detailed. Not surprisingly, their use in regions of high seismic activity has increased considerably in recent years, even though prediction of their strength and deformation capacities remain complex. Difficulties exist because the hysteretic behavior of R/C walls under severe seismic conditions is still not fully understood. Considerable advances have occurred in the last decade, however. Since 1966, experimental and analytical studies have been carried out at several institutions in the United States (Portland Cement Association, University of Illinois, and University of California, to name a few) [38], at the University of Canterbury in New Zealand [39], and at numerous research laboratories in Japan [40].

In predicting strength capacity, as well as its overall hysteretic behavior, it is convenient to classify walls as "squat" or "low-rise," and "tall," "high-rise," or "flexural."

Squat walls. Squat shear walls are usually defined as those having a height-to-horizontal length ratio of 1.0 or less. Maximum strength (ultimate shear) is controlled by the interaction of shear and flexure. The observed shear strength of the test specimens was substantially greater than that predicted using the suggested ACI code equations [41]. Thus new equations have been suggested for obtaining maximum shear strengths that are in reasonable agreement with the experimental values [42].

Tall walls. Efficient seismic design of tall walls requires that all brittle modes of failure be suppressed. Maximum strength must therefore be controlled by flexure, which can be predicted accurately with present methods. It appears then that strength capacity can also be predicted accurately, as illustrated in Fig. 35, although this is not generally the case. Strength prediction for the case of Fig. 35 was facilitated by knowing the lateral loading pattern (i.e., the distribution of horizontal inertial forces through the height of the wall), which was the same as that selected for determining the test loading. Furthermore, the wall was designed so that behavior under these loads would be controlled by the maximum flexural strength capacity of the critical wall region. However, the loading pattern that can be developed and the resultant state of internal forces at certain wall regions during any ground motion response generally vary continuously between broad bounds, making it difficult to establish the loading pattern. Therefore, a *unique strength*

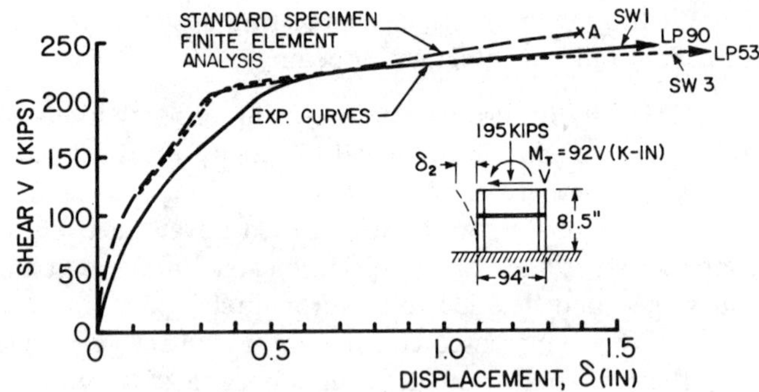

Figure 35 Comparison of experimental and analytical load-displacement curves obtained for specimens of R/C walls [43].

capacity cannot exist because strength will depend upon how the wall is loaded.

It is thus clear that specifying the strength capacity of a tall wall for possible severe seismic ground motion necessitates prediction of its time-history response. Once this is determined, it is possible to obtain the loading pattern that induces maximum bending moment at a specific critical region, as well as that which induces the largest shear force. While flexural strength can be predicted with sufficient engineering accuracy, more research on predicting shear strength is needed. This is another reason why tall walls should be designed so that flexure rather than shear controls maximum strength.

R/C Frame-wall System. The main problem in predicting the strength capacity of a combined frame-wall system subjected to severe seismic ground motions is the determination of the critical loading pattern. The interaction between frame and wall makes prediction of the critical loading pattern more complicated than that for an isolated cantilever wall. An illustration of the time variation of the moment through the height of the wall, as well as of the moment-shear ratio at the base of the wall, is illustrated in Fig. 36 [43].

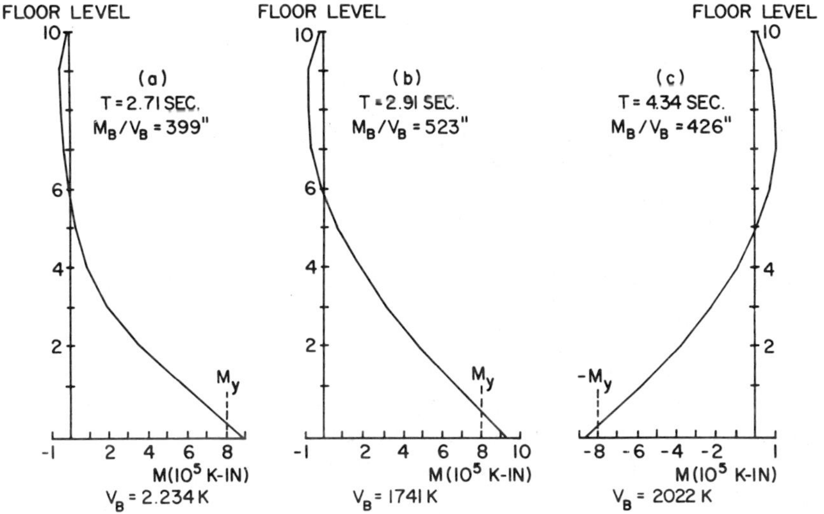

Figure 36 Variation of moment diagram and moment-shear ratio at the base of a wall in a wall-frame system [43].

Prediction of deformation capacity

While strength capacity can be predicted in most cases, particularly for bare structural systems, with sufficient engineering accuracy, the prediction of deformation capacity is more difficult. A number of problems are encountered: (1) the meaning of "deformation capacity" remains unclear; (2) the amount of usable structural deformation depends on the type of loads (excitations) acting on the structure—static (sustained) or dynamic; and (3) the hysteretic behavior of structures, particularly those of reinforced concrete, is very sensitive to the time history of loads (e.g., monotonic versus cyclic with reversals of forces and deformation).

The first problem can be illustrated by considering tests carried out at Berkeley on the behavior of a R/C subassemblage under seismic loading conditions [44]. The behavior of that subassemblage is shown in Fig. 37. While

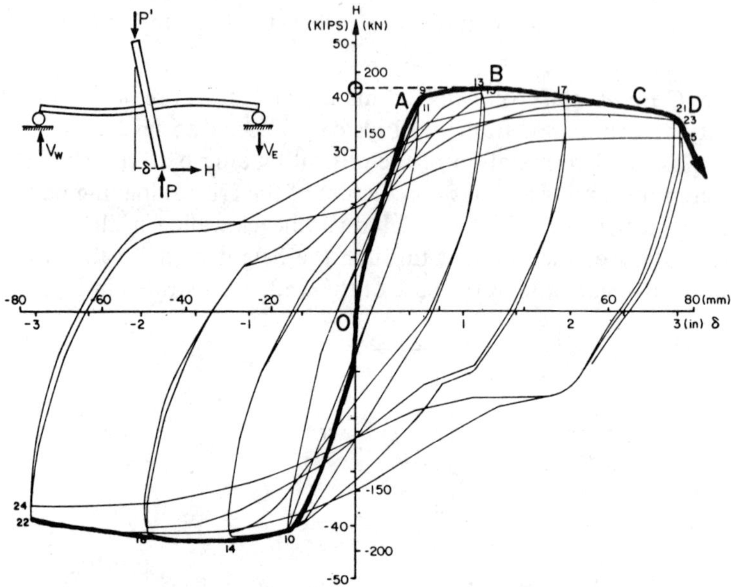

Figure 37 Lateral load-displacement relationship for a R/C beam-column subassemblage [44].

lateral strength capacity is clearly defined by the peak value of the lateral load, H (point B), the point in the curve $OABCD$ at which deformation capacity is reached is less clearly exhibited—is it the displacement corresponding to point B, the strength capacity, or is it the maximum deflection obtained just before a sharp drop in resistance takes place, (i.e., point D)? Usually neither of these points is used to define the deformation capacity of the specimen, and an intermediate point C (obtained by limiting the drop in strength to a certain percentage of the peak strength) is selected for measuring the usable deformation capacity of the specimen.

The lack of uniqueness in what constitutes the deformation capacity is closely related to the second problem. It is clear from Fig. 37 that if the lateral load applied to the subassemblage is of a sustained nature when deformation reaches the value corresponding to peak strength (point B), that value is all that can be used. On the other hand, if H represents the strength developed by a dynamic excitation (such as an earthquake ground motion), usually all the deformation up to point D can be used without the danger of collapse. Thus it is not surprising that no single definition has been formulated for deformation capacity.

The sensitivity of the deformation capacity to time-history loading is illustrated in Fig. 38 [22]. As shown in that figure, the maximum deformation

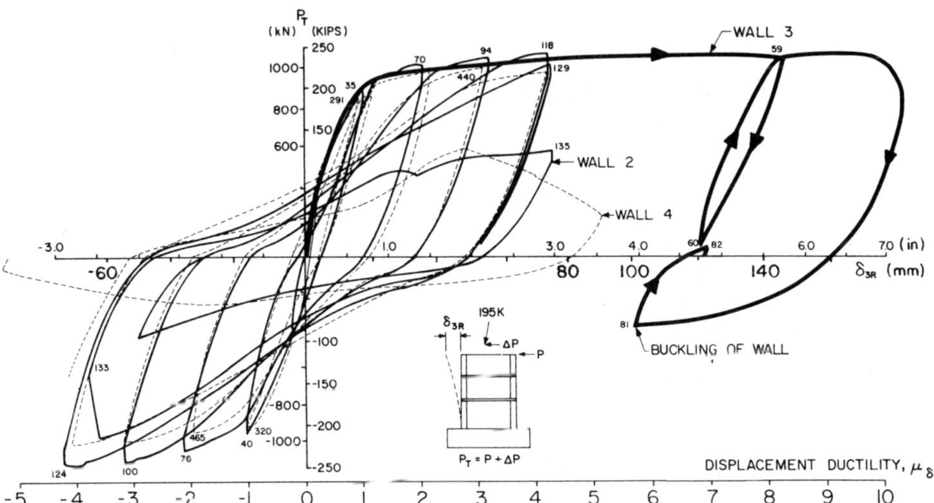

Figure 38 Comparison of behavior under monotonic loading (wall 3) with hysteretic behavior under increasing displacement reversals (walls 2 and 4) [45].

of the wall reached a value of about 175 mm (7 in.) when it was subjected to monotonically increasing lateral load (or deformation). On the other hand, when an identical specimen was subjected to cyclic loading with full displacement reversals, the maximum deformation was only 76 mm (3 in.) (i.e., a decrease of about 60%). These results point out that it is not possible to predict the behavior of structural wall systems under severe seismic ground motions based only on load-deformation relationships obtained under monotonically increasing loads. The experimental results described in [22] show that although it was possible to attain, under monotonically increasing load, a maximum displacement of 175 mm (7 in.) corresponding to a displacement ductility of about 10, this large ductility is not recommended for seismic-resistant design. When the load was reversed, the wall buckled under a lateral load of only 356 kN (80 kips), about one-third of the maximum strength developed (Fig. 38). The experiments show that if seismic response had required one or more cycles of full reversals at maximum deformation, the maximum usable ductility would have been about 4. On the other hand, if seismic response had required only a few full reversals of load without significant reversals of displacement, a ductility of about 6 could have been used (Fig. 39).

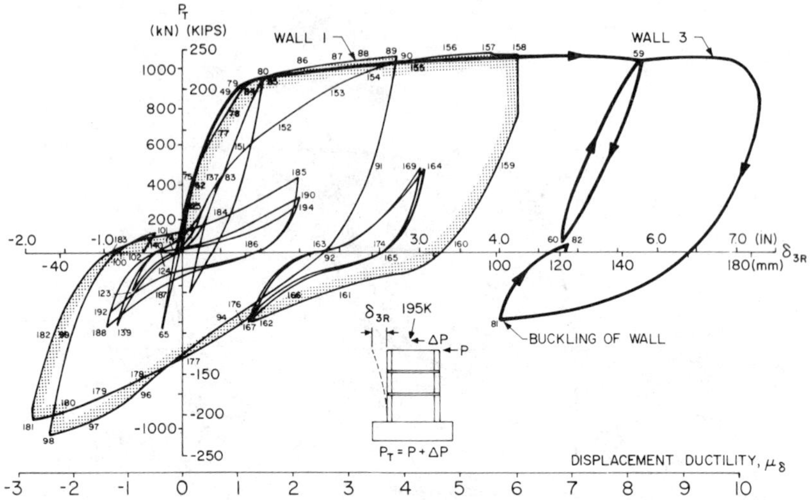

Figure 39 Comparison of behavior under monotonic loading (wall 3) with hysteretic behavior including partial reversals of displacements (wall 1) [22].

ANALYTICAL METHODS.   In general, no reliable methods are presently available for predicting the deformation capacities of real buildings. In discussing this topic it is convenient to distinguish buildings that undergo a brittle type of failure from those that experience ductile behavior. Current methods permit more accurate prediction of the deformation capacities of brittle buildings, which are undesirable for extreme environments. Prediction of deformation capacities for ductile buildings requires dynamic analysis using computer programs based on nonlinear models of the building. In real buildings deformation capacity is usually controlled by the reduced deformation capacity, which is very difficult to predict, of one or more regions that have been designed, detailed, or constructed improperly due to poor quality control of materials and/or workmanship. Only in the case of buildings that can be modeled quite accurately [i.e., very simple (trivial) buildings, properly designed, detailed, and constructed] is it possible to obtain results that can be considered reliable guidelines for estimating their deformation capacity.

Although a few three-dimensional computer programs are available for predicting the response of building structures, most predictions of deformation capacity for buildings under severe seismic excitations are obtained using nonlinear dynamic programs based on two-dimensional models with either infinite or limited ductility.

Infinite Ductility Models. If P–Δ effects are neglected, the analysis can lead to an unrealistic prediction of infinite deformation capacity. If the model includes these effects, maximum deformation can be defined as the point at which the structure becomes unstable under service loads (Fig. 5) or at the stage where strength drops a certain percentage of its peak value. In most cases the structure will fail because of the limited deformation capacity of the critical regions, which would appear to invalidate analyses based on infinite ductility models. Despite their limitations, the results of these analyses, together with some experimental information, can be used effectively in judging the behavior of the analyzed structure. If the rotational time history (demand) at the critical regions can be computed, the deformation capacity of the whole structure can be obtained by comparing the demand with the available rotation capacity, which is sensitive to the detailing of the critical regions. The deformation capacity obtained can thus be conservatively estimated as that corresponding to the deformation at which the most critical region starts to fail.

In the seismic response of a real structure with a high degree of redundancy, when the most critical region reaches its deformation capacity, other parts of the structure may not experience their maximum deformation because of the effect of high modes. Thus the start of failure of just one critical region does not usually lead to sudden collapse or to a decrease in lateral load resistance. Consequently, the lateral displacement at which failure of the most critical region occurs can be considered a conservative measurement of the deformation capacity of the structure.

The discussion above stresses the importance of predicting the deformation (in most cases, rotation) capacity of the possible critical regions of a structure. In this sense it is necessary to distinguish regions of ductile moment-resisting frames from those of wall structures. While the rotation capacity of flexural critical regions in the beams of ductile moment-resisting frames can be predicted with sufficient engineering accuracy (except for very short beams in which shear stresses can reach high values), the deformation capacity of critical wall regions is more difficult to determine, particularly if high shears develop. Integrated analytical and experimental research is needed to determine the rotation or shear deformation capacity of these critical regions under generalized loading conditions.

Limited Ductility Models. Direct prediction necessitates that analyses be performed using models that include limited deformation capacity (ductility) based on actual detailing of the critical regions. In these models, when computed ductility requirements at the critical regions exceed available ductility, a gradual rather than sudden drop in resistance to zero is introduced. The lack of adequate data regarding this decrease in resistance of critical regions in real structures and the difficulties in modeling the actual decay in strength have necessitated the use of simplified models. Mahin and Bertero used simplified, limited ductility models to investigate the seismic behavior of buildings that suffered significant structural damage during recent earthquakes [45]. Although the nonlinear analyses based on the model used correctly identified the main sources of observed damage, they were incapable of simulating observed deformations. Efforts should be directed to obtaining reliable data on the complete (i.e., up to structural collapse) hysteretic behavior of the critical regions of different components of a structure, and then to developing computer programs based on realistic idealizations of the observed behavior.

A reliable computer program that accounts for deterioration of strength as well as stiffness of critical regions is needed to enable the designer to determine progressive failure (i.e., collapse) of the complete structure. From analysis of available experimental data on behavior of structural elements and their subassemblages, the most reliable indices or parameters on which to base prediction of deformation capacity appear to be the inelastic rotation of critical regions whose behavior and failure is controlled by flexure and the shear distortion of those regions whose behavior is controlled by shear. These parameters can be measured directly and some experimental information is already available, mostly for the case of critical beam regions [46].

STATE-OF-THE-PRACTICE IN PREDICTING STRENGTH AND DEFORMATION CAPACITIES

The most desirable and practical method for evaluating strength and deformation capacities should combine simplicity of execution with acceptable levels of reliability. In practice it is not surprising that strength and deformation capacities are currently predicted on the basis of linear response analyses. Proper interpretation of those analytical results can be used to identify deficiencies and, perhaps, lead to acceptable estimates of the strength of buildings, particularly those of a relatively brittle type, or those in which inelastic deformations develop simultaneously at enough regions to convert the structure into a mechanism with enough strain hardening to compensate for the P–Δ effect. However, such behavior is more often the exception than the rule in actual buildings.

Comparison of results obtained for the behavior of complete structures and subassemblages from laboratory and field experiments with those from analyses using linear-elastic methods indicates that the latter usually lead to very conservative predictions of strength. The main reasons for this conservatism are: (1) the low minimum values of code-specified strength of materials; (2) the effect of strain hardening of the steel used in construction; and (3) the redistribution of internal forces after first yielding or first plastic hinge formation due to ductility of structural materials. The actual strengths of buildings that have been designed by codes are considerably higher than those predicted not only when linear-elastic methods are used, but even when methods based on elastic–perfectly plastic models of critical regions whose yielding strengths are based on code-specified strengths are used. Figure 40 illustrates the considerably higher strength observed in a static test of a R/C frame with respect to the strength obtained from linear-elastic and elastic–perfectly plastic analyses. This frame had previously been subjected to a series of dynamic tests on a shaking table, in which base shears up to 124.5 kN (28 kips) were recorded [47].

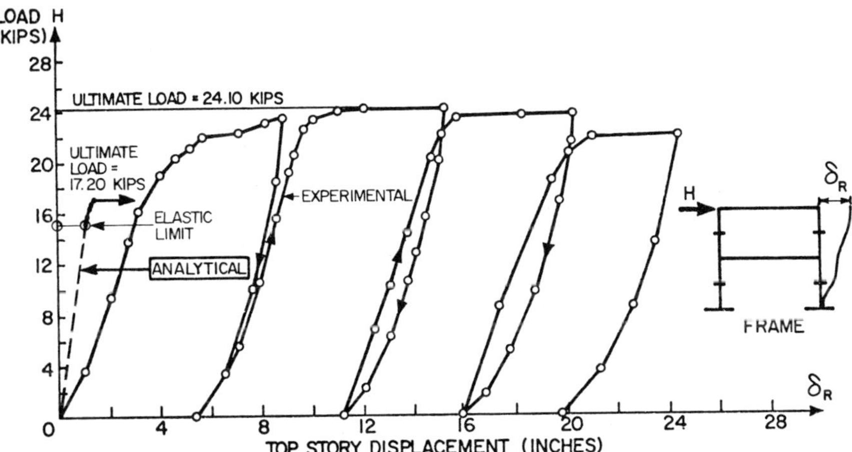

Figure 40 Comparison between analytical and experimental (static test) later load-deformation relationships for a R/C frame [47].

Linear-elastic methods, except when applied to brittle types of structures, generally fail to give information adequate for predicting deformation capacity. The studies carried out in [10,11,35,45] prove that computed nonlinear displacement response histories can differ substantially from elastic predictions. Because linear-elastic methods are relatively simple and inexpensive to use, studies should be carried out to correlate computed elastic response and actual inelastic response of standard types of structures. Guidelines should then be formulated for interpreting results from elastic analysis to obtain an idea of the actual strength and deformation capacities of such structures.

EDUCATIONAL, RESEARCH, AND DEVELOPMENT NEEDS

In this paper the difficulties encountered in predicting the strength and deformation capacities of buildings have been reviewed, as have their implications for achieving optimum seismic-resistant design and identifying and retrofitting existing (hazardous) buildings, which have been shown to constitute one of the major threats to life and loss of property from natural hazards. Emphasis has been placed on improvements needed for predicting the complete hysteretic behavior (of which strength and deformation capacities are part) of buildings under the combined effects of the normal and extreme environments to which they may be subjected during their service life. Such improvements can be effected only through integrated educational, research, and developmental efforts. Some of these needed efforts are summarized below.

Educational needs

Despite many unresolved problems in predicting the complete hysteretic behavior of buildings under the combined effects of normal and extreme environments and gaps in our knowledge of environmental hazards, our understanding has advanced significantly in the last two decades. There is at present a significant body of knowledge regarding the problems created by extreme environments. However, comparison of the state-of-the-art with the state-of-the-practice indicates that such knowledge has generally not been incorporated in practice. It is therefore crucial to impart this knowledge to structural engineering students through their curriculum in universities and professional schools, as well as to professionals through refresher courses that could be offered by continuing education programs. Implementation of such knowledge could be encouraged by making it a requirement for professional registration. Among the topics that should be included in the curriculum are:

1. Sources and mechanisms of possible extreme environments. This should include the proper way of measuring the severity of the excitations created by the environment on the building.
2. Actual mechanical behavior of soils and building materials under conditions imposed by the combination of excitations induced by normal and extreme environments.
3. Three-dimensional mechanical behavior of actual soil–building systems up to collapse when subjected to loads or deformations due to the combined actions of normal and extreme environments. This includes damage accumulated during the service life of buildings from variations in loading and environmental histories. Knowledge of cumulative damage will enable

the designer to estimate the building's condition (e.g., its residual stresses, cracking, deterioration in strength and stiffness from aging, and corrosion) at the time of occurrence of the extreme environment.

4. Procedures and criteria for establishing the strength and deformation capacities of existing buildings. This would require knowledge of the complete hysteretic behavior of the building.

5. Use of probability and statistics to develop sounder methods of predicting strength and deformation capacities under extreme environments.

6. Socioeconomic aspects of hazards created by extreme environments.

Research and development needs

Research needs are necessarily related to educational needs since increased knowledge is brought about through research. However, as Holley pointed out, research alone is not enough; analytical and experimental studies must be augmented by developmental work [48]. Among the different research and development needs discussed here, the following are emphasized:

1. Develop more precise methods for predicting the characteristics of the possible extreme environments that can occur at any given building site.

2. Conduct comprehensive, integrated experimental and analytical studies on the mechanical behavior of materials, single and composite, used in the construction of soil–building systems (both structural and nonstructural components) under loads and/or deformations induced by different extreme environments.

3. Carry out investigations of new materials that could improve the performance of buildings under abnormal environments.

4. Develop more accurate methods for nondestructive testing of materials in the field.

5. Develop more reliable procedures for determining strength and deformation capacities of soil–building systems under load or deformation conditions that can be imposed by the combined effect of normal and extreme environments. Required are integrated field, laboratory, and analytical studies and development of (a) new techniques for testing buildings up to collapse; and (b) nonlinear three-dimensional computer programs for predicting the progressive collapse of buildings.

ACKNOWLEDGMENTS

It is not possible here to thank all the individuals who contributed to the research reviewed in this paper, much of which was developed in course of studies sponsored by the National Science Foundation, whose support is gratefully acknowledged, but a few should be mentioned. The work conducted by my former students, Drs P. Hidalgo, H. Kamil, R. Klingner, S. Mahin, and T. Y. Wang, and more recently by graduate students Mssrs. J. Vallenas, S. Viwathanatepa, and S. Zagajeski, must be acknowledged. My deepest appreciation goes to the man whom we are honoring in this symposium. Professor Popov brought me to Berkeley nearly 20 years ago and during this time has exerted a tremendous influence on my own growth as a teacher and researcher. The continuing research collaboration and encouragement to use that research for improving and developing new courses and to participate in professional activities is acknowledged with gratitude.

In closing, I would like to express my admiration for Egor Popov's devotion to both teaching and basic and applied research and to praise him for the excellence of his numerous contributions in the field of engineering.

REFERENCES

[1] SAWYER, H. A., JR., "Comprehensive Design of Reinforced Concrete Frames by Plasticity Factors," 9th Planning Session of CED Symposium: Hyperstatique, Ankara, September 1964.

[2] TICHY, M., AND N. VORLICEK, "Flexural Mechanics of Reinforced Concrete Frames by Plasticity," *Proceedings*, International Symposium of Flexural Mechanics of Reinforced Concrete, ACI SP-12, Miami, November 1964.

[3] POPOV, E. P., AND R. MCCARTHY, "Deflection Stability Frames Under Repeated Loads," *Journal of the Engineering Mechanics Division, ASCE*, vol. 86, no. EM1, January 1960, pp. 61–78.

[4] POPOV, E. P., P. SHARIFI, AND S. NAGARAJAN, "Inelastic Buckling Analysis of Pipes Subjected to Internal Pressure, Flexure, and Axial Loading," *Pressure Vessels and Piping: Analysis and Computers, ASME*, 1974, pp. 11–23.

[5] *International Recommendations for Design and Construction of Concrete Structures*, F.I.P. 6th Congress, 2nd edition, Prague, 1970.

[6] Technical Committee 26, "Limit State Design," *Proceedings*, International Conference on Planning and Design of Tall Buildings, ASCE, Lehigh University, August 1972.

[7] STEVENS, L. K., "Limit State Philosophy and Application," *Proceedings*, Australian and New Zealand Conference on Planning and Design of Tall Buildings, Sydney, 1973.

[8] BERTERO, V. V., "Research Needs in Limit Design of Reinforced Concrete Structures," *Report EERC 71-4*, Earthquake Engineering Research Center, University of California, Berkeley, 1971.

[9] COHN, M. Z., "Optimal Limit Design of R/C Structures," *Inelasticity and Nonlinearity in Structural Concrete*, University of Waterloo Press, 1972.

[10] BERTERO, V. V., AND H. KAMIL, "Nonlinear Seismic Design of Multistory Frames," *Canadian Journal of Civil Engineering*, vol. 2, no. 4, December 1975.

[11] ZAGAJESKI, S. W., AND V. V. BERTERO, "Computer-Aided Optimum Seismic Design of Ductile Reinforced Concrete Moment-Resisting Frames," *Report UCB/EERC 77-16*, Earthquake Engineering Research Center, University of California, Berkeley, 1977.

[12] BERTERO, V. V., AND B. BRESLER, "Design and Engineering Decisions: Failure Criteria (Limit States)," panel discussion paper presented at the 6th World Conference on Earthquake Engineering, New Delhi, January 10–19, 1977.

[13] NEWMARK, N. M., AND E. ROSENBLUETH, *Fundamental Earthquake Engineering*, Prentice-Hall, Inc., Englewood Cliffs, N.J., 1971.

[14] YAMADA, M., AND H. KAWAMURA, "Resonance-Fatigue Characteristics for Evaluation of the Ultimate Aseismic Capacity of Structures," presented at the 6th World Conference on Earthquake Engineering, New Delhi, January 10–19, 1977.

[15] BERTERO, V. V., AND E. P. POPOV, "Hysteretic Behavior of Ductile Moment-Resisting Reinforced Concrete Frame Components," *Report EERC 75-16*, Earthquake Engineering Research Center, University of California, Berkeley, 1975.

[16] *Uniform Building Code*, International Conference of Building Officials, Whittier, Calif., 1976 edition.

[17] Seismology Committee, *Recommended Lateral Force Requirements and Commentary*, Structural Engineers Association of California, San Francisco, 1976.

[18] ACI-ASCE Committee 352, "Recommendations for Design of Beam-Column Joints in Monolithic Reinforced Concrete Structures," *Journal of the ACI*, Proceedings, vol. 73, no. 7, July 1976.

[19] *Final Review Draft of Recommended Comprehensive Seismic Design Provisions for Buildings*, Applied Technology Council, ATC-3-05, Palo Alto, January 1977.

[20] MAHIN, S. A., AND V. V. BERTERO, "Nonlinear Seismic Response of Coupled Wall Systems," *Journal of the Structural Division, ASCE*, vol. 102, no. ST9, September 1976.

[21] BERTERO, V. V., "Establishment of Design Earthquakes—Evaluation of Present Methods," *Proceedings*, International Symposium on Earthquake Structural Engineering, St. Louis, August 1976.

[22] BERTERO, V. V., ET AL., "Seismic Design Implications of Hysteretic Behavior of Reinforced Concrete Structural Walls," presented at the 6th World Conference on Earthquake Engineering, New Delhi, January 10–19, 1977.

[23] *Building Practices for Disaster Mitigation*, Building Science Series 46, National Bureau of Standards, U.S. Department of Commerce, February 1973.

[24] FERVER, G., "Existing (Hazardous) Buildings," AIA Research Corporation, Report to National Science Foundation on Architects and Engineers: Research Needs, New York, December 1976.

[25] "The Los Angeles County Earthquake Commission Report," March 1972, 8 vols.

[26] BRESLER, B., ET AL., "Developing Methodologies for Evaluating the Earthquake Safety of Existing Buildings," *Report UCB/EERC 77-06*, Earthquake Engineering Research Center, University of California, Berkeley, 1977.

[27] DRENIK, R. F., P. C. WANG, AND W. WANG, "Case Study of Critical Excitation and Response Structures," *Report POLY EE/EP-75-DID*, Polytechnic Institute of New York, November 1975.

[28] GUENDELMAN, I. R., AND G. H. POWELL, "DRAIN-TABS, A Computer Program for Inelastic Earthquake Response for Three-Dimensional Building Systems," *Report EERC 77-08*, Earthquake Engineering Research Center, University of California, Berkeley, 1977.

[29] BATHE, K. J., E. L. WILSON, AND R. H. IDING, "NONSAP, A Structural Analysis Program for Static and Dynamic Response of Nonlinear Systems," *Report SESM 74-3*; University of California, Berkeley, 1974.

[30] MONDKAR, D. P., AND G. H. POWELL, "ANSR-I General Purpose Computer Program for Analysis of Non-linear Structural Response," *Report EERC 75-37*, University of California, Berkeley, 1975.

[31] BATHE, K. J., "ADINA, A Finite Element Program for Automatic Dynamic Incremental Nonlinear Analysis," *Report 82448-1*, Massachusetts Institute of Technology, Cambridge, May 1976.

[32] MAC NEAL SCHWENDLER CORPORATION, "NASTRAN, the NASA Structural Analysis Program," 1977.

[33] POWELL, G. H., "Computer Programs for Analysis of Seismic Response of Reinforced Concrete Buildings," Workshop on Earthquake-Resistant Reinforced Concrete Building Construction, University of California, Berkeley, July 1977, *Proceedings*, 1977, Vol. II, pp. 969–980.

[34] GATES, W., "The Art of Modeling Buildings for Dynamic Seismic Analysis," Workshop on Earthquake-Resistant Reinforced Concrete Building Construction, University of California, Berkeley, July 1977, *Proceedings*, Vol. II, pp. 857–865.

[35] MAHIN, S. A., ET AL., "Response of Olive View Hospital Main Building during the San Fernando Earthquake," *Report EERC 76-22*, Earthquake Engineering Research Center, University of California, Berkeley, 1976.

[36] SUDHAKAR, A., "Computer Program for Small Displacement Elasto-Plastic Analysis of Plane Steel and Reinforced Concrete Frames," Master's thesis No. 561, University of California, Berkeley, 1972.

[37] KLINGNER, R. E., AND V. V. BERTERO, "Infilled Frames in Earthquake-Resistant Construction," *Report EERC 76-32*, Earthquake Engineering Research Center, University of California, Berkeley, 1976.

[38] FIORATO, A., AND W. CORLEY, "Laboratory Tests of Earthquake-Resistant Structural Wall Systems and Elements," Workshop on Earthquake-Resistant Reinforced Concrete Building Construction, University of California, July 1977, *Proceedings*, Vol. III, pp. 1388–1430.

[39] PAULAY, T., "Earthquake Resistant Structural Walls," Workshop on Earth-

quake-Resistant Reinforced Concrete Building Construction, University of California, Berkeley, July 1977, *Proceedings*, Vol. III, pp. 1339–1365.

[40] TAKEDA, T., "Design of R/C Frame-Wall Structures," preprint from Workshop on Earthquake-Resistant Reinforced Concrete Building Construction, July 1977.

[41] ACI Committee 318, *Building Code Requirements for Reinforced Concrete*, ACI-318-71, Detroit, 1971.

[42] BARDA, F., J. M. HANSON, AND W. G. CORLEY, "Shear Strength of Low-Rise Walls with Boundary Elements," in *Reinforced Concrete Structures in Seismic Zones*, ACI Publication SP-53, Detroit, 1977, pp. 149–202.

[43] WANG, T. Y., V. V. BERTERO, AND E. P. POPOV, "Hysteretic Behavior of Reinforced Concrete Framed Walls," *Report EERC 75-23*, University of California, Berkeley, 1975.

[44] POPOV, E. P., ET AL., "On Seismic Design of R/C Interior Joints of Frames," presented at 6th World Conference on Earthquake Engineering, New Delhi, January 10–19, 1977.

[45] MAHIN, S. A., AND V. V. BERTERO, "An Evaluation of Some Methods for Predicting Seismic Behavior of Reinforced Concrete Buildings," *Report EERC 75-5*, Earthquake Engineering Research Center, University of California, Berkeley, 1975.

[46] BERTERO, V. V., AND E. P. POPOV, "Seismic Behavior of Ductile Moment-Resisting Reinforced Concrete Frames," in *Reinforced Concrete Structures in Seismic Zones*, ACI Publication SP-53, Detroit, 1977, pp. 247–292.

[47] HIDALGO, P., AND R. W. CLOUGH, "Earthquake Simulator Study of a Reinforced Concrete Frame," *Report EERC 74-13*, Earthquake Engineering Research Center, University of California, Berkeley, 1974.

[48] HOLLEY, M. J., "Not Research Alone, but Research and Development," *Proceedings*, Workshop on Simulation of Earthquake Effects on Structures, San Francisco, September 1973, National Academy of Engineering, Washington, D.C.

THE PROFESSIONAL NEEDS
OF THE PRACTICING ENGINEER

———— *M. S. Agbabian* and R. G. Johnston*** ————

Dedicated to Egor P. Popov, whose teaching and research continue to benefit the engineering profession. His work has influenced analytical techniques of static and dynamic response of structures, and has opened the way to improved design procedures, construction details, and building codes.

INTRODUCTION

Practicing engineers have always made a significant impact on the economic, environmental, and social progress of their communities, but only in the last decade have engineers come to an intimate awareness of their influence. In the past, they assumed that the professional engineering societies, armed with the power of the group, would tackle the social issues. If properly motivated, engineers might work for the committees of their professional societies and consider themselves sufficiently involved. It is now apparent to them that their daily work as designers, analysts, or consultants influences decisions

---

*President, Agbabian Associates, El Segundo, California.
**Vice President, Brandow and Johnston Associates, Los Angeles, California.

reaching far beyond the immediate assignments. Today's powerful drive to assure public safety and enhance the environment must be satisfied on the design board.

The first task faced by practicing engineers is to assure the integrity of the system being designed, while meeting all safety and cost guidelines. The second task is to convince the public that they have indeed achieved these goals without damaging the environment. Their accomplishments will have serious setbacks each time they plan, design, or construct a major facility that fails to meet the standards. Catastrophic failures, escalating costs, environmental pollution—all will have repercussions beyond the projects for which they are responsible. Failure of a dam makes the safety of all dams suspect. Uneconomical design of a desalination plant makes all similar plants unattractive. Leaking radioactivity detected outside a containment shell gives credence to claims that all nuclear power plants are dangerous to the environment. The responsibilities of practicing engineers are indeed awesome.

Fortunately, most facilities are safe, are economically designed, and may even enhance the environment. Practicing engineers can usually be satisfied that they have bettered their communities. But each new assignment makes him/her vulnerable. Whether a new graduate or an expert with 25 years of experience, every engineer is daily challenged to perform at the highest levels in engineering history—not only high levels of technical competence, but equally high levels of social, political, legal, and environmental wisdom.

For instance, what has happened to practicing engineers in the field of structural engineering? More specifically, what has changed in the field of earthquake engineering?

THE PRACTICING ENGINEER IN EARTHQUAKE ENGINEERING

In the last decade, practicing engineers have become involved in earthquake engineering at a spirally increasing rate. Earthquake engineering is no longer confined to the western states. Consider these milestones:

The U.S. Congress has taken note of the need to support earthquake engineering research, by introducing legislation on earthquake hazards reduction.

The National Science Foundation has broadened its sponsorship of research to include societal aspects.

The Seismic Engineering branch of USGS is extending its reliable network of strong-motion instruments.

ASCE has formed a technical council on lifeline earthquake engineering.

EERI has enlarged its mission by becoming a national organization.

The State of California has appointed a Seismic Safety Commission.

The Nuclear Regulatory Commission has made seismic safety one of the most critical requirements for nuclear power plants.

The Structural Engineers Association of California, through its Applied Technology Council, is producing a comprehensive document of provisions for the development of seismic regulations for buildings. The document will be applicable in every state of the United States.

These activities center around practicing engineers. Let us consider the Applied Technology Council activity known as ATC-3. Eighty-eight prominent engineers and scientists from the academic and practicing fields have formed 5 task groups, 14 committees, and 3 review groups; through these groups they are formulating design provisions that may very well be the guidelines of practicing engineers for the next decade.

Does this mean that such guidelines will automatically chart the course of action of practicing engineers? That research engineers may now take a well-deserved rest? On the contrary; it is clear that the technology of earthquake engineering is burgeoning. Much research and many years are needed to evolve increasingly rational approaches. During the growth of this science, the practicing engineer's judgment must span the gap between the unknown and the known. Engineers will have to be more creative in their designs, yet their judgments must withstand the sharpest scrutiny.

We will be more specific. Practicing engineers should look at a building code or a regulatory guide as their point of departure. Satisfying the code is the first step in good design. Building-code officials are becoming as conscious of this change in outlook as are practicing engineers; creative thinking is being encouraged.

We will draw a brief portrait of the type of professional challenge faced by today's practicing engineers. Our examples are taken from actual earthquake engineering projects. Each project posed problems that required creative solutions.

Example 1. Dam safety

The task was to determine the probable structural behavior of a concrete gravity dam subjected to strong earthquake ground motion. A three-step study was planned:

1. Review the behavior of a similar dam.
2. Perform equivalent static or dynamic elastic analyses.
3. Perform a nonlinear dynamic analysis if the results of steps 1 and 2 warrant it.

A similar dam subjected to an earthquake input had shown cracks in the concrete; leaks were observed, but the reservoir of water was contained. The elastic analysis showed very large tensile stresses, in excess of the tensile strength of the concrete, at several critical sections. If the earthquake was considered an equivalent static force, instability was indicated. If the earthquake input was dynamic, using the time history of motion at the bedrock foundation, large relative motions of concrete monoliths were indicated, which would compromise the ability of the dam to contain the reservoir.

The engineer at this point could either declare the dam unsafe or pursue his investigation with a more realistic model of the material properties. When the tensile stresses in the dam reach a critical level, there will be a redistribution of the stresses within the structure. Moreover, beyond the time when a crack is formed, owing to the cyclic nature of the earthquake input, some rebonding will occur when the concrete section is in compression. Under the earthquake input, the sections with tensile cracks would still perform a function during the stress reversal stage when they would be in compression. Finally, while all the complexities of this phenomenon occur there will be energy dissipation that will increase the damping of the structural system.

The engineer decided that a nonlinear dynamic analysis was necessary before a final decision could be made. When such an analysis was performed, he observed that nearly all external elements of the dam cracked, but interior cracking was restricted; and at any given instant of time, enough areas within critical sections remained in a state of compression to maintain overall stability.

The engineer had to make a "go/no-go" decision. The elastic analysis indicated a "no-go" decision. The nonlinear analysis indicated a "go" decision. Relying on his study of the behavior of a similar dam that had experienced a moderately severe earthquake, and relying even more heavily on the results of the nonlinear analysis, he declared that for the conditions investigated the structure was safe—in other words, the impounding of the reservoir would not be compromised. But this may not be the end of the story. What if the selected earthquake input is not the maximum credible earthquake? What if the nonlinear material properties he has selected from laboratory data are not representative of the mass concrete? By quantifying these and other uncertainties, he makes a probabilistic assessment of the survival of the concrete dam for a prescribed period of time.

Example 2. Hospital rehabilitation

The San Fernando earthquake damaged or collapsed a group of buildings at the Sylmar Veterans Hospital. Damaged buildings were demolished and the site was abandoned. It became necessary to investigate the sites and structures of other VA hospitals.

The criterion for hospital safety included the provision that key activities should continue after an earthquake. Abandoning a hospital would involve evacuation and relocation of patients. The investigation showed that certain hospital buildings in California would not resist the assumed level of ground shaking. The question was: Should the buildings be reinforced, should they be declared safe, or, in some cases, should they be abandoned?

There were uncertainties at each stage in the investigations. Quantifying each of these uncertainties would have provided a rational basis for a decision, but implementing such a procedure was hampered by severe limitations. The characteristics of old, existing structures were not well defined. Statistical data were lacking. Some aspects of building response were not susceptible to mathematical analysis. Moreover, there was always uncertainty in the definition of the maximum credible earthquake.

In this example, the earthquake input was estimated by a number of independent geotechnical consultants. Structural performance was evaluated on the basis of experience augmented by analysis. The consensus of experts—that is, their best judgment—was considered to be the most reliable basis for the decision-making process. This method has also been referred to in the past as "the common sense of the profession." In one case in West Los Angeles, where the existing building was demolished and a new building erected, it was clear that this was the only possible decision.

Example 3. Office buildings

An eight-story, steel-frame office building that was built in accordance with the 1970 building code did not comply with current code requirements. A potential buyer was concerned that he would acquire a nonconforming building whose safety was suspect.

In reviewing the design calculations and drawings, the structural engineer initially arrived at the conclusion that the potential buyer should be warned that the building did not have the ductile moment resistance capacity currently required by codes and accepted as good practice. It occurred to the engineer, prior to issuing his report, that he should investigate whether the as-constructed building differed from the design. When construction deviates from design, the engineer typically expects that those deviations will be a step away from good practice. In this case, the "as-built" drawings and the "as-constructed" details of the building frame connections showed the contrary. The contractor had elected to provide 100% welded beam-to-column connections in place of bolted connections that were not designed to resist moments. He had apparently elected to do this for cost considerations after establishing that he had provided a substitute design detail that was accepted by the design engineers as being equal to the original design. In fact, his proposed change had inadvertently improved the original design, resulting in a ductile moment-resisting frame. The structural engineer's final report declared the building to be capable, within acceptable limits of ductility, of resisting the lateral forces specified for that seismic region.

These three examples illustrate some of the capabilities engineers should possess: analytical capability using the newest engineering techniques, a knowledge of what constitutes good design, and even some detective work to produce all the evidence that will help them form judgments. They also illustrate the environment for today's practicing engineers. They can no longer confine themselves to a quiet corner, concentrating only on satisfying the minimum design requirements of a building code or a regulatory guide.

The entire spectrum of their responsibilities as earthquake engineers includes much more than the examples of the dam, the hospitals, and the office building have revealed. They must first develop seismic criteria, because the science of earthquake engineering is so new that a dependable data base does not exist; regulatory guides and reliable reference criteria have not been established for many situations and sites. Therefore, they

1. Review the geologist's and seismologist's assessment of the earthquake hazard.
2. Evaluate the effect of the soil properties at the site.
3. Study strong-motion records that match the geology and seismicity of the site.
4. Evaluate the sensitivity of the criteria to changes in values of critical parameters.
5. Attempt to quantify the uncertainties in the information at their disposal.
6. Apply judgment in selecting the seismic criteria.

Next, earthquake engineers design the structures; they

1. Select a structural system. (Most architects respect creative thinking and good judgment during this most important step of the design process.)
2. Conceive a structure/foundation system that remains stable as it responds to ground motion.
3. Design the structural elements and connections to resist cyclic loads without degradation.

But they are not finished. They must analyze the structures to verify the adequacy of the designs for earthquake resistance. Since the analysis is critical, engineers

1. Include a sufficient number of degrees of freedom to describe the seismic response within the critical frequency range.
2. Account for the damping and ductility of the structure/foundation system.
3. Evaluate torsional effects, discontinuities, directional effects, coupling, soil/structure interaction, relative displacements, and other response characteristics that were not explicitly considered during design.
4. Recognize the possible variations in material properties and workmanship.

Examples were given from the field of earthquake engineering to point out the breadth of professional competence that the practicing engineer should possess. Similar developments have imposed requirements in other fields that are comparable in complexity to those faced by earthquake engineers. In a general sense, the responsibilities of all practicing structural engineers are the same, regardless of their areas of specialization.

What, then, are the professional responsibilities of today's practicing engineers? What new technologies have become their engineering tools? What are their needs? What are their resources for professional development? How can their needs be better satisfied?

Answers to these questions will give us the perspective to chart new directions.

WHAT ARE A PRACTICING ENGINEER'S PROFESSIONAL RESPONSIBILITIES?

The duties of all practicing engineers are essentially the same, whether they are private consultants, members of engineering teams, or employees of large A/E organizations. At some time in their careers, engineers may work in a number of areas—conceptual development; design and design criteria, studies, review, verification; site selection; construction management. Table 1 describes these activities. To carry out their duties, they should have analytical, experimental, design, and management capabilities.

Analytical

All engineers should be capable of (1) defining problems in a manner that accounts for significant physical phenomena and provides for amenable solutions, (2) reducing systems to idealized mathematical models appropriate for analysis, and (3) using suitable techniques to solve the problems.

Experimental

All engineers should be capable of (1) developing empirical solutions from experimental data, (2) verifying analytical procedures by correlation with experiments, and (3) demonstrating that facilities, a system, or a component, subjected to a specified environmental input, will function in accordance with the design requirements.

Design

All engineers should be capable of (1) visualizing the function of each structural element as part of the total structure and the deformed shape of the structure under the applied loads, (3) joining the structural elements into a

Table 1

TECHNICAL RESPONSIBILITIES OF PRACTICING ENGINEERS

Area	Activity
Design criteria	Define facility objectives.
	Formulate requirements for meeting objectives.
Site characterization	Make beneficial-siting analyses.
	Select optimum site.
	Verify site adaptability to facility objectives.
Concept development	Develop several design concepts based on functional requirements of facility.
	Select candidate concept for final design.
Design	Design facility to criteria.
	Prepare construction drawings, specifications, cost estimates.
Design studies	Investigate special design features or limitations, such as: Cost optimization
	New construction materials or methods
	Hazards due to explosion, radiation, earthquake, wind, floods
	Failure modes and safety margins.
Design review	Evaluate the design against the operational and environmental requirements.
	Assure compliance of design with criteria.
Design verification	Plan, design, and direct test programs simulating operational conditions.
	Construct mathematical model.
	Compare test and analysis results.
Construction management	Maintain schedules, budgets, and quality of workmanship.
	Evaluate field changes.

stable structure that efficiently transmits loads from element to element, and (3) understanding the properties of the materials and how they influence the behavior of the structure.

Management

All engineers should be capable of (1) seeing a project as a subsystem within a system, (2) determining the time in the sequence of events when final decisions are to be made, (3) appreciating the interaction with socioeconomic and other technical disciplines, and (4) completing the project on schedule and within the budget while performing at a level of technical excellence.

The tasks performed by practicing engineers have become more complex as they interact not only with the owners of various projects but with society. The environment has now been added explicitly to structural safety and cost effectiveness, making a three-way demand on talents. Safety measures against earthquake, fire, wind, and floods have added a new dimension. More rational methods of analysis and evaluation are demanded from engineers. Design-to-cost and value engineering, procedures first developed for national defense programs, are finding their place in civil-engineering systems. Environmental impact requirements force structural engineers to interact with other professions, to become active members of multidisciplinary teams. The public recognizes this change and expects practicing engineers to work within this broader framework.

WHAT NEW TECHNOLOGIES HAVE BECOME TOOLS?

Just since the 1950s there have been four major technological advances in structural engineering:

1. Ultimate design methods have received general acceptance.
2. Probabilistic design methods have come to complement engineering judgment.
3. The computer has become a tool in the design office.
4. The finite element method has moved ahead of other numerical methods.

Probabilistic design methods, which generate quantified values of uncertainties in load input, material properties, and mathematical idealizations, are providing a more rational basis for selecting the design load factors in ultimate design. Solutions by the finite element method, implemented on the computer, help provide a better understanding of the behavior of complex structures. It demonstrates the sensitivity of structural behavior to variations in parameters. These new tools have, without question, helped the engineering profession reach pinnacles of accomplishment. They have also created transitional problems.

These four technological advances have occurred within the career lifetimes of many practicing engineers. For example, the transition from hand calculation to computer solution was also a transition from moment distribution to finite element analysis. To earlier generations the abbreviation F.E.M. meant "fixed-end moment." To the new generation of engineers it means "finite element method." Practicing engineers went through a brief period when they were unable to visualize the deformation of a structure which they had been so well able to perceive while designing by moment distribution. This was partially corrected later when computer plots supplemented data printouts with pictures of structural deformations. We say "partially" because engineers, once active participants in the moment distribution process as they followed a structure's effort to reach its state of equilibrium by successive relaxations, now became passive observers as the computer performed these steps for them.

The new technologies have in many instances widened the gap between recent graduates and practicing engineers who earned their degrees some 20 years ago. Recent graduates have learned the new techniques as a coherent part of their college engineering studies. Yet more experienced engineers who direct the work of recent graduates have had to add those same new techniques to their capabilities by various uncoordinated methods, short courses, seminars, and self-teaching, which are not necessarily designed to be comfortably integrated with their previous studies.

WHAT ARE THE NEEDS?

From the foregoing it may appear that the needs of practicing engineers are numerous. This is not the case. Although the technology is constantly evolving, the nature of engineers' needs has remained the same:

Basic principles.
Creative thinking.
Good judgment.

For a large number of practicing engineers, training in basic principles more or less ends at graduation. Unfortunately, some engineers are not trained in creative thinking and good judgment until they have worked in the field for several years. Such a demarcation, where the university provides the principles and engineering practice develops creativity and judgment, shows a lack of interaction between the university and career experience. The principal need of practicing engineers is therefore a balanced program in which all three areas are developed during undergraduate training and continue to be developed during their careers. Figure 1 shows the ideal dynamic interaction between university training and research, and professional practice.

WHAT ARE THE RESOURCES
FOR PROFESSIONAL DEVELOPMENT?

The professional development of practicing engineers is strongly dependent on the published literature, technical conferences, university extension courses, and collaboration with other professional engineers.

Most engineers take advantage of the published literature. We wish to make here a positive statement in favor of the pressures on university faculty members to publish the results of their research. Their publications have an enormous influence on practicing engineers, provided that the research can be translated into applied procedures. The needs of practicing engineers

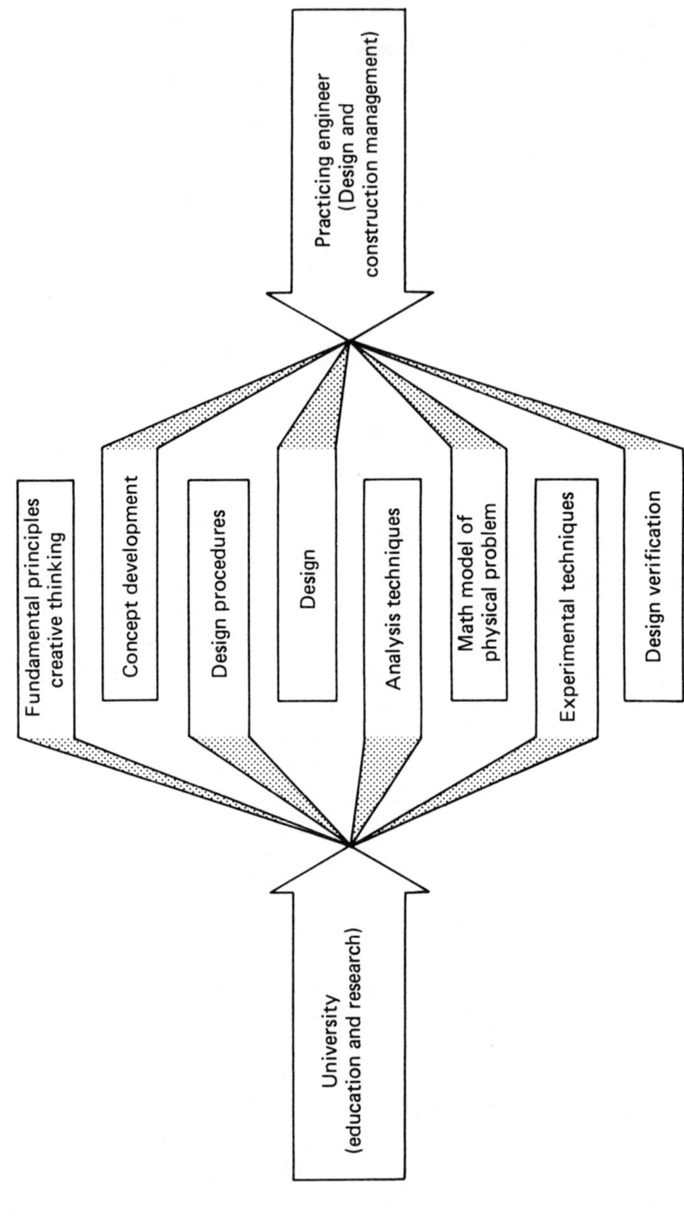

Figure 1 The dynamics of university and engineer interaction.

248

will not be overlooked if researchers have the foresight to see the potential applications of their research and do not seek merely the acknowledgment of their peers.

Technical conferences and symposia will continue to be an excellent educational resource. We have observed that there has been a conscious effort to tailor such conferences and symposia to the needs of practicing engineers.

Practicing engineers have often been criticized for not doing enough to further their professional development after graduation. This is a misconception. A survey sponsored by the California State Board of Registration and conducted by the UCLA Institute for Social Sciences Research has shown that three out of four engineers spend one or more hours per week in professional development courses. University extension courses generally cover state-of-the-art subjects. Our review of the courses offered has shown that they are generally well organized, and that they succeed in presenting the state-of-the art to practicing engineers.

Collaboration with other engineers is another important resource for professional development. The transfer of knowledge, the exchange of ideas, the critical reviews, and the conceptual trade-offs that take place in teamwork prepare engineers to accept increased responsibility. In engineering offices where creative work is done, line supervision is limited to administrative matters, and engineering decisions take shape from the interaction of collaborating professionals.

HOW CAN NEEDS BE BETTER SATISFIED?

Improvements are both possible and desirable in the four resources we have cited. Publications, technical conferences, university extension courses, and collaboration among engineers can certainly be carefully planned to make them more responsive to the needs of professional engineers. If there were an attempt at the national level to better coordinate these resources, practicing engineers would have clearer directions in their professional development.

However, we believe there is a more basic improvement that should be seriously considered: a change in the university program that prepares engineers for their professional careers. As the demands on professional engineers have increased in the past 25 years, university undergraduate programs have attempted to meet these demands by increasing the requirements for nonengineering courses at the expense of engineering courses. It is our observation that the four-year program has indeed become a pre-engineering curriculum. The required engineering courses we once studied in our undergraduate years are now available only in graduate programs. It is paradoxical that as the demand for more preparatory work in the university has increased, more and more of the requirements have shifted to the post-graduate course level. Yet universities continue to claim that they graduate engineers ready for professional practice after four-year undergraduate programs.

The state boards of registration, recognizing the need for more training, require that engineers' work be fully supervised for at least two years before they are allowed to take the board exams. Only after passing the exams may they practice autonomously. The board's mandate obviously implies that a four-year engineering course is not sufficient preparation for professional engineering. However, registration is required only for the protection of public health and safety, and the jurisdiction of the state boards does not extend to those who work as members of a team, as long as one of its team members is a registered engineer and is willing to accept responsibility for the work of all other members.

The question is: Can a four-year college-engineering curriculum even provide the necessary formal education in basic physical principles, let alone the fundamentals of creative thinking and good judgment? It is not enough to answer that students can take advanced studies in graduate school, when those students can be certified by the university as engineering graduates after finishing only the four required years. The time has come for the engineering profession to require five years, even six years, of mandatory training. This can only be achieved if the universities declare the four-year curriculum to be a preengineering program, similar to the four-year prelaw and premedicine requirements. We believe that this comparison to law and medicine is not unreasonable. In some respects, demands on the engineering professional are greater than those on lawyers and doctors, since most engineering projects affect large groups of people rather than simply a few individuals.

Assuming that our educational policies can be directed toward a mandatory five- or six-year program, the cornerstones of creative thinking and good judgment should not be more specialization but an expansion of the engineer's outlook on the multidisciplinary programs incorporating public policy, economics, business administration, and management. Figure 2 compares the conventional program with the proposed program for preparing professional engineers.

Specialization has a unique place in structural engineering. It should, however, begin with a foundation of multidisciplinary training. Here again, the similarity to law and medicine is obvious. Radiologists, for example, first receive degrees in premedicine (B.S.) and medicine (M.D.), and then become diplomates in radiology. It may sound revolutionary in concept, but the Ph.D. specialization program (including the dissertation) may be reduced to two years instead of the current three- to four-year program, provided that candidates have completed a five- or six-year mandatory general program. Their total residency in college will still be seven to eight years, but their

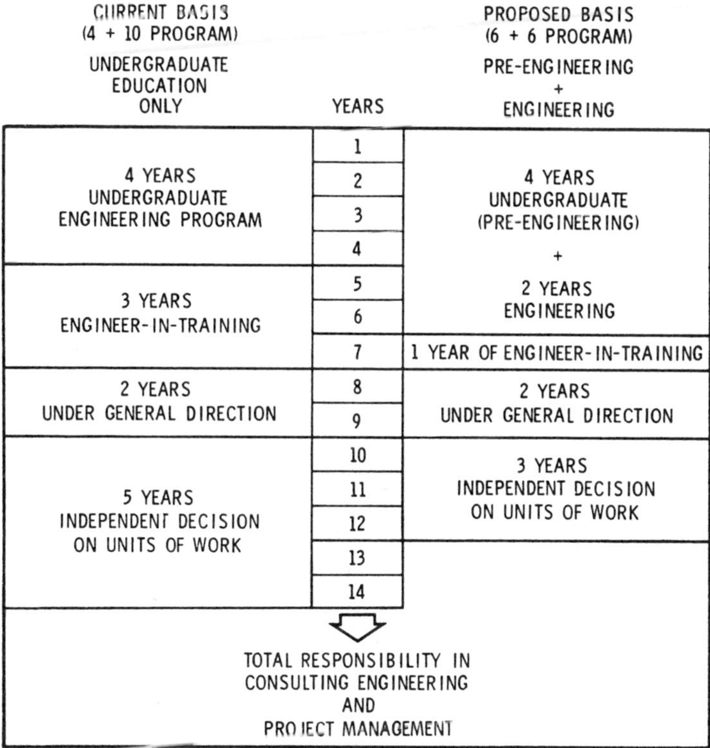

Figure 2 Preparation for total responsibility as a professional engineer.

specialist talents, together with the broader mandatory program, will prepare them to be more effective professionals. The schedule for a specialization program is outlined in Fig. 3 and compared to the current program and schedule.

Some would argue that professional engineering practice is the best training ground. Requiring college residence for one or two years longer would appear to remove these precious years from the training period. This is not the case. The training period during practice will become shorter as a result of improved educational background. For example, the greatest challenges practicing engineers face are in project management and consulting engineering. In project management, the engineer directs a group of engineers and technicians to complete a project on time, within the budget, and with technical excellence. In consulting engineering, the optimum solu-

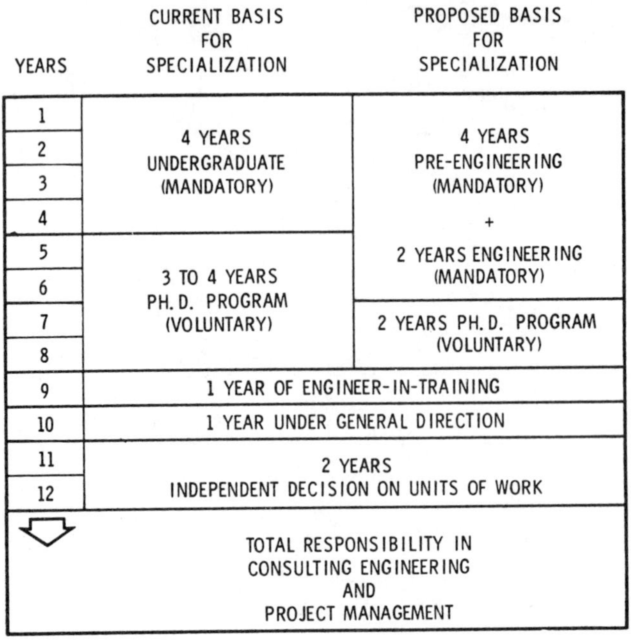

YEARS	CURRENT BASIS FOR SPECIALIZATION	PROPOSED BASIS FOR SPECIALIZATION
1	4 YEARS UNDERGRADUATE (MANDATORY)	4 YEARS PRE-ENGINEERING (MANDATORY)
2	4 YEARS UNDERGRADUATE (MANDATORY)	4 YEARS PRE-ENGINEERING (MANDATORY)
3	4 YEARS UNDERGRADUATE (MANDATORY)	4 YEARS PRE-ENGINEERING (MANDATORY)
4	4 YEARS UNDERGRADUATE (MANDATORY)	+
5	3 TO 4 YEARS PH.D. PROGRAM (VOLUNTARY)	2 YEARS ENGINEERING (MANDATORY)
6	3 TO 4 YEARS PH.D. PROGRAM (VOLUNTARY)	2 YEARS ENGINEERING (MANDATORY)
7	3 TO 4 YEARS PH.D. PROGRAM (VOLUNTARY)	2 YEARS PH.D. PROGRAM (VOLUNTARY)
8	3 TO 4 YEARS PH.D. PROGRAM (VOLUNTARY)	2 YEARS PH.D. PROGRAM (VOLUNTARY)
9	1 YEAR OF ENGINEER-IN-TRAINING	
10	1 YEAR UNDER GENERAL DIRECTION	
11	2 YEARS INDEPENDENT DECISION ON UNITS OF WORK	
12	2 YEARS INDEPENDENT DECISION ON UNITS OF WORK	
⬇	TOTAL RESPONSIBILITY IN CONSULTING ENGINEERING AND PROJECT MANAGEMENT	

Figure 3 Preparation for total responsibility as a specialist professional engineer.

tion to a technological problem is developed. Such responsibilities are not usually given to engineers until 10 to 15 years after graduation, a relatively long apprenticeship felt to be necessary in disciplines that are not covered by the educational program. We believe that a project management or a consulting engineering responsibility can be assumed in six years after graduation from the engineering education program we propose. Although these statements are made as a result of our experience as professional engineers in the structural engineering field, we feel they are equally applicable to other fields of engineering.

CONCLUSIONS

The professional needs of practicing engineers are met through a two-step process: academic training followed by experience. During the practice period, engineers also pursue voluntary professional development programs, benefiting from opportunities provided by advanced education courses, technical symposia, and collaboration with their colleagues. Their needs at the fundamental level have not changed: basic principles, creative thinking, good judgment. Changes have occurred, however, in practicing engineers' environment. They are now members of multidisciplinary teams, and their work overlaps social and economic issues. They are also expected to base their decisions on rational rather than empirical criteria because their technical assignments are becoming more complex. Practicing engineers today interact with social scientists and public officials and are in continuous touch with the technical research of academic colleagues if they are to perform at the level of competence that is demanded of them.

It is our conclusion that the professional standing of practicing engineers has been improved in past decades primarily by the initiative shown by engineers to meet the new challenges. They have used results from research conducted at academic institutions to sharpen their technical skills. Their needs in a multidisciplinary environment have, however, been met only incidentally, depending on their exposure to situations that developed in their practices. We must recognize that practicing engineers of the future will need a more formal program of preparation to enter the profession. Such a program, which can be developed at universities, should consist of a balanced curriculum of multidisciplinary training, followed by specialization when specialization is desired. It is time to admit that the education of engineers should be raised to the level of other professions. It is time to declare the four-year college program a preengineering program. Following this preparatory work, a mandatory professional engineering program should be required, followed by a voluntary specialization program. Practicing engineers will then start their career better prepared to accept their responsibilities as principal contributors to public safety, technological improvements, and economic well-being.

UNDERGRADUATE EDUCATION FOR PROFESSIONAL PRACTICE

*Don O. Brush**

INTRODUCTION

Engineering undergraduate education changed profoundly after World War II. Wartime experience with new developments such as jet propulsion, atomic energy, and radar revealed that the undergraduate programs of the day provided engineering undergraduates with an inadequate foundation in the sciences and mathematics. Engineering curricula subsequently were revised to emphasize mathematics, basic science, and engineering science at the expense of technical skills and engineering design, and faculty recruitment criteria were revised to emphasize the Ph.D. degree at the expense of professional experience. The changes have been permanent.

Not everyone agrees that the changes have led to a net improvement in preparation for practice in structural engineering. Members of small consulting firms frequently are critical of present undergraduate programs. Four criticisms are common: some graduates are unfamiliar with structural terminology, such as girders and gusset plates; they have little intuitive feel

---

*Associate Dean of Engineering and Professor of Civil Engineering, University of California, Davis.

for quick-and-dirty methods of analysis; they are incompetent draftsmen; and they have trouble with graphical communication. Generally speaking, the criticisms are serious and accurate.

On the other hand, members of large engineering organizations frequently are critical of undergraduate preparation for an entirely different set of reasons. Some new graduates, they say, have too little skill in the use of matrix methods of structural analysis, have had too few courses in advanced mechanics, and have poorly developed writing skills. These criticisms, too, are serious and accurate.

Can we have the best of both worlds? Can we design a four-year undergraduate program for structural engineers that will assure coverage of quick-and-dirty methods *and* matrix methods, technical skills *and* advanced mechanics? Of course we can. It is necessary only to require undergraduates to specialize early enough and narrowly enough. Would such a specific focus serve the best interests of the student and the profession? In most cases it would not. Education for professional practice is a lifetime occupation, and what graduating seniors know is less important than what they have been prepared to learn. Furthermore, we seldom know which undergraduates will in fact become structural engineers, and early specialization can turn out to be counterproductive.

Proponents of continuing education like to compare undergraduate program design with the problems of a traveler who is packing for an extended journey into a desert region. Perhaps you have heard the analogy. If no supplementary supplies are to be available along the way, he must take with him all that will be needed for the entire journey, and he must carefully provide for the inevitable problems of spoilage and breakage. If supplementary supplies *are* to be available along the way, however, the set of supplies needed at the outset is entirely different. Analogously, if learning is to take place after graduation, an effort to design an undergraduate program that will provide everything an engineer will need to know for a lifetime of practice is effort misdirected.

Continuing education is provided by schools, employers, and professional societies. Approximately half the engineers in this country enroll in formal graduate work after completion of B.S.-degree requirements. About half of those with an M.S. degree completed graduate work before beginning practice. The other half returned to school after a period of full-time employment for full- or part-time graduate work. Other graduate engineers enroll in university-sponsored certificate programs designed to introduce practicing engineers to new technology or to provide refresher courses in long-unused fundamentals. Less formal still are university-sponsored symposia such as the present one.

Employers are also a principal source of continuing education for practicing engineers. Larger employers maintain elaborate training programs for new employees and for old employees who have been assigned to new jobs. In-house short courses and seminars also are provided, especially in high-technology industries.

One of the primary functions of the professional societies is to assist their members in continuing professional development. The major societies have developed impressive programs to encourage participation and to provide appropriate opportunities. Examples include technical publications, national meetings, postmeeting short courses, and specialty conferences. The June issue of *Civil Engineering* listed 17 short courses to be offered in conjunction with the ASCE national meeting in San Francisco in October. Topics ranged from timely subjects such as "Offshore Structures: Design and Analysis" to perennial problems such as "Management of Project Paperwork."

A recent bill passed by the Iowa legislature requires professionals to participate in a continuing education program as a condition of license renewal. Although such legislation focuses on form rather than substance, it is clear evidence of the importance placed on continuing learning.

The second argument against a narrowly focused undergraduate program is that we seldom know which undergraduates will become structural engineers. In the nation, 16% of the engineering B.S. degrees are in civil engineering. The fraction of engineering graduates whose major interest is structural engineering probably is about 3%. The number who focus on structural engineering prior to the senior year may be less than 1%.

Selection of a career objective for most people is an evolutionary process that extends over a period of years. Many engineering freshmen have only the most tentative commitment to engineering, and transfers into and out of engineering undergraduate programs occur in large numbers. Engineering undergraduates must slowly come to perceive the differences among the fields of engineering, such as civil engineering and mechanical engineering, and among functions of engineers, such as operations and design. They must simultaneously develop a sense of their own aptitudes in relation to respective career alternatives. Throughout the decision-making process, evolving choices are strongly influenced by real or perceived changes in the state of the job market.

Most important, an individual engineering undergraduate's ultimate field of interest will be shaped more by employment and advancement opportunities after graduation than by pregraduation choices. Some of the 3% whose major interest at the time of graduation was structural engineering will become manufacturing managers, and some of the individuals who become structural engineers will have graduated in mechanical engineering. A student who specializes too soon and too narrowly may be less able or less willing to respond to a changing job market.

The analogy to the traveler who is packing for an extended journey into a desert region again is appropriate. What he should take with him depends on whether the desert is tropical or arctic. He should not plan too exclusively for either region if he is among the majority who will settle upon a destination only after packing is well advanced. If he changes his destination after departure, narrowly focused planning is even less useful.

If early specialization is counterproductive, what kind of undergraduate program does provide the best foundation for professional practice in structural engineering? Let us look briefly at four aspects of undergraduate education: the curriculum, the faculty, campus services, and support by professional societies and practicing engineers.

CURRICULUM

General objectives

The commonly accepted objectives of an engineering curriculum are to develop in students (1) an ability to define and solve in a practical way the problems of society that are susceptible to engineering treatment, (2) a sensitivity to the socially related technical problems that confront the profession, and (3) an ability to maintain professional competence through continued learning. To achieve these objectives, engineering curricula today commonly include a half-year of mathematics beyond trigonometry, a half-year of basic science, a year of engineering science (e.g., thermodynamics), a half-year of engineering design, and a half-year of humanities and social sciences. These components add up to three years of course work.

In addition, engineering curricula also should strongly emphasize written, oral, and graphical communication skills. Such skills should be fostered by requiring specific courses in English, speech, and drawing, and by emphasis on communications skills in engineering courses. Many entering freshmen still believe the engineering profession to be an appropriate haven for people who want to avoid learning to write.

It is beneficial to students and to the profession for students with appropriate aptitudes to select engineering as a career, and for those without appropriate aptitudes to select something else. Freshman-level orientation courses that describe the role and function of professional engineers in society should be provided to support and reinforce career guidance and counseling services. The courses should attempt to answer the common and difficult question: What does an engineer do?

A program intended to provide a foundation for professional practice also should include a senior-level course in professional responsibilities.

The course should provide instruction in topics such as the professional aspects of engineering practice, ethical issues, contracts and specifications, and law as it affects engineering.

Subdivision into majors

The traditional subdivisions in engineering curricula are electrical engineering, civil engineering, and so on. In some programs, disciplinary names, such as systems, mechanics, and so on, are used for the undergraduate majors. Use of such names reflects the fact that, for example, a structural designer may have more in common with a machine designer than with, say, a transportation systems planner. The most logical arrangement of all may be a unified engineering major with no formal subdivisions. Of course, each of these academic arrangements has strong advocates, and each is represented around the country by undergraduate programs of high quality.

For the long-term health of the profession, a diversity of organizational arrangements probably is desirable. Nevertheless, I believe that the best preparation for professional practice is provided by the traditional majors. Organization into disciplinary majors is indeed logical, but the logic is more apparent to professors than to undergraduates, especially younger undergraduates. The traditional structure fosters closer identification with professional practice. Students who graduate in civil engineering are more likely to think of themselves as civil engineers, to join the ASCE, and to seek registration as civil engineers.

Civil engineering

If the undergraduate program *is* subdivided into traditional majors, what should be the characteristics of the civil engineering curriculum? Because the various fields of engineering share a common foundation in mathematics, basic science, and engineering science, and because most students do not know for a year or two which engineering major interests them most, a common core of courses that extends over the first two years or so should be required in most programs. It should be possible for students to delay formal declaration of a major within engineering until the end of their sophomore year.

The future interests of graduates and future needs of the marketplace are largely unpredictable. For these reasons, the last two years of the civil engineering curriculum should include required exposure to all the fields of civil engineering, such as environmental engineering, water resources engineering, and structural engineering. Under this curricular arrangement, undergraduate courses in, say, building design, advanced methods of structural analysis, and advanced mechanics can be offered, but not required. It follows, inevitably, that some future structural engineers will miss such courses at the undergraduate level. This flexibility emphasizes the importance of an effective system of advising.

As much latitude as possible should be provided for the election of technical courses to suit the interests of individual students. Those who *have* identified their personal preferences tend to learn more from courses they want to take than from courses they do not want to take. Curriculum requirements should assure that every student's personal set of choices provides an understanding of the kinds of insights, information, and procedures used by professional engineers in the solution of the problems of society that are susceptible to engineering treatment.

FACULTY

Professional orientation

The atmosphere in which the undergraduate program is presented is determined by the faculty. The program is professionally oriented if the faculty is professionally oriented, and the program is not if they are not. The majority of faculty members in a program should have had full-time experience in engineering practice. Lack of such experience limits the faculty member's professional perspective as an instructor and diminishes his/her value as a career adviser. It may also diminish his/her appreciation of the essential role of quick-and-dirty methods in engineering analysis.

Faculty consulting and industrial sabbaticals also contribute to the professional orientation of faculty members. In terms of net benefit to the undergraduate program, however, both are mixed blessings. Consulting can provide undergraduate instructors with an incentive for keeping up with technological change, and it can lead to beneficial interaction with practicing engineers. Unless closely limited, on the other hand, it can divert the instructor's attention from his/her teaching and research functions. An industrial sabbatical may provide much-needed experience, but the academic advancement of a young faculty member may benefit more from sabbatical time spent in research. In most cases, administrative emphasis on dollar volume of funded research has a negative impact on the quality of the program.

Professional registration of faculty members is another indicator of a professionally oriented program. The narrow specialization that characterizes engineering graduate work at the Ph.D. level is not compatible with the broad, practical knowledge required for professional registration. The extensive effort required for a Ph.D. to become registered serves to remind him or her of the difference between Ph.D. training on the one hand and the needs of typical undergraduates on the other.

An atmosphere of professionalism is also fostered by active participation by faculty members in programs of the professional societies. Service with practicing engineers on local and national committees has a beneficial effect on the quality of the faculty member's teaching and advising.

Teaching

An individual faculty member can make his/her most important contribution to the quality of the program through good teaching. Learning is an active process, and good teaching consists in facilitating that process. By carrot or stick, a teacher must engage each student in the active pursuit of his/her own individual knowledge and understanding.

The most useful method for improvement of teaching may be the use of instructor-rating questionnaires for direct feedback from class members. Every instructor should be required to obtain such feedback. A campus-wide teaching resources center should be provided to assist faculty members in improving their teaching. Such a center can furnish equipment and personnel for the voluntary videotaping of a class meeting, and for playback for the instructor in the presence of an experienced teaching evaluator. No other experience is quite so informative to an instructor. The center can also provide effective-teaching seminars and information on technological aids, such as videotapes for individualized instruction.

Advising

Flexibility in the undergraduate programs to meet the needs of individual students is essential. The key to successful flexibility is an effective system of advising. Every undergraduate student should have an engineering faculty adviser. Every full-time engineering instructor should serve as an adviser. A close, personal association between a student and his or her faculty adviser can be one of the most important factors in a successful educational experience. Such an association is commonplace among Ph.D. candidates. It is widespread, although less close, among engineering seniors who are surveying the job market. Freshmen, however, tend to avoid faculty advisers altogether. The slowness with which freshmen begin to "join up," that is, to participate in the educational process outside the classroom, should be a continuing concern in every undergraduate program.

Faculty obsolescence

To provide good teaching and advising, engineering instructors must keep abreast of constantly changing technology. For many faculty members, research is the most effective means for avoiding obsolescence. Others include consulting, active service on technical committees, including refereeing and review service on publications committees, and active participation in short courses and symposia. Such activities are essential ingredients of a successful program.

CAMPUS SERVICES

I want to mention briefly four campus services that contribute to the education of engineering undergraduates: the placement office, the cooperative work-study program, advising services, and the office of continuing education.

Placement office

Perhaps the most familiar campus service for most graduate engineers is the placement office. Most of us obtained our first professional employment through an on-campus interview arranged by such an office. The placement office can also help undergraduates in the evaluation of their individual interests, talents, and values in relation to employment. A central feature of a placement office is a career-resources library that contains up-to-date information on careers, employers, and job vacancies. The placement center may provide workshops on topics such as use of information in the career-resources library, identification of personal talents and career interests, résumé writing, and how to act during and what to expect from an interview with a prospective employer.

Work-study

Provision of a cooperative work-study program contributes immeasurably to the professional growth and development of undergraduates. In co-op programs, periods of full-time study are alternated with periods of full-time employment in work that is related to the student's educational objective. Such programs have been an important part of engineering education for more than 50 years.

The purpose of cooperative education is to provide an opportunity for students to observe and engage in engineering practice before completion of undergraduate work. The experience can help a student in the formulation of clear educational and career goals. It provides an opportunity for the individual to assess his/her skills and capabilities, and it permits a more realistic appraisal of academic alternatives. It provides an opportunity for the application of academic education to practical problems, as well as an opportunity for an exercise of initiative and an assumption of responsibility that might not be provided on campus. It permits a student to work with professional people who may serve as role models. It provides an opportunity for employers to correct what they may perceive as an educational deficiency in graphical communication, familiarity with terminology, and the like. The experience may affirm a student's tentative career choice, or, of equal value, it may convince him/her that some other career is a better choice.

The cooperative education office can arrange for co-op positions, provide on-campus publicity about available positions, seek student applicants, and provide the administrative procedures through which interested students and employers are brought together. It can also establish guidelines to distinguish between jobs that provide a suitable learning experience and thereby qualify as co-op jobs, and those that provide only income.

Advising services

Advising and counseling by nonfaculty staff members is a third campus service that contributes to the professional education of students. This service supplements the advising provided by faculty members. Nonfaculty advising and counseling generally are of two kinds, those furnished by full-time counseling staff members and those furnished by part-time, paid, undergraduate students. Staff members in counseling services generally are trained counselors, not engineers. They know less than engineering faculty advisers about the profession of engineering, but they may be more skillful than faculty advisers in helping students to formulate their thoughts and articulate their questions. Counselors also can provide standardized interest-identification tests for students who have lost headway in the process of career selection. The early identification of students who have little aptitude for engineering is as beneficial to the individual and the profession as the identification of those who do have appropriate aptitudes.

Most freshmen and a few upper classmen are intimidated by faculty members. Many make special efforts to avoid contact with faculty advisers. For such students, peer advisers can be a useful source of advising information. Because of their structured job training, peer advisers sometimes are more knowledgeable than faculty advisers about available campus services such as placement, counseling, and financial aid, and therefore may be better sources of referral information. Furthermore, peer advisers see the academic program from a different perspective and can complement the advice offered by faculty members.

Office of continuing education

The influence on the undergraduate program of the campus service called the office of continuing education, or the engineering extension center, is indirect but not insignificant. Through the development of short courses, workshops, and symposia, the service brings about interaction between faculty members and practicing engineers and contributes to the professional development of both. Campus support of a continuing education office is overt evidence of the understanding that the engineering undergraduate program can provide only one part of the lifelong learning process.

PROFESSIONAL SOCIETIES AND PRACTICING ENGINEERS

The last factor to be considered is support by professional societies and practicing engineers.

Intersociety organizations

The fact that the professional societies share the responsibility for the quality of undergraduate education is commonly recognized. The most obvious expression of the societies' efforts is the Engineers Council for Professional Development (ECPD). The ECPD is an agency of 14 participating professional societies, including the ASCE. It is recognized by the National Commission on Accrediting as the sole agency responsible for accreditation of educational programs leading to a degree in engineering.

A second intersociety organization that has a direct influence on undergraduate education is the Engineering Manpower Commission of the Engineers Joint Council (EJC). More than any other body, the EJC represents the engineering profession as a whole. The Engineering Manpower Commission provides authoritative information on nationwide engineering enrollments, manpower supply and demand, and salaries. Recent *Engineering Manpower Bulletins* have been entitled, "Job Prospects for Engineering and Technology Graduates" (December 1976), "Engineers Earning More and Enjoying It Less" (January 1977), and "Engineering Degree Statistics and Trends" (March 1977). The publications of the Engineering Manpower Commission provide the most authoritative data available on employment supply and demand for graduating seniors.

American Society of Civil Engineers

For undergraduates in civil engineering, the professional society that has the greatest impact on education is the ASCE. One of the official purposes of the ASCE is to "develop professional consciousness among civil engineering students." At the national level, educational activities are administered through the office of the Manager of Student Services. One of the services is the preparation and general distribution of guidance material for high school students and college freshmen. Recent publications include, "Is Civil Engineering for You?" and "Careers in Civil Engineering—The Structural Engineer." A statement in the former accurately reflects the state of mind of many undecided students: "If you are like most of us, your interests are constantly changing, your life's ambition varies, and the world around you is forever in a state of flux."

The national office provides information to ASCE student chapters in large volume. Examples are copies of the new ASCE Code of Ethics and the ASCE Guide to Employment Conditions. The latter contains guidelines and recommendations on terms of professional employment, salary ranges, and fringe benefits. The national office also encourages members of student chapters to subscribe to *Civil Engineering* and provides a reduced subscription rate.

Most of the effort at the national level is devoted to the organization and support of activities in the local sections. Each local section is encouraged to establish a Committee on Integration of Education and Practice. Tasks suggested for local committees include the following: "Prepare a listing of firms that are willing to employ faculty during summers or sabbatical leaves. Prepare a listing of firms which can hire students during summers or part-time during the school year. Provide a list of civil engineers who are willing to speak to student meetings or classes, together with their topics and qualifications. Provide engineering schools with a list of engineers who are willing and qualified to teach courses on a part-time basis, together with their areas of expertise. Aggressively promote several meetings a year at which engineering faculty and students could meet and talk with practicing engineers." Needless to say, the success of such proposals depends on the enthusiasm of local chapter members, and it varies widely.

The local section appoints two "contact members" for each student chapter in its region. The contact members attend chapter meetings and provide liaison between the chapter and the section. The section also provides speakers for meetings of student chapters, sponsors annual student-night dinners for the chapters in its region, sponsors annual awards for outstanding student-chapter members, encourages student attendance at national meetings by providing for partial reimbursement of travel expenses, and provides interest-free loans to engineering undergraduates for educational expenses.

One of the educational functions of student chapters is the organization of frequent field trips to engineering projects or offices such as building or bridge construction sites, rolling mills, fabrication shops, and consulting offices. On many campuses, review sessions for the Engineer-in-Training examination are sponsored by student organizations. The provision of such a review not only helps students pass the first part of the examination for professional registration, but also greatly increases student awareness of the professional registration process.

Many other professional organizations contribute directly and actively to the quality of the undergraduate program. Services include sponsorship of field trips, seminar speakers, design contests, and an imaginative variety of other contributions. Such services attest to the strong interest in and support of undergraduate education by the professional societies.

Individual engineers

Practicing engineers contribute to the quality of undergraduate education both as society members and as individuals. Individual engineers serve as members of ECPD accreditation teams, as members of advisory panels for engineering deans, as employers of co-op students, and in other ways. Undergraduates benefit from on-campus contact with practicing engineers who appear as classroom guest lecturers and as seminar speakers. Practicing engineers who teach courses as part-time lecturers can contribute significantly to the professional orientation of a program. For maximum benefit the courses should be practice-oriented ones such as building design, construction management, and water resources planning, rather than core courses such as statics and mechanics of materials. Such lecturers bring to the course not merely an up-to-date set of case histories, but also an attitude toward professional practice that may not be adequately represented among the faculty. Of course, not all practicing engineers are effective teachers. Some are excellent, and a few turn out to be thoroughly ineffective. To locate and engage appropriate people usually requires a major effort, but it is an effort that leads to substantial benefits for the program.

SUMMARY

We have discussed four aspects of undergraduate education that contribute to preparation for professional practice: the curriculum, the faculty, campus services, and the support of professional societies and practicing engineers. We have said that the best foundation for professional practice is provided by a curriculum that furnishes extensive guidance information for the benefit of undecided students, provides a common core of freshman and sophomore courses for all engineering majors, requires civil engineering majors to take a broad spectrum of courses in civil engineering, and offers specialized structures courses as electives. The best foundation is provided by faculty members who are professionally oriented, who are effective teachers, who enthusiastically support the advising system, and who are themselves actively engaged in continuing learning. The best foundation is provided by a campus that offers an effective career planning and placement service, a well-developed cooperative-education service, a program of advising services to complement the faculty advising system, and an active program in continuing education. Finally, in engineering undergraduate programs there is a quality of richness that is a result not of curricular requirements, faculty qualifications, or campus services, but of extensive interaction by students and faculty with practicing engineers. This interaction occurs through field trips, cooperative-education employment, local section meetings, classroom guest lectureships, and in other ways. The quality of the undergraduate program benefits immeasurably from the support of the professional societies and practicing engineers, and the best preparation for professional practice is provided in those programs in which that support is most active.

GRADUATE EDUCATION AND RESEARCH FOR STRUCTURAL ENGINEERING PRACTICE

*Daniel C. Drucker**

INTRODUCTION

The topic of graduate education and research for structural engineering practice in the context of natural hazards is one of the more interesting elements of the general subject of graduate education and research in engineering. If there were consensus on the broad issues, we should move immediately to structural engineering. There is no consensus, but that's not bad. Diversity can make for vitality. What is bad is the growing pressure we all feel to return to what its proponents believe is "real" engineering, a pressure to turn back the clock to those imagined good old days of this century, days that never were and never should have been, when engineers did not do research or otherwise waste their time learning esoteric topics and others' ways of thinking. Engineering students at the upper division and master's level were isolated from students of other disciplines in this imagined utopia except perhaps for a few measured and efficient doses of training in areas of known value to engineers. Law and medical schools are cited as models to be emulated, actually schools of earlier decades and of traditions now aban-

---

*Dean, College of Engineering, University of Illinois at Urbana-Champaign.

doned at many leading universities. Those of us who lived through the good old days, or have talked with those who did, know that each decade has had its proponents of a return to "real" engineering who viewed the current professorial staff as impractical or worse.

Why do I trouble you with these thoughts? You are researchers as well as teachers and practitioners of engineering. You are keenly aware of the fine research traditions of the engineering experiment stations or their equivalent established in this country at the start of this century and of their counterparts and predecessors abroad. You understand that much of the engineering of the future will be based on the research of today, just as the engineering of today owes much to the research of yesterday. The progress of engineering practice is in large measure the result of the work of people like our colleague Egor Popov, whose contributions we recognize through this symposium. You know that unless the education of our students is primarily in the hands of a research-oriented faculty, a faculty with its eyes on the future and its feet on the ground, we do our students a grave disservice. We must point our students forward, not backward. We must prepare them for the engineering of the future far more than for the engineering of today. Today becomes yesterday by the time they leave our tender, loving care.

THE NARROWING OF GRADUATE EDUCATION

What is there to discuss at this level of generality among friends who share the same fundamental tenets of academic faith? Some years ago there certainly would have been little to say, but now I am not so sure. Does it startle you to learn that, when I look over the academic transcripts listing the graduate courses taken by prospective assistant professors, more and more I find no courses at all outside engineering, no mathematics taught by a mathematician, no science taught by a scientist? The great concern for the appropriate interaction of technology and society, expressed by many individual engineers and by all the major engineering societies as well as this symposium, finds almost no reflection in the taking of single social science course at the graduate level taught by a social scientist. Often the transcript of the graduate years shows no courses outside the chosen discipline of civil, electrical, or mechanical engineering. Sometimes there are hardly any formal courses outside the speciality within the discipline. An extreme example that I have not seen, but do not doubt exists, would be a graduate program composed solely of structures courses. Day by day, we and the rest of the profession express the need for interdisciplinarity; year by year, some of us seem to be permitting our students to become less broadly educated in an effort to teach them as much as we can of our own specialty.

Should there be a marked difference in our approach to the needed breadth and depth of formal education at the bachelor's, master's, and doctor's level? I would suggest that there should not be. At each level the student should be given the minimum of needed expertise to acquire confidence and to function effectively in an engineering position. At the same time, the student should be provided with the maximum breadth and depth in adjacent areas to avoid overconfidence and overdependence on a limited set of tools and concepts of transient value at best. Learning is a life-long process made far easier by an appropriate formal education.

The balance I would like to suggest to students at the graduate levels, both master's and doctor's, is roughly what is best for the undergraduate years: about $\frac{1}{4}$ mathematics and science; $\frac{1}{4}$ engineering science; $\frac{1}{4}$ analysis, synthesis, and design; $\frac{1}{6}$ humanities and social studies; and $\frac{1}{12}$ anything of interest and value. Forced to choose between another design course and an advanced area of mathematics or science or engineering science, for a graduate student with an accredited undergraduate engineering program, I would opt for the advanced area (any advanced area) but would not insist if the student felt otherwise.

PRACTICE ORIENTATION
AND RESEARCH ORIENTATION

You will recognize that I deplore the existence of the present ECPD accredited advanced-level master's degree programs that permit a year devoted fully to design, and insist upon one-third of a year. I consider them to be unprofessional rather than professional, because they severely restrict the horizons of their students and set an apparent seal of approval by the profession on one of the least desirable of the many acceptable educational paths to the practice of engineering. Equally unprofessional and equally destructive for the future of the engineering profession is the attempt to divide faculty into "practice-oriented" and "research-oriented," with the practice-oriented faculty responsible for the education of future professional engineers in isolated professional schools of engineering. How soon are the lessons of history forgotten! How easy it is to fall into the trap of believing that in the future we shall experience only slow evolutionary change with ample time to learn what is needed rather than recognizing the increasing frequency of revolutionary change in all branches of engineering. Students and staff alike must maintain a life-long research attitude toward learning and professional practice.

Yet, perhaps in all these matters we also are in agreement. Most branches of civil engineering involve large questions of public policy and much uncertainty as to how the public can best be served. In the mitigation of the hazards of earthquake, wind, water, and fire, it is the combination of our painful awareness of the limitations of our knowledge and our concern for public safety that motivates both teaching and research.

HARD CHOICES OF TOPIC COVERAGE
AND LEVEL OF UNDERSTANDING

A structural engineer needs a fundamental understanding of the continually changing spectrum of individual and societal values and behavior to develop appropriate designs. Certainly anyone entering structural engineering practice and worrying about the natural hazards of wind and water and fire and earthquake must know a great deal about a wide variety of materials, about wave propagation through rock and soil, about combustion and heat transfer, about aerodynamics, hydromechanics, and the dynamics of elastic and inelastic continua and structures, as well as about people and their reactions to the possibility and the actuality of disaster. New mathematical concepts, new mathematical techniques, new discoveries of science and the applications of science, along with new experimental and field information, are absorbed continually into practice. Problems of public policy, trade-offs between safety and cost involving individual perceptions and willingness to take risks, round out the need for a broad interdisciplinary pattern of education and research. So much needs to be learned by the student, so much is yet unknown to the experts, that the problem of choice is far from trivial.

There is no limit to helpful courses in traditional and in new interdisciplinary areas. There is no problem in seeing how well the knowledge gained through research will aid in design and enrich the curriculum. Taking a good course or sequence of courses is a very efficient way of learning, of getting to the forefront of existing knowledge as quickly as possible. The problem in research is to select from the wide scope of the unknown those topics whose solution if obtained will have some receptive audience for application or for future useful research. Of course, the theoretical, experimental, or analytical difficulty should be commensurate with the ability and level of the participants. An analogous problem is faced in designing curricula for the broad purpose of preparing a student for the practice of structural engineering. Two full years of course work or 16 formal semester courses is about the maximum any of us require our properly prepared graduate students to take on their way to a doctorate. That's not very many in the context of the needed breadth, depth, and interdisciplinarity, but I would not require more if I could. Hard choices must be made at all levels of education so that our students can begin to be productive at a reasonably early age.

A PARENTHETICAL COMMENT AND QUESTION

A parenthetical comment on the earlier years of education is relevant here. Undergraduate engineering students, the major fraction of whom are in the research universities of this country, are very bright people on average. They are intellectually capable of reaching much higher levels of understanding and knowledge than they do in the many years set aside for their precollege education.

Over most of the last 100 years, with the usual local and national fluctuations that must be expected over time, we have seen an enormous increase in the average level of knowledge of high school graduates who have entered engineering schools. It is this increase that has permitted engineering educators to prepare their students minimally but adequately for the ever-increasing challenges of entry-level engineering practice within the four-year baccalaureate time span. It is the great margin for further increase still available that makes me so strong an advocate for continuation of the tradition of the four-year baccalaureate as the first engineering degree.

Our present students and all of us are suffering the severe consequences of a substantial change that has taken place over the past dozen years or so in the elementary and secondary schools. There seems to have been a marked lowering of teacher expectation of both the qualitative and quantitative performance of college-bound students. The evidence is clear in the records of the combination of the high school percentage rank (HSPR) and the SAT or ACT aptitude and achievement scores of our entering feshmen. Those of us who maintained HSPR standards in the early 1970s in the face of decreasing applications for admission to engineering nevertheless found the scores on a downward trend. To our surprise, this same downward trend was observed for the students electing the then tremendously popular fields of the liberal arts. Over the past four years, the admission pressure on engineering has increased enormously. Those of us who have raised our admission standards because of the limited numbers we can accept have found that although the median HSPR of our entering students has gone up appreciably, the ACT or SAT scores are doing barely better than just leveling off. Despite the uncertainty of the meaning of test scores, it does seem to be true that any given reasonably high level of student performance is now achieved by a far smaller fraction of the graduating class than 10 years ago in most or possibly all high schools.

Most of this decline in standards is a natural accompaniment to a very permissive, largely affluent, society which mistakenly assumes that the need for hard work and intellectual excellence is past, an egocentric society that places finding oneself above helping one's neighbor. Part is an equally mistaken response, or possibly simply a lazy response, to the valid need for equality of educational opportunity and proper access to higher education by poor minority groups. Disadvantaged groups deserve far better opportunity than society yet provides. The answer, however, is not to lower the standards for all and our expectations for our best students. It is to provide the disadvantaged with an adequate environment, with the educational opportunities, the special attention and considerable resources needed to upgrade the quality of education now offered from kindergarten on up, and the additional education needed to bring those already in the existing terribly inadequate system to the level of understanding they must achieve.

The majority of the high schools, the junior high or middle schools, and the primary schools have evaded the hard choices they should make. Their students talk with their teachers about how to solve the problems of the environment, ecology, energy, and materials without obtaining the beginnings of the basic knowledge needed to understand, and later to contribute to, the solutions needed. Incisiveness and clarity of thought in written or oral communication are put aside in favor of encouragement to speak out and communicate with each other and the world as best they can in the language of their childhood years.

Have we done so much better at the undergraduate level? Do we insist upon upholding the standards that too many of our colleagues in the preparatory levels of education have abandoned? Are we incisive in our use of mathematics and science in our engineering courses? Have the levels of our proposed undergraduate and graduate courses and the general quality of our educational offerings in engineering been maintained? If the answer to each of the rhetorical questions were a clear yes, I would not, of course, have diverted our attention from graduate education and research in this parenthetical statement about earlier education. Do we, too, skim over or avoid the fundamentals and skip to the solutions?

A TREND TOWARD LESS DEPTH AND GREATER SUBDIVISION OF EACH DISCIPLINE

We have made choices, but in too many instances they may have been the wrong ones, driven by the understandable desire to have our students not reflect discredit on us, their educators. Often the choice is to acquaint our students with all aspects of the spectrum of current engineering activity they may be called upon to perform on graduation so that they are able to make an immediate start on any assignment they are likely to be given. If we are civil engineers, all other branches of engineering are pushed out. Structural engineers within civil will push many other areas of civil engineering to an optional category. The options rarely are selected by a student because they are in competition with the dazzling array of structural engineering and closely related courses offered.

In the best of these undesirable arrangements at the graduate level there will be year sequences such as structural analysis and design, or mechanics of continua and structures, or soil mechanics and foundations. Even then, the level the doctoral student reaches in formal course work rarely is higher than that of the master's degree candidate; the breadth alone is greater. In the worst of these undesirable arrangements there are few or none of the year sequences. A wide variety of courses are listed and the number grows larger each year: structural optimization, matrix analysis of structures, structural design, structural reliability, approximate methods of structural analysis, plates, shells, dynamics of structures, plastic design, soil mechanics, foundations, systems approach, construction, steel structures, concrete structures, wood structures, ocean structures, earthquake engineering, wind engineering, and so on.

Such a delectable smorgasbord of one-semester courses *all* at the beginning graduate level attracts students from many diverse undergraduate programs in civil engineering. Therefore, each tends to be a little more like an advanced undergraduate course than a true graduate course. Since each course is offered at the beginning level, any math or science prerequisite at the graduate level would be superfluous. Do I exaggerate? In some instances only a little. What saves us from disaster is that many of our advanced undergraduate courses across the country are at quite a good level—not a graduate level but definitely at the fourth-year level.

Are most of our master's candidates educated broadly through the fifth-year level and doctor's degree recipients through the sixth? Far from it; midway through the fourth year would be closer! Should they be so broadly educated? Yes! Depth as well as breadth is needed at all levels of engineering education.

WHAT OF THE FUTURE?

If we do not educate our students deeply enough, not only will the fundamental advances in structural engineering come from others, but our students will not be able to appreciate and use the advances in their practice of structural engineering. As an example, I would guess that 20 years from now a structural engineer will not be in the mainstream without a firm grounding in the statics and dynamics of the complex inelastic response of structural materials, structures, and continua to complex loadings. There is no way such a fundamental understanding can be achieved without three or four genuine graduate courses in sequence and far more mathematics and knowledge of materials than most of us now require.

Speaking of mathematics, I would guess that 20 years from now a structural engineer unfamiliar with functional analysis will not be able to read the current structural engineering literature. I may be wrong about functional analysis, but think back to the largely ignored status of matrix algebra in engineering education just 25 years ago. Without some good fundamental mathematical understanding, achievable only at the advanced level, no matter what branch of mathematics becomes the current coin of the realm, our students who learn what to do by cookbook methods or canned computer routines will be at a terrible disadvantage and likely will become either managers or technicians. They will be able only to talk about the difficult new problems, not to solve them.

In the era for which we educate our students, the back-of-the-envelope calculations they will need in order to make preliminary designs or check the computer printout results of others will take for granted a full understanding of the upper- and lower-bound plastic theorems of limit analysis and design, shakedown, maximum displacement bounds for inelastic structures, and a host of other and yet-to-be-developed general theorems. Detailed calculations and analysis will be ever so much more realistic in the degrees of freedom, complexity of loading, and the response of structural elements and complete structures.

Corresponding advances will be made in the techniques of experimental structural analysis and inspection. Interpretation of unusual readings and observations will, in many instances, require a fundamental understanding of physical phenomena now solely in the domain of solid-state physics and chemistry.

Beyond all these conventional technical advances will come the integration of the hierarchy of individual and societal values into the development of alternative design proposals. The depth of understanding and the technical competence needed here far exceeds the limits of present-day knowledge.

FORMAL COURSE PATTERNS
FOR DOCTORATE EDUCATION

I would suggest that the formal course structure of an appropriate curriculum should look more like an imposing multistory building in block form than a friendly ranchhouse with all courses at the entry level of the senior year or the first graduate semester (Fig. 1).

Although a sequential building of level of understanding is essential, and the most careful attention must be paid to the inclusion of prerequisite knowledge, the detailed choices of topic coverage for each level are mainly a matter of taste and local option. For example, a sequence of courses can go from framed structures to plates to shells to continua, or the starting point can be the mechanics of continua or soil mechanics. Probability and reliability can be intimately interwoven with the structures courses or taught separately and used subsequently. Similarly, mathematics or materials courses can be

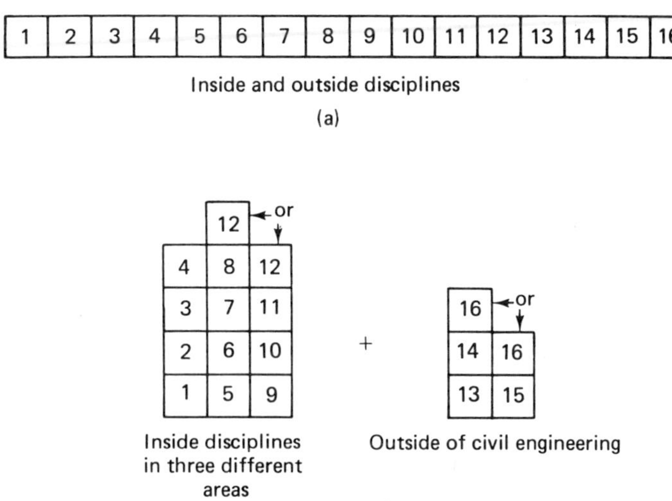

Figure 1 The more demanding multistory curriculum is preferable to the friendly ranch style. (a) Ranch-style friendliness: all graduate students welcome in all courses; no graduate course prerequisites for any other graduate course. (b) A four- or five-story pattern with genuine prerequisites; each successive level of graduate course requires one or more graduate courses (or equivalent knowledge) at a lower level as prerequisite.

taught early or the concepts brought out in cohesive chunks when interest will be high. Although it can be enormously helpful to students to learn the fundamentals in context from a teacher who understands the applications, it is essential that the teacher be genuinely expert in the fundamentals. What must be avoided is the tendency, often encouraged by students, to omit the necessarily extensive discussion of the fundamentals in order to give more time to the obviously applicable. Continual feedback is necessary no matter what choices are made. The teacher as well as the student must be made explicitly aware of the level, coverage, and prerequisites for each course offered.

ROLE OF RESEARCH AT UNIVERSITIES

Much has been said already of the role of research in clarifying and generating concepts and information that continually upgrade the curriculum and the practice of structural engineering. The major role is to educate students through the experience of doing, or at least trying to do, what has not been done before. Care must be taken in the selection of the topic and the approach to extend the capability of each student as far as possible with success within the constraints of the time available. Much of the effort at the master's level will be routine by the standards of those at the forefront of the field, but interesting and valuable both to the participant and the profession. A great amount of testing and simple experimental investigation, of analytical and computer studies, based on known principles, can be carried out at universities simultaneously with the effort to develop new experimental and analytical or numerical techniques and new theoretical and design concepts. Effective use is not always made of this rich diversity at individual schools and within the worldwide university community. When all goes well, as in the development of plastic analysis and design in the 1950s and late 1940s, progress is awesome. The development of fracture mechanics provides a much more recent example which included many nonuniversity research laboratories in the fraternity of joint effort because of an enormous need for almost immediate answers by the nuclear industry. The needs of the energy industry, nonnuclear as well as nuclear, provide similar opportunities for this broadened cooperation in addressing natural hazards. However, it is likely to be university research primarily that will develop the basic principles for mitigation of the tragic consequences when major public works, or large buildings and other privately built large systems, are subjected to severe natural hazards.

NEED AND PROSPECTS FOR RESEARCH
ON NATURAL HAZARDS

Turning now to the research needs and prospects, the talk "The Future of Earthquake Engineering" given by Professor Nathan M. Newmark on the occasion of the Inaugural Symposium of the John A. Blume Earthquake Engineering Center provides an excellent starting point. Building on his years of experience and his recent activity as Chairman of the Advisory Group on Earthquake Prediction and Hazard Mitigation, Nate said:

> Structural engineering and structural design are old and established fields of learning. Yet, buildings designed in accordance with recent building codes have suffered damage and in some cases even collapse in earthquakes, and we cannot yet say that we can design to resist earthquakes without paying the severe economic penalty of the extreme conservatism that we find necessary in the case of special structures such as nuclear power plants.
>
> Regardless of our limited understanding of earthquake mitigation methods, investments are being made, structures are being built, land is being developed, earthquake motion precursors are being observed, and interpretations based on our current inadequate understanding are being made.

However, because of the fact that the public in general, and responsible public officials, are rapidly becoming aware of the problems of earthquake hazard and the necessity for hazard mitigation, I feel confident that within the next ten years there will be substantially increased funds available for those areas of research and study that need to be developed in order to attain the goal of the earthquake engineer and his brother scientists in the areas of geotechnology.

Surely similar statements can be made about the engineering and science applicable to the mitigation of the hazards of wind and water and fire. So also are there the direct equivalences of the classifications of "earthquake prediction, modification and control, hazard assessment, design improvement, and utilization including social and behavioral response."

CONTINUALLY CHANGING ATTITUDES TOWARD NATURAL HAZARDS

It is impossible to guarantee absolute safety even with unlimited knowledge, funds, and resources. In our world of large but limited resources, at any given state of knowledge and independent of our intentions, the higher the general level of safety required, the less available to mankind will be the necessities of life. This trade-off holds true country by country over long times. Consequently, all of engineering should be very responsive to the attitudes of individuals and the public at large toward the danger of injury or death. Structural engineering is especially exposed to public view and so has no choice but to be especially sensitive. Large structures are designed for a long life so that likely future attitudes as well as current views of acceptability of risk should be taken into account as best they can be. Attitudes and economic needs and strengths vary from country to country and from time to time. The more that is known about hazards, structures, and people, the more pleasing, the safer and the more economic the design can be; but the most suitable design for a given hazard is both time and location dependent.

In the days of prehistory and early history, of the voyages across unknown oceans, of the early settlers and the great western migration in this country, leaders and followers risked enormous dangers. Both officers and crews of clipper ships, for example, knew that their chance of surviving as many as 10 voyages was negligible, yet these terrible odds were faced with equanimity because there was no great expectation of survival for 20 years in the safest of occupations close to home. Disease, infection, and accident made life terribly uncertain under the best of circumstances. Fire was an ever-present danger. Any added hazard of earthquake or wind made but a minor perturbation.

Then great engineering achievements in the public health area provided many areas of the world with a safe water supply, a sanitary disposal of waste, and drained swamps. Later came the development of miracle drugs and great advances in medical practice. Now almost everyone in the affluent societies of this world has come to expect a long and comfortable life and is unwilling to be subjected to natural hazards that are not self-selected. With ample food and adequate provision of individual and public health care, many of the risks imposed by others or by unexpected forces of nature are viewed as intolerable. The driver of an automobile cheerfully takes the high risk of personal error but wants the manufacturer to guarantee zero defects. Occupants of a commercial building assume that engineers have designed that building to keep them completely safe from internal or natural hazard.

The average person in a developing country is far more like our early settlers in this country who willingly traded away considerable safety for a more satisfying existence. There is acceptance of a somewhat unsafe means of transportation rather than no transportation at all. Structures of many types are needed to help increase the standard of living and the quality of life. More structures of lower cost and lifetime safety are likely to be favored over fewer of high cost and safety. Risks and dangers unacceptable to us today will be assumed willingly in return for adequate food, clothing, and shelter.

Such current and likely future attitudes here and elsewhere must be factored into graduate education and research and the process of design. There is a pressing need for the development of clear design principles that exhibit cost versus risk trade-offs in understandable terms that permit public decision making for schools, hospitals, bridges, and all public works, as well as private structures. Not even we are so wealthy that we can evade trade-offs. The lack of success in achieving rational dialogue on nuclear power plants shows how difficult the job can be. However, there is more hope because the usual public reaction to natural hazards is very different from that toward nuclear plants. Most often there is the ostrich-like denial reaction, a pretense that disaster will never strike. Summer homes and complete summer colonies on some ocean or gulf beaches are swept away regularly every 30 years or so and are rebuilt optimistically. Subdivisions of expensive as well as inexpensive homes occupy flood plains. Fires periodically do great harm when simple precautions would preclude their occurrence or at least preclude the entrapment of people. Yet if the structure is not their structure, or when the need to be in or on the structure is job-related or otherwise not a free choice with explicit knowledge of the hazard, the reaction of the public will be closer to that toward nuclear power plants. Somewhat similarly, records and observations of the performance of structures subjected to natural hazards are very limited, as is our ability to predict what will happen from first principles.

FROM DESIGN AGAINST FAILURE TO DESIGN
FOR WORKING LOADS AND BACK AGAIN

The theory of elasticity, including its elementary form for beams, frames, and trusses, is a relative newcomer on the scene of structural design. Spectacular structures of the ancient world, the Greek and Roman eras, the Middle Ages, and the beginning of the industrial revolution all were built without concern for working stresses and working loads. Through trial and error, plus simple but effective concepts of static equilibrium, structures were designed not to fail under the loads and conditions to be experienced. Temples and cathedrals withstood winds, bridges withstood floods, but the infrequent strong earthquakes sometimes proved troublesome.

Static and dynamic elastic response to repeated application of live loads superposed on dead loads is an important and fascinating aspect of structural analysis and design. It is necessary to limit deflections and to guard against fatigue or progressive failure in this range. Otherwise, however, the elastic response to working loads is not very important for its own sake, although needed for many other calculations.

So, once again, after a long hiatus, we return to the concept that it is design against failure that is the engineer's key to success. A spectrum of overloads rather than working loads constitutes the crucial design conditions. The theorems of plastic limit analysis and design were a first step for static or quasi-static loading of most civil engineering structures. Developed for structures of steel, other metals and alloys, and reinforced concrete, they proved to be directly applicable to stone structures as well and helped to unify the engineering of today with that of the distant past. Along with plastic buckling and shakedown theorems, they provided a basis for rational preliminary design against excessive deflection or low-cycle fatigue due to occasional expected overloading. In the analysis of the preliminary design, geometric changes in the structure under these overloads have to be taken into account and the design modified appropriately.

Strong winds or waves or earthquakes to be expected several times during the life of the structure would be approximated by static loading and included in this overload analysis, but the truly unusual, less-than-once-in-a-lifetime wind or waves or earthquake requires a much more realistic approach. Dynamic inelastic response to the most severe dynamic loading chosen for the design should be contained safely. Much remains to be learned before an accurate analysis of a complete structure can be performed; far more still is needed before a useful design procedure can be established.

The open question of crucial importance is not so much the degree of damage to be tolerated under the most severe design loading but the choice of that loading. In this return to design against failure there are unlimited opportunities for both short-range and long-range useful and instructive research at the graduate level.

RESPONSE OF LIGHTLY
AND HEAVILY DAMAGED STRUCTURES

Mitigation of natural hazards in structures requires analysis of the response of damaged structures and the design of structures to continue functioning after suffering severe damage. Civil engineering was an offshoot of military engineering. Now in relation to natural hazards, structural engineering for the civilian sector returns to such analogous concerns of military structural engineering as the danger and containment of fire and designing for the ability of airplanes to fly or naval vessels to function or bridges to carry load after being hit one or more times.

Commercial aircraft are also designed with fail-safe features. However, civil engineering structures, unlike aircraft, have little or no maneuverability. They cannot avoid hazards by not flying, or flying around storms, or not landing at the scheduled time. Also, aircraft are not designed against collisions, a more likely occurrence for a plane than a terribly severe earthquake for most buildings.

The extension of analysis and design to structures with varying degrees of damage or failure requires research and synthesis of field experience that is far from trivial. Specifications for design of civil engineering structures surely have built into them through decades of experience many imperfections of structure and material and some estimates of tolerable damage caused by common overloads. Yet it is quite unlikely that specifications based on experience will properly comprehend design for those extreme conditions rarely or not yet encountered in recent times.

Within the quite severe limitations of our knowledge of the behavior of materials, connections, assemblages of structural and nonstructural components, and so on, we can assume that computers will soon be able to follow the static and dynamic response of a structure over time as it continually degrades. Much can be learned by running typical examples on a computer, by studying structures in the field, and by carrying out thoughtful experimental investigations in the laboratory on lightly and heavily damaged components and structures.

However, a comprehensive theoretical approach is essential to develop a fundamental understanding and useful set of design concepts and principles to cover the range from no damage under likely hazards, to easily repairable damage for the rare occurrences that are likely several times during the life of the structure, to repairable with considerable difficulty and expense for the once-in-a-lifetime hazard, to irreparable but life-preserving damage for the hazard predicted at some multiple of the design life. In our return to design against failure in the context of natural hazards, the spotlight must now be turned on the behavior of materials and structures in the range labeled as failed and therefore discarded in other contexts.

POSSIBLE RESEARCH AREAS RELEVANT TO ANALYSIS, DESIGN, AND RETROFITTING

The key step in design is an appropriate preliminary design choice of geometry, materials, structural and nonstructural components, connections, and other details. Subsequent analysis leads to considerable refinement but rarely outside the domain of the original preliminary design.

It is well worth looking for, and most difficult to find, general theorems such as the plastic limit theorems or maximum deflection theorems and their corollaries to guide preliminary design. Any such valid and useful theorems for the dynamic response of damaged structures or for bounds on the damage are likely to be at a rather abstract, high conceptual level. Sufficient understanding must be developed of principles of optimum design for extreme dynamic loadings with comfortable performance at static working loads and with fail-safe features. General principles are needed for well-designed connections and details, as well as main members, including their response when damaged lightly and severely.

In experimental structural analysis and experiments on the behavior of materials, as well as in the theoretical development, attention should be concentrated on the static and dynamic response of lightly and heavily damaged structures and materials. The cross-sectional shape of steel beams, the placement and the type of reinforcing in reinforced concrete beams columns and slabs, the different configurations of walls and floors and the materials of which they are made, all need and warrant study in the damaged range. Alterations in damping or in frequency response with friable and frictional material, with construction details that slip or let go or catch, provide ample scope for research. So also does the possibility of active as well as passive response to the dynamic action of wind and water and earthquake. Internal structures or supplemental components that maintain their integrity as the main structure fails are another direction that might be suitable initally or in the retrofitting of existing buildings and other structures.

Part of the basic approach found so useful in elastic fracture mechanics will carry over to the response of damaged structures and components. Most of the structure is likely to remain elastic with progressive or complete failure occurring in discrete local regions. Therefore, the details of each process of local failure can be duplicated and the overall response can be determined with a laboratory specimen. With more efficient designs, more of the structure behaves inelastically and the analog is to plastic fracture mechanics. Duplication of local conditions is less certain but still reasonable. Overall modes of failure become more likely and are much more troublesome to deal with analytically or experimentally. However, the needed or desirable improvements in material or component behavior can be identified and tested.

You are far more knowledgeable than I on the most promising of the areas of useful research, so it is inappropriate for me to continue with a list of topics. It is best to close the portion of this paper on research with a reminder to look closely at the research, field repair, development, design, and retrofitting activities of the aerospace and defense establishment. Much of their sponsored unclassified university rescarch is directly relevant to the mitigation of natural hazards despite the great disparity between the public unwillingness to accept an imposed risk and the great dangers necessarily taken for granted in military operations. It is worth remembering that ONR was the major sponsor of the development of plastic analysis and design in this country following World War II.

SUMMARY AND CONCLUSION

An effort is made to exhibit the promise and some of the problems within the broad scope of graduate education and research for structural engineering practice. Retrograde steps in education at the graduate as well as the elementary level are castigated. Hope is expressed that the profession will support forward-looking, research-oriented graduate and undergraduate education as the preferred path to the practice of engineering. Engineering educators are urged to make the needed hard choices of topic coverage at the graduate level, to abandon the friendly ranchhouse curriculum style, with all courses at the entry level, and return to curricula with courses in sequence that are the analog of imposing multistory buildings (Fig. 1). In the discussion of research in the context of natural hazards, brief attention is devoted to the differing and continually changing attitudes of individuals and groups throughout the world toward imposed and self-selected risks. Major emphasis is placed on the need to shift attention from the response of structures to working loads to the analysis of and design for the later stages of progressive failure. The research spotlight should be thrown on the principles determining the response of slightly and heavily damaged structures and structural components to subsequent static and dynamic loading. Structural civil engineering can take advantage of the design, repair, and retrofit experience and the research sponsored by the aerospace and defense establishments, including the approach so successful for elastic and plastic fracture mechanics. Much remains to be done. The research universities individually and collectively, worldwide, have the interest and intellectual competence to make great contributions to the well-being of mankind through the intimate combination of their research and the education of the practitioners of the future.

TECHNOLOGY AND PUBLIC POLICY
FOR EARTHQUAKE SAFETY

*Robert A. Olson**

INTRODUCTION

The translation of technological knowledge into effective public safety programs is a difficult, complex, controversial, and somewhat opportunistic process involving the dissemination and implementation of that knowledge throughout the social system. It should not be forgotten that the social system has the major voice in determining whether or not this knowledge shall be put into practice, how it shall be done, and if the resulting programs shall survive. By its very definition, the social system is composed of many elements, and for purposes here these include a multitude of interest groups, professional associations, interested and influential professionals, various public agencies, the legislative bodies of government, and numerous other actors who appear periodically, and often very influentially, in the public decision-making process.

It is no wonder that people who put their ideas forward lament what

*Executive Director, Seismic Safety Commission, State of California.

The comments in this paper are those of the author and do not necessarily represent those of the Commission.

happens to them and how they are expressed when the policy process is complete, assuming those ideas get through the process at all. On the other hand, there are those who are drawn into the process because these ideas are put forward and they respond to them. Some are pleased with the outcome because it is the best that can be done, it is workable, or their particular interests have not been jeopardized to any significant degree. Others perceive their losses more severely.

As far as public policy for the reduction of earthquake hazards is concerned, the relationship between technical expertise and the policy-making process is a relatively new phenomenon in California. By no means, however, can it be isolated from the strong movement since the close of World War II to have the scientific and research community more involved in the development and implementation of public programs. In relation to earthquake hazard reduction, there were notable exceptions prior to this, but, as will be noted in detail later, the broad, continuing involvement in the policy process of much of the expertise in earthquake hazard reduction had its origin only in about the last decade.

The result of this process of acquiring knowledge and translating it into public policy inevitably brings about social change. These changes show up in organizations, the educational process, professional practice, decision making and legislative behavior, public opinion and attitudes, and in the general level of public knowledge of the subject.

The realization by members of the scientific and technical community involved in earthquake hazard reduction policy that they are involved in a process of social change is also a relatively new phenomenon. A few key members of this community have become well aware of this, but it appears that this realization is not felt as strongly by many other professionals who have not been as closely involved with the development of public programs.

In this paper I hope to acquaint the reader with the history of seismic safety policy, the complexities of the public policy process, the influences and pressures that shape the development and implementation of public policy, and what the future evolution of public policy related to earthquake hazard reduction may look like.

It is important, too, to recall that the ultimate goal of earthquake engineering, as well as other technical and professional specialties contributing to seismic safety, is the saving of lives and to a lesser extent the mitigation of damage to property.

As a point of departure, it should also be noted that public safety generally has been the responsibility of government. Changes in public policy related to earthquake safety are to a large extent, therefore, aimed at modifying policy-making and administrative processes so as to "improve" the performance of the responsible public institutions.

HISTORICAL OVERVIEW

In outlining the development of public policy dealing with earthquake hazard reduction in the State of California, it has been traditional to follow a chronological sequence of events, usually beginning with the development of local codes following the 1925 Santa Barbara earthquake, the Field Act of 1933, and other examples through the years. Depending upon the particular viewpoint, this leads to such observations as: "Look at how much we have been able to do, particularly in the recent past" or "Look at how little we have learned and applied from past earthquakes." Although beyond the scope of this paper, it would seem that a valuable area for research could be to select a few key public policy decisions that were expressed as specific programs, such as the Field Act, the Hospital Act, the Seismic Safety Element program, and others to identify the forces at work at the time that influenced the creation of the program.

When the chronology of damaging earthquakes is separated from the chronology of major public policy developments, there is an interesting split. Some historical earthquakes have been followed by major policy changes. Examples include the Santa Barbara earthquake of 1925, which generated the beginning development of local codes for the earthquake-resistant design of structures, and the Long Beach earthquake of 1933, which produced the Field Act and all its subsequent amendments that regulate the construction of public elementary and secondary schools in the State. However, the biggest cluster of policy changes followed the 1971 San Fernando earthquake. Some of the key programs resulting from that earthquake included the establishment of state-wide construction practices for hospitals; the creation of a State Strong Motion Instrumentation Program, which is or soon will be the largest in the world; the requirement that cities and counties consider earthquake hazards in their general land use planning programs; that inundation maps and evacuation plans be prepared for communities vulnerable to the failure of large dams upstream in the event of an earthquake; and others.

Thus, there have been significant policy changes following damaging earthquakes, but this discussion must also account for other damaging earthquakes after which nothing significant in terms of public policy occurred. The real question is: Why not? Examples include the 1906 San Francisco "fire," the 1952 Kern County earthquakes, the earthquakes in Eureka and Dixie Valley, Nevada, in 1954 (the latter inflicting damage in Sacramento), the 1957 Daly City earthquake, the 1969 earthquake in Santa Rosa, and the 1975 Oroville earthquake.

In the last two cases, however, there have been some policy implications worth noting. Following the Santa Rosa earthquake, the city, largely because it was already involved in the renewal process, made extensive use of urban redevelopment programs to expedite the removal and replacement of damaged buildings. In the case of the 1975 Oroville earthquake, it is a little early to tell what the policy implications of that earthquake were, but it seems clear that its biggest effect has been on the construction of Auburn Dam in the Sierra foothills. The earthquake brought into focus the potential activity of faults in the Sierra foothills, whether this major dam was located on a "safe" or "unsafe" site, and if it was designed to take an adequate amount of earthquake loading. It was concern expressed by knowledgeable professionals that led to the current reevaluation.

INFLUENCES AND PRESSURES
THAT SHAPE PUBLIC POLICY

In a recent paper, "Earthquake Hazard and Public Policy in California," Steinbrugge and Johnson noted that [11]: "For many years, however, the California public's reaction to this hazard through their elected representatives to government, as well as the governmental bureaucracy at all levels, was to ignore the problem as if in the hope that the peril would quietly disappear. It is only in the last several decades that public policy has changed, and the hazard is now receiving increasingly heavy attention, particularly at the state level."

This section attempts, in a crude way, to identify those forces that have acted to facilitate or impede the design and implementation of earthquake hazard reduction programs. Some valuable research could be done by studying in detail the policy-making environment surrounding a few public programs for the purpose of identifying the pressures and influences that led to the success or, just as important, to the failure of programs or proposals. There is very little known about these factors, and those who share the responsibility for the translation of knowledge into effective public policy would benefit greatly from a more thorough understanding of the forces that affect those decisions.

A number of factors seem to be influential in promoting earthquake safety. It is beyond the scope of this paper to go into these in great detail and others could be added, but a brief summary seems in order. First, the occurrence of damaging earthquakes itself presents opportunities for public action. This is due primarily to heightened public interest and consequent motivation of public decision makers to act. These bodies face many pressures and crises in normal times, and it is easy to understand why problems related to earthquake hazard reduction usually receive priority attention only when significant events occur.

A second factor is the activity of advocate organizations. They offer ideas, proposals, support, and influence necessary to achieve seismic safety objectives. This includes many organizations, professional societies, and groups such as the former Joint Legislative Committee on Seismic Safety, which capitalized on the San Fernando earthquake to achieve a higher level of seismic safety by initiating several major programs. Closely allied to the presence of advocate organizations is the presence and influence of opinion leaders. Legislators, members of city councils and boards of supervisors, and other vocal experts play a key role in designing, leading, and negotiating the development of new public policy.

Another factor of more recent origin which seems to be supporting improved seismic safety is the general heightening of awareness throughout society of environmental quality. The concerns about air and water quality, conservation of resources, growth management, and similar problems have supported increased attention to environmental safety, particularly when it is related to natural hazards or those created by man interacting with the natural environment. The rapidity of communications continues to be a major factor in heightening the awareness of seismic safety. One need only recall the vividness of the earthquake damage communicated to the American public as a result of the recent earthquakes in Nicaragua, Guatemala, northern Italy, Romania, and China. Rapid communication has meant that American viewers and readers can understand the effects of disastrous earthquakes almost immediately after they have occurred. These countries are no longer remote in terms of affecting public opinion.

A fourth factor has been the growth of financial and human resources devoted to the analysis of earthquake problems and the expansion of the number of people, particularly researchers and practicing professionals, who are concerned about earthquake hazards. It seems apparent that the results of this investment are that the earthquake problem is better understood; knowledge of its implications has entered the fields of practitioners who must be concerned about ground motion in their work; college and university curricula have expanded to include courses dealing partially or entirely with earthquake hazards; and expanded research programs, undertaken within or financed by government and other organizations, have contributed significantly to the understanding of earthquake hazard mitigation. This has produced a larger community of knowledgeable people, answers—or at least approaches to answers—to problems that needed study, and support of some action programs, such as the creation of the Seismic Safety Commission, which is concerned about the effectiveness of existing programs and the design and implementation of new ones.

The advancement of knowledge regarding earthquake hazards and the additional financial resources available for earthquake research have produced a vast amount of relatively new knowledge that needs to be utilized more effectively and rapidly. In addition, in very recent years there has been growth in research and study oriented toward the public policy and social science aspects of earthquake hazards. This knowledge, coupled with advances in technical fields, has helped those who are concerned with designing and implementing programs understand what some of the major social, economic, and political problems are and what the impacts of such changes might be on the interests affected.

A last and relatively new factor that may be facilitating the development and implementation of improved public policy regarding earthquake safety has been the role and the publicity surrounding earthquake prediction research. As noted earlier, the greatest advancements in earthquake safety seem to be in response to damaging earthquakes. The emergence of earthquake prediction as a major research effort in the United States and elsewhere has provided the subject of earthquake hazards with a continuing popularity. Although members of the public may get confused about the validity of predictions, sources of information, and other factors, they are continually reminded that the earthquake hazard is present, that prediction is only part of the problem, and that people are continually working toward solutions.

These few factors seem to be the major ones that have facilitated the improvement of public policy, particularly in California, with regard to seismic safety. To a large degree, these factors have probably been very influential in raising the level of consciousness about earthquake problems in other states.

It is also fairly easy to identify some factors that appear to impede the development of seismic safety policy. First, the absence of damaging earthquakes does effect the receptivity of decision-making bodies to enact or support new earthquake programs. Another closely related factor is the problem of other priorities that demand attention. Public policy making is a dynamic process that tends to be crisis-oriented. One need only look at the hundreds of pieces of legislation submitted during any regular session of the California Legislature to understand the pressure on that body. The California Legislature, for example, has been concerned recently with property tax reform, public school financing, medical malpractice insurance, and other major issues. Should a big earthquake occur, one effect would be to change the priorities of this body, and more attention would be given to earthquake programs. This was clearly demonstrated following the 1971 San Fernando earthquake. Nevertheless, the press of other priorities impedes the advancement of seismic safety during "quiet times." One example will suffice. An assembly committee held public hearings in late 1975 on the implications of earthquake prediction for earthquake insurance. Some weeks later, the committee staff consultant was asked what was going to be done as a result of those hearings. His reply simply was, "Probably nothing. It is not a crisis yet."

One factor that has impeded the progress of seismic safety policy has been the inability to define the threat in precise enough terms so that people perceive that there is a high probability that they will be affected. This seems to be a strong factor today, and the development of a reliable and effective prediction system will almost certainly erode its influence. California has been fortunate in that damaging earthquakes are relatively infrequent and, from a life hazard viewpoint, have occurred at fortunate times of the day. This contributes to the general belief that it probably "won't happen here," particularly during "my term of office."

An additional factor that has appeared over the years to impede further seismic safety policy has been the reaction to existing programs, which has required their strong defense, particularly as the time between earthquakes becomes longer. Many people have spent time over the years going back to the California Legislature to defend the standards for school construction enacted after the 1933 earthquake, and since the 1971 earthquake there have been such occasions with regard to programs which were initiated following that earthquake. At these times it is very hard to initiate new programs, because the proponents are often asked questions that put them on the defensive about existing programs. Fortunately, there has been a marked decline in recent years of the fatalistic attitude expressed in such phrases as, "We know we experience earthquakes, but we can't do anything about them."

Another factor that has been influential is the inability, in most cases, to demonstrate the effectiveness of many of the existing programs. Partly, this is a function of the relative infrequency of damaging earthquakes after which the performance of existing programs can be judged. It is only in recent years, for example, that enough data have been accumulated about the behavior of public school buildings built according to the Field Act of 1933 to show that the program is basically sound.

Stanley Scott, of the Institute of Governmental Studies, University of California, Berkeley, touched on this issue when he stated [8]:

> All goals of society have values that must be weighed against the costs of achieving different objectives. Increased earthquake safety—or reduced risk— can be attained with today's technology, but only at a price. On the other hand, data for reliable cost–benefit ratio calculations are often unavailable, and many costs and benefits can be expressed only in non-quantifiable terms. Accordingly, the evaluation of cost–benefit ratio considerations, and the definition of acceptable risks, requires the exercise of informed judgments.

In sum, the factors that facilitate or impede the development of seismic safety policy may help account for the different responses to historic damaging earthquakes. An extremely valuable research project, carefully done and based on a refinement of these factors, tested against the public records of previous earthquakes, might show that certain combinations of them have produced actions, such as the Field Act or the several programs following the

1971 San Fernando earthquake, and the absence of these factors might help explain the absence of action following other earthquakes.

POLICY PRINCIPLES
IN PROGRAM ADMINISTRATION

When one looks at public policy it is important to recognize the basic principle underlying a given program. Most difficulties that result from the enactment of programs involve operational questions, such as the standards required, the performance characteristics, the rules of the organizations involved, and others. A few examples of basic policy principles will suffice. The Field Act governing public school construction in California stated essentially that school children were a special population deserving additional safeguards and that it was proper for the state to preempt enforcement to help ensure uniformity. The Hospital Seismic Safety Act set statewide standards for the construction of new hospitals and other medical facilities. The underlying principle was that hospitals are critical community facilities whose importance following disasters is increased, and it is important for these facilities to survive and remain as functional as possible. It has also reinforced the principle of state preemption of certain responsibilities and introduced into public policy the concept of damage control, which relates directly to the need to keep these facilities functioning in some basic manner.

The state requirement that local emergency organizations have maps showing the areas that might be inundated should dams fail under earthquake conditions is based on the simple policy that "fail safe" does not exist. For planning purposes it has to be assumed that absolute safety is impossible, and this information will facilitate an effective emergency response. The amendment to the State Planning Act which requires cities and counties to have seismic safety elements in their general plans is based on the belief that better land use decisions will be made if communities are aware of seismic and geologic hazards in their planning processes. The Special Studies Zones Act, which requires geologic reports for most structures planned for construction in the major fault zones in the state, is also based on a relatively simple principle; that is, it is permitted to build in the fault zones, but the builder and the local jurisdiction granting the permit must be aware of the geologic conditions of the site as part of the building permit process. In other words, it makes information available that potentially effects a decision. There are other examples, but these illustrate some of the principles involved when public policy is developed and new proposals are debated.

Public policy is expressed by the way programs are administered and by the types of decisions that are effected. Given the governmental structure, one can look at programs that are developed and administered locally. Examples include the adoption of the Uniform Building Code, local siting and grading ordinances, parapet ordinances, and programs dealing with earthquake hazardous buildings. Programs may in some cases be developed by the state but administered primarily by local government. The requirement to have seismic safety elements in general plans is one example. Another is the Special Studies Zones Act, in which the decision to issue a building permit is primarily a local one. Other programs are both state-developed and administered. This includes the Field Act, the Hospital Seismic Safety Act, and the professional registration and licensing programs. In certain other cases, there is almost no role for local and state government, since the jurisdiction primarily belongs to the national government. The siting, design, and construction of nuclear power plants is the most prominent example. Others include federally owned dams and other facilities.

Another way of looking at seismic safety programs is by the type of decision that a proponent of the new program wishes to effect. These fall into a few general categories, including land use, building systems, lifeline systems, and support programs. In the area of land use planning and regulation, there are programs such as the Seismic Safety Element requirement, the Special Studies Zones Act, the requirements of some jurisdictions for detailed site studies, and the procedures of specific organizations such as the Bay Conservation and Development Commission, which has permit jurisdiction over development around the perimeter of San Francisco Bay. Programs affecting building system decisions are those such as the Field Act, the Hospital Act, the Uniform Building Code, the elevator regulations of the Department of Industrial Relations, and other requirements related to electrical, mechanical, architectural, and foundation design. One area that is emerging rapidly in terms of public policy concern is the area of lifeline systems and critical facilities. The types of decisions to be effected include those that relate to the entire array of public utilities, communication systems, transportation facilities, and perhaps dams and other critical facilities. The final set of decisions that may be affected by changes in seismic safety policy relates to support programs such as professional licensing and registration requirements.

THE FUTURE OF SEISMIC SAFETY POLICY

It is worth venturing into the future to see what influences there may be that will affect the policy process with regard to the continued development of improved seismic safety. In the July 1976 issue of the *Smithsonian Magazine*, Daniel S. Greenberg may have set the tone for this work in the future. He states [3].

During the next quarter of a century, research will thrive in terms of expenditures, scientists and technicians at work, ambitious· projects, and costly facilities. But it will do so as a tightly reined captive of the surrounding society, and in an atmosphere of caution and thrift. Most important, opportunities for scientific research will no longer be judged in the traditional terms of knowledge being a desirable goal in itself regardless of the purposes to which it might be applied.

Scientists and research administrators will find that whole regions of research and technological application have been declared off limits, shunned as economically unworthy of pursuit or legally prohibited as too dangerous to be entrusted to frail humans and their unpredictable social and political institutions.

Research in many, if not most, fields will be actively encouraged, for the thickening problems of mankind on a crowded globe will demand scientific and technological solutions. But the choice of what to research will not be left entirely to the people who perform the research. Rather, it will receive the scrutiny and require the approval of a wide-ranging assortment of nonscientists who are concerned with cost, priorities, social impact, public safety, and political significance.

Some of the major policy issues that will be present in the foreseeable future include a continuing effort to try to find new ways to reduce the hazards from non-earthquake-resistant buildings; the continuing work on earthquake prediction research and its parallel social, economic, and political implications; concern about the adequacy of siting and design standards for lifeline and critical facilities and the role of public agencies in governing those systems and facilities; the role of the Uniform Building Code as a basis for minimum standards and the possibility of using the code to achieve important public ends, such as the recognition of important structures; and the increasing role of intergovernmental coordination of local, regional, state, and federal agencies in achieving earthquake hazard reduction.

It can be expected that the factors identified earlier that facilitate or impede the development of public policy will continue to be present and new ones may appear.

There will also be increasing emphasis on more effectively utilizing knowledge and speeding up the process of implementing it into effective and practical programs. Organizations such as the Seismic Safety Commission may take on added significance as the link between the research community and the practitioner and governing authorities. There will continue to be a strong need for policy-oriented research that will help answer some of the difficult questions related to the design, acceptance, and administration of strengthened or new public programs.

CONCLUSION

In conclusion, it might be useful to touch on the role of the Seismic Safety Commission in the State of California.

The purpose of the Commission is to strengthen earthquake safety by improving public policy. The Commission provides an overall policy framework, is a means for coordinating various programs, and is a focal point for proposals, recommendations, and implementation actions to deal with the continuing earthquake threat. In some ways it can be seen as a link between science and technology, professional practice, and politics.

The Commission has a broad range of responsibility. These include setting long-range goals and priorities, requesting state agencies to prepare standards and criteria, recommending changes in existing programs to reduce hazards further, reviewing reconstruction efforts after damaging earthquakes, helping to coordinate the seismic safety activities of all levels of government, and providing information, encouraging research, and sponsoring training.

The Commission also advises the Governor and the Legislature on earthquake programs and needed legislation. It may hold hearings on matters important to seismic safety. In addition, it advises the Division of Mines and Geology on the Strong Motion Instrumentation Program and the Special Studies Zones Act. The Building Safety Board, which is responsible for earthquake standards for hospital construction in the state for the Department of Health, reports annually to the Commission on its program.

REFERENCES

[1] CAMPBELL, I., "The Influence of Geologic Hazards on Legislation in California," *Bulletin of the International Association of Engineering Geology*, No. 14, 1976, pp. 201–204.

[2] DYE, THOMAS R. (Editor), *The Measurement of Policy Impact*, Conference Proceedings, Florida State University, 1971.

[3] GREENBERG, DANIEL S., "Scientists Wanted—Pioneers Needn't Apply; Call A.D. 2000," *Smithsonian*, Vol. 7, No. 4, July, 1976, pp. 60–67.

[4] Institute of Governmental Studies, *In the Interest of Earthquake Safety*, University of California, Berkeley, 1971.

[5] National Academy of Engineering, *Public Safety: A Growing Factor in Modern Design*, Symposium Proceedings, Washington, D.C., 1970.

[6] Panel on Earthquake Prediction, *Predicting Earthquakes: A Scientific and Technical Evaluation—with Implications for Society*, National Research Council, Washington, D.C., 1976.

[7] Panel on the Public Policy Implications of Earthquake Prediction, *Earthquake Prediction and Public Policy*, National Academy of Sciences, Washington, D.C. 1975.

[8] SCOTT, STANLEY, *Earthquake and Geologic Safety Policies and Objectives* (unpublished paper), Institute of Governmental Studies, University of California, Berkeley, April, 1976.

[9] State of California, Seismic Safety Commission, *Annual Report to the Governor and the Legislature for 1975–76*, Seismic Safety Commission, Sacramento, 1977.

[10] State of California, *Meeting the Earthquake Challenge: Final Report to the Legislature*, Joint Committee on Seismic Safety, Sacramento, 1974.

[11] STEINBRUGGE, KARL V., AND CARL B. JOHNSON, "Earthquake Hazard and Public Policy in California," *Engineering Issues, Journal of Professional Activities*, American Society of Civil Engineers, Vol. 99, No. PP4, Proc. Paper 10105, October, 1973, pp. 513–519.

EARTHQUAKE-RESISTANT DESIGN CRITERIA FOR MAJOR PROJECTS

G. W. Housner and P. C. Jennings**

INTRODUCTION

Industrialized societies rely for economic well-being upon a wide variety of major facilities, including large dams, nuclear power plants, oil refineries, petroleum and liquefied natural gas tank farms, offshore drilling platforms, high-rise buildings, chemical process facilities, and other similarly complex and costly installations. Failure of these would entail large losses and, in addition, some could be hazardous in the event of catastrophic failure. When such facilities are to be located in seismic regions, the provision of appropriate earthquake resistance is often the most critical engineering problem faced in their design. The earthquake design of major projects should be given special consideration, as distinguished from ordinary structures which are designed according to a code. In the case of a code design, the only question is whether the design satisfies the requirements of the code. In the case of a major project, however, more fundamental questions are at issue, such as: How will the structures perform during future earthquakes? What is acceptable, infre-

---

*California Institute of Technology.

quent damage? How much should be invested in the provision of earthquake resistance? and Will the design be approved by an outside review panel?

Problems of design

In the earthquake-resistant design of major projects, two types of problems are encountered. The first are problems of a purely technical nature, which include the determination of the desired strength of the structure, the choice of structural type and material, the method of framing, the allowable stresses and strains, and the many details that comprise the engineering design process from its inception to the final structure. The second kind are more managerial in nature; these include the coordination of the contributions of consultants, such as geologists, seismologists, and earthquake engineers, and the presentation and defense of the project and its earthquake-resistant design before various governmental bodies and regulatory agencies, including the preparation of backup documentation. The second type of problem was at one time unimportant, but in recent years activity in this area has increased greatly and some large projects must now receive approval from as many as 50 different political or regulatory bodies. A sizable fraction of the attention of senior project engineers is devoted to this aspect of the project; and in some instances, it seems to assume even greater importance for the success of the project than does the engineering design itself.

The prime technical problem in the earthquake-resistant design of a major project is the formulation of the design criteria, although the subsequent engineering analysis and design may also be difficult. When formulating the criteria it is necessary to keep in mind that, fundamentally, they are a means of specifying the desired aseismic capacities of the structures and facilities. The objectives of the criteria are twofold: first, to provide levels of earthquake resistance for the various parts of the project that are consistent relative to each other; and second, to provide an absolute level of earthquake resistance that is appropriate to the desired performance of the project.

On the nontechnical side, the requirements of coordinating the technical specialists and the involvement with regulatory agencies and political bodies can place a heavy burden on the project manager. To do the job effectively, he/she must ensure good communication among geologists, seismologists, earthquake engineers, and designers, that is, among those who contribute information upon which the design criteria are based and those who will utilize the information. In the past, difficulties have arisen because of misunderstandings, particularly between engineers and seismologists, whose training and experience predispose them to look at the earthquake problem differently. It is also essential for project managers to have a good overall grasp of the various aspects of the earthquake design problem, for they must assess the conservatism, or lack thereof, of the final design and must arrange efficient interaction with regulatory and political groups.

Function of design criteria

The primary function of the design criteria is to restate a complex problem, which has unknowns and uncertainties, into an unambiguous, simplified form having no uncertainties. The design criteria should provide clearly stated guidelines for the designers. For example, when designing a structure, an engineer needs to know the forces and deformations that the structure should be able to resist. Some of these forces, such as gravity dead loads, are known, but forces that result from the transient action of nature or man, such as earthquake, wind, or live loads, are not. This lack of knowledge must somehow be circumvented and a precise, unambiguous statement of the design conditions must be given to the design engineer. This is accomplished by means of the design criteria. The designer also needs to know the properties of the materials and structural elements that will be used, but as these are not precisely known, mainly because of imperfections in materials and workmanship, the design criteria must also take this into account. In the preparation of the design criteria, allowance must be made for the uncertainties, and it is necessary to be cognizant of all the unknowns for which allowance must be made.

The traditional engineering design criteria for gravity and live loads, for example those in the Uniform Building Code, specify design loads that are greater than the actual loads typically encountered and specify allowable design stresses that are appreciably less than the expected ultimate strength of the material. The purpose of this procedure is to ensure strength sufficient for unforeseen variations in loads, material properties, and workmanship. These criteria, in effect, tell the design engineer: "If you design according to these requirements, the structure will be considered adequate." A similar approach can be taken for earthquake-resistant design if the conditions are more or less the same for all projects. However, if the seismic hazard varies markedly from place to place, and if structures vary in importance, cost, length of life, ease of repair, and consequences of failure, the formulation of seismic design criteria cannot, in general, be reduced to a simple rule of thumb, for then special knowledge and judgment are required.

USE OF SEISMOLOGICAL AND GEOLOGICAL DATA

When designing for a seismic region it would be very helpful to know the exact ground shaking that a structure under consideration will experience during its lifetime. This knowledge, however, is unavailable, so the recourse is to estimate what might happen in the future. Seismological and geological data form the bases for estimating future ground shaking at a site. The seismic history of a region, by indicating what has happened in the recent past, provides a clue as to what might be expected in the near future. In this sense the past predicts the future, but the reliability of the prediction depends upon the quality and quantity of available data. By definition, earthquake data of high quality have instrumentally determined magnitudes and epicenters of all significant events. Earthquake data of satisfactory quantity would, by definition, include a sufficiently large number of events so that enough earthquakes of larger magnitude are included.

Earthquake magnitude

In practice, the size of an earthquake is denoted by its assigned magnitude. In its original sense, the magnitude of an earthquake is the logarithm to the base 10 of the maximum amplitude of the response of a standard seismograph with a natural period of 0.8 and 80% of critical damping. at 100 km from the center of the earthquake.[1] In a more useful but less precise sense, the magnitude is a number that describes the size of an earthquake; that is, it describes the seismic energy released by the fault rupture as well as the size of the area affected by strong ground shaking. (The imprecision of the magnitude scale when used in this sense is described in the Appendix.) In practice, the magnitude of a large earthquake is based on a measurement made at a distance of hundreds of miles and which, therefore, contains no direct information about the nature of strong ground shaking near the causative fault. However, it is customarily assumed that the ground shaking of two earthquakes assigned the same magnitude number will be similar, other things being equal, but it should be remembered that other things are seldom exactly equal.

The adequacy of seismological data depends upon having a sufficient number of data points in the historical record, with magnitudes and locations determined, so that large-magnitude events are also included. For example, if the data include only earthquakes having $M \leq 5$, the probability distribution would not be defined and the reliability of an extrapolation to the probability of earthquakes of $M \geq 8$ would be questionable. Lacking sufficient data to define a probability distribution, it is customary in the United States to assume a frequency distribution for M that is consistent with the seismic history of California, even though this introduces a degree of uncertainty.

In the less seismically active regions of the United States, the seismolog-

---

[1]This definition now applies only to the local magnitude, M_L. There are currently several magnitude scales in use, and in the case of large earthquakes the commonly reported magnitude is not M_L but the surface-wave magnitude, M_S, or the magnitude determined in some other way.

ical data are relatively scarce and of poor quality. For example, in the eastern part of the country, information available on damaging earthquakes seldom includes the instrumentally determined magnitude of the event but instead gives modified Mercalli intensity (MMI) numerals. The MMI is of lower statistical quality than the magnitude, not only because it is based on personal observations of earthquake effects instead of instrumental records, but also because the interpretation is often unreliable. For example, a review of the effects of the August 12, 1929, Attica, New York, earthquake indicates a maximum MMI of VII instead of the VIII originally assigned to it [1]. The uncritical use of MMI data introduces a degree of uncertainty, which may lead to an overestimation of seismic hazard.

Required seismological and geological information for design

The seismic history of the United States is not very long—two hundred years more or less, depending on location—and this is a short time for earthquake occurrence. This relatively short-time information can be supplemented by geological information about long-time tectonic processes that are measured in many hundreds or thousands of years. For example, faults that can be identified as having slipped during the past hundreds, or past thousands, of years can be taken to contribute to the seismic hazard of the region, but it is difficult to quantify this contribution. Various procedures have been employed to interpret the seismic hazard posed by such faults. The crudest approach is that which assigns a Maximum Capable Earthquake (MCE) to the fault. (The MCE is sometimes called the Maximum Credible Earthquake, but the latter name is so ambiguous and poorly defined that it is best avoided.) For example, a fault whose discernible length is approximately 40 miles might be assigned a MCE of magnitude 7, or one with a discernible length of 15 miles might be assigned a MCE of magnitude 6.5. The MCE by itself is not a very informative number, for it does not distinguish between a fault that will have events of the approximate size of the MCE once per 200 years and one for which the return period is once in 500,000 years, even though this information would be very important to engineers preparing seismic design criteria.

A project manager should require that the geological and seismological consultants address the question of probability of occurrence. He should not accept a report that merely states "the recommended design earthquake is magnitude 7.5," not only because it does not indicate frequency of occurrence, but also because it implies that the geologist–seismologist has made a decision about engineering "design" which is outside his/her area of competence. Geological and seismological consultants are experts on geologic and seismic hazards, and their reports should describe possible earthquakes together with estimates of probability of occurrence, or the possible intensity of ground shaking together with estimated probability of occurrence. Persons who understand engineering design and performance of structures should be responsible for incorporating such information into the design criteria.

Strong-motion accelerograms illustrate the kind of ground motions to be expected in the future, and the ground motion to be considered in the design can be exemplified by three components of ground acceleration which are consistent with recorded accelerograms. The recommendations of a seismological consultant should, preferably, present ground accelerations in the form of appropriate recorded accelerograms from particular earthquakes, or synthesized accelerograms that have appropriate intensity, duration, and frequency characteristics. The seismological consultant should also give the estimated probability that ground shaking in excess of this accelerogram in severity will occur. For example, it would be appropriate for him/her to give either the ground motion that would be exceeded once in 50 years, or the motion that would be exceeded once in 200 years, as long as it is properly identified, but it would not be acceptable to give a ground motion without identifying the expected frequency of exceedance.

Sometimes seismological consultants do not present accelerograms but instead give a less complete description of the ground motion. Properly, this less complete description should include (1) the intensity of ground shaking, (2) the duration of strong ground shaking, (3) the frequency characteristics of the expected motion, and (4) the frequency of occurrence. The intensity of the ground shaking indicates to the engineer how severely a structure will vibrate; the duration of strong ground shaking indicates the degree of damage to be expected if the structure is stressed beyond the elastic limit; the frequency characteristics of ground motion should be identified, for earthquakes in different parts of the world may have different frequency characteristics and therefore effects on structures can differ; and the estimated frequency of occurrence of ground shaking indicates the conservatism of the recommended ground motion.

Often the intensity of ground shaking is described by giving a value of peak acceleration, which by itself is an ambiguous description, because the intensity of two ground motions with the same peak acceleration can differ appreciably as far as structural response is concerned. A much better method of describing ground motion would be to compare it to a known accelerogram, such as Taft 1952 or a synthesized accelerogram. The description could thus be phrased as: 1.5 times as intense as Taft 1952, duration of strong shaking 1.2 times as long, and frequencies of motion all greater by a factor of 1.3. When the information is presented in this manner, the engineer will understand what the seismologist means. More information can, of course, be given, but if any less information is given, the meaning will be ambiguous. Sometimes the seismological consultant describes the ground motion by recommending a smooth "design spectrum." This, however, is not proper because a design spectrum is not the same as a "response spectrum" of actual ground motion or a smoothed "average spectrum," and it is precisely this difference that involves engineering judgment. For example, if the top 25% of the highest peak on an accelerogram is lopped off, it would, in general, have very little effect on the response of structures; therefore, when an engineer selects a smooth design spectrum based on an acclerogram, the zero-period spectral acceleration (sometimes called "effective acceleration") may, with justification, be smaller than the peak ground acceleration. If the structure to be designed is highly ductile, the project manager may set the entire design spectrum at a level lower than the response spectrum. The validity of thus specifying the design spectrum depends on how well the spectrum is correlated to the properties of the structure to be designed.

ENGINEERING CRITERIA

Design spectrum

The central feature of most earthquake-resistant design criteria is the design spectrum, an example of which is shown in Fig. 1(b). The jagged response spectrum [Fig. 1(a)] describes the computed response of different oscillators to a particular ground motion, whereas the smooth design spectrum is a specification of the level of seismic design force, or displacement, as a function of natural period of vibration and damping level. Implicit in Fig. 1(b) is the condition that the level of force prescribed by the design spectrum is to be associated with a specified level of material resistance, for example the allowable design stresses or strains. The resultant effect is thus a specification of the required earthquake resistance of the structure and its elements. If the material resistance is stated in terms of allowable stresses, the design spectrum is a specification of the strength of the structure; if the material resistance is expressed in terms of permissible ductile strains, the design spectrum becomes a specification of the capacity of the structure to deform, that is, its ductility.

Three factors—the level of the design spectrum, the specified spectral damping, and the allowable stresses and strains—together specify the capacity of structures to resist earthquakes. They combine to form the basis of the earthquake-resistant design criteria. For example, an increase in allowable damping is equivalent, in effect, to a decrease in the level of the design spectrum or to an increase in the allowable stresses. Owing to this, the specification of all three items should be correlated so that the degree of conservatism in each can be taken into consideration when establishing the overall conservatism of the design. To specify these independently without

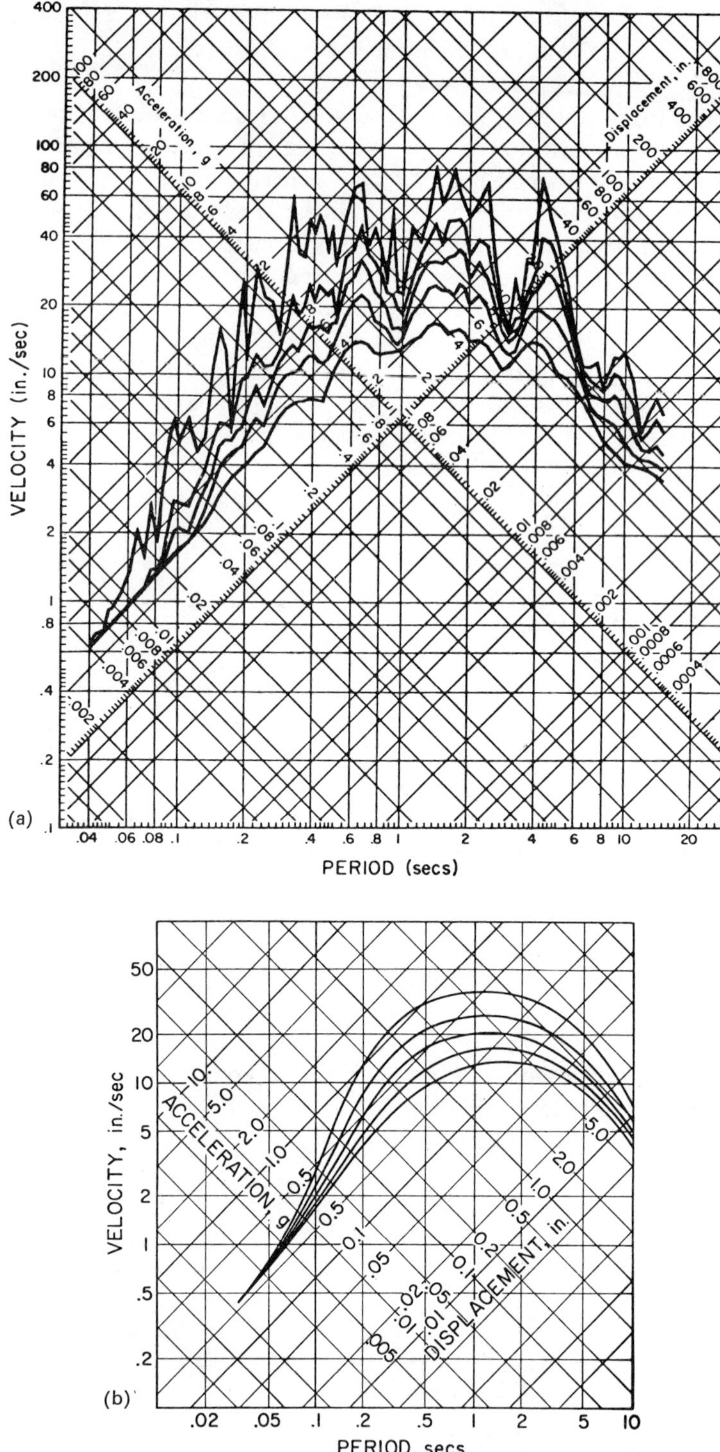

Figure 1 (a) Response spectrum of north-south ground motion recorded at Holiday Inn during the February 9, 1971, San Fernando earthquake; (b) example of smooth design spectrum based on Fig. 1(a).

identifying the individual conservatisms makes it impossible to assess the margin of safety of the final result.

In setting design spectra, project managers should take into account acceptable degree of damage and likelihood of occurrence. They must also consider the actual capacity of a structure that results from the use of the design spectra and the specified capacities of the materials of construction. This is obviously a problem that requires both engineering knowledge and judgment and, because of the complexities and uncertainties, considerable reliance must be placed on knowledge of how engineered structures have performed in past earthquakes. For example, the San Fernando earthquake provided evidence that the level of design, together with the material resistances and the quality control specified in California's Field Act, ensure the successful performance of typical one- and two-story school buildings during very strong ground shaking [2]. The buildings performed successfully even though the nominal levels of design accelerations were much lower than the actual ground accelerations.

Capacity of structures

The apparent paradox that the code value of acceleration for which a structure is designed is much smaller than the recorded peak acceleration of the ground motion that the structure successfully survived can be explained without recourse to such terms as "effective peak acceleration" and "sustained peak acceleration," which are smaller than the peak acceleration itself. The explanation is that the allowable design stresses and strains in the building code are not indicative of the material and structural resistances under dynamic conditions. To clarify this, it is necessary to establish the true relation between the actual dynamic capacity of engineered structures and the levels of the basic components of the design criteria: spectrum level, damping, and allowable material resistance.

An example of this relation is shown in Fig. 2, which is derived from the San Fernando earthquake, an accelerogram of which is shown in Fig. 3. In Fig. 2, three sets of data are shown for multistory reinforced concrete structures constructed since 1964, plotted as a function of the fundamental periods of the buildings measured during the earthquake. Two of the buildings are shown in Figs. 4 and 5. The triangular data points, the top group, are the maximum acceleration measured on the roof of the buildings. The circular data points represent the maximum base shears experienced during the earthquake by the fundamental modes of vibration of the structures. The base shears were determined from the computed maximum displacements reported by Hudson and others (1968–1976), and from a knowledge of the distribution of mass in the buildings and the shapes of the fundamental modes. The ticked circles indicate that this level of base shear was associated

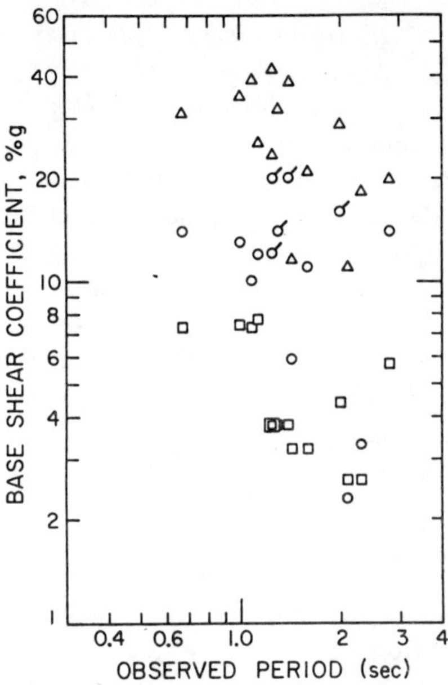

Figure 2 Base-shear coefficients of multistory buildings that experienced the San Fernando earthquake. The triangles represent the maximum acceleration recorded on the roof; the circles represent the maximum first mode base shears experienced during the earthquake; the ticked circles represent structures experiencing some structural damage; squares represent the nominal code values of base shears for which the buildings were designed. The periods are those of the fundamental modes of vibration.

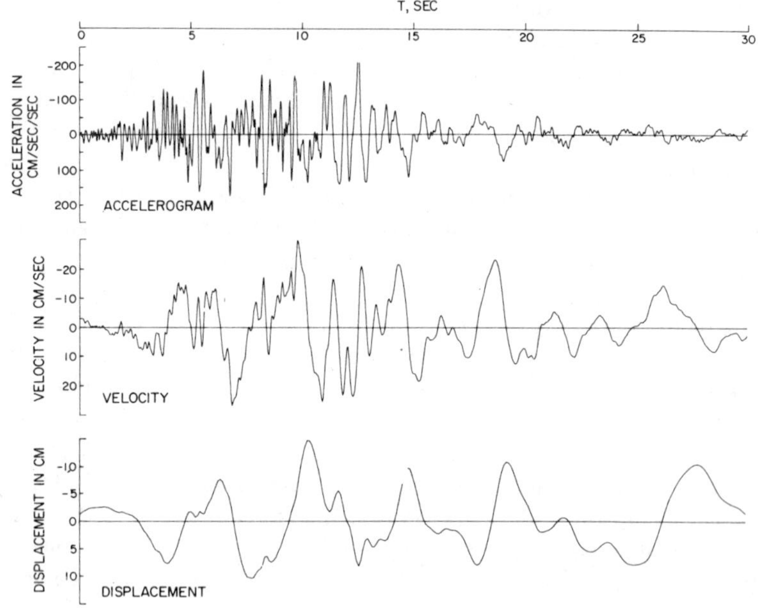

Figure 3 Recorded ground accelerations and computed velocities and displacements at Holiday Inn during the San Fernando earthquake.

Figure 4 Holiday Inn after the earthquake. This concrete building received some structural damage: cracking of beams and columns, and so on.

Figure 5 Bank of California after the earthquake. This building received some structural damage: cracking of beams and columns, and so on.

with some structural damage. It should be noted, however, that none of the structures represented in the figure were dangerously damaged and all could have resisted significantly stronger shaking without collapse. The square data points in the figure are the base shear values employed in the designs; these were determined by the designer in accordance with the applicable building code. The significance of Fig. 2 is that it indicates that such structures, on the average, can be expected to resist base shears that are two to three times larger than the code design values without severe structural damage. The margin of safety against collapse of these structures was not tested by the San Fernando earthquake, but the data suggest that, on the average, responses equivalent to five or more times the design base shear could have been resisted without collapse, although severe damage would probably have resulted.

The capacity of buildings to resist strong ground shaking is illustrated from another viewpoint in Table 1. The table describes, for California conditions, the expected performance of buildings of different types to the poten-

Table 1

EXPECTED DEGREE OF DAMAGE[a] VERSUS INTENSITY BASED ON
SAN FERNANDO EARTHQUAKE SHAKING FOR SOUTHERN CALIFORNIA

	Intensity of Ground Shaking			
Class of Building	*15–20 % g*	*20–30 % g*	*30–40 % g*	*40–50 % g*
A: above-average modern building	None	Minor or none	Moderate	Severe
B: average modern building	Minor or none	Moderate	Severe	Major
C: below-average modern building	Moderate	Severe	Major	Partial collapse
D: old, precode building	Moderate to severe	Major	Partial collapse	Partial collapse

[a]Minor damage: Can be repaired without appreciable interference with normal operations.

Moderate damage: Can be repaired with small interference with normal operations; perhaps equivalent to closing down for several days.

Severe damage: Significant damage to structural members; repairs will require closing for at least several weeks.

Major damage: Extensive damage to structural elements; repairs require closing down for several months.

Partial collapse: Repairs require closing down for an extended period, from 5 months to a year. For class D structures, the building may have to be abandoned.

tially damaging shaking that can occur in major earthquakes. The table is not meant as a quantitative guide to the assessment of hazard, but rather as a first approximation to the expected effects of strong ground motion.

If the observed ability of structures to resist earthquakes is not taken into account when formulating the design criteria, inconsistencies may result. For example, as shown in Fig. 6, several concrete buildings with 8-in.-thick shear walls survived, without damage, the severe ground shaking at the Veterans Administration Hospital during the San Fernando earthquake. However, the seismic design criteria for the new, postearthquake, Olive View Hospital building, at a site adjacent to the VA hospital, were so stringent that these VA buildings could not satisfy them.

Figure 6 Two concrete buildings at the Veterans Administration Hospital that survived the San Fernando earthquake without structural damage. An old, weak building between these two collapsed during the earthquake.

Modifications in the shape of the spectrum

To specify consistent levels of structural capacity for buildings with different natural periods, the shape of the design spectrum should reflect the relative intensities of expected motions at different frequencies. Because the energy in the site ground motion at shorter periods would be dominated by nearby earthquakes of moderate size rather than by more distant, larger shocks, and because the longer-period energy would be dominated by large earthquakes even if these were to occur on more distant faults, the shape of the design spectrum may depend to a degree on the expected occurrence of earthquakes in the region of the site. The present state of knowledge does not, however, warrant going to great lengths to tailor the shape of the design spectrum to fit hypothetical earthquake hazards. It is recommended that such modifications be limited to simple and relatively minor alterations to a spectrum of standard shape.

A related problem arises concerning the adjustment of the design spectrum to accommodate possible influences of local geology and soil conditions. Unfortunately, data bearing directly on this problem are scarce and can only be accumulated slowly because the necessary instrumentation is so costly and strong earthquakes occur infrequently. To throw light on the possible effects of local soil conditions, special computations are often made, which involve estimating the ground motion at depth (bedrock or firm soil) and then propagating this motion to the surface through linear, nonlinear, or iteratively linear models of the overlying soils. The seismic waves are assumed to be planar, horizontal shear waves that propagate vertically. Such analyses are often made for major projects sited on relatively soft soil. They can give useful insight if the actual geological and seismological conditions do not differ greatly from the conditions postulated by the computational procedures, as was the case, for example, of the well-known recorded behavior of the soft soil in Mexico City. However, it is very difficult to assess the differences between the assumed and the actual conditions, and in practice this approach has sometimes been misused. The inconsistencies that result in trying to assess potential site effects could be reduced by carefully comparing predicted results with accelerograms recorded under similar geological and seismological conditions. Such relevant accelerograms, if available, should be used as the primary guide in adjusting the shape of the design spectrum, and the results of analytical and computational studies should be used as secondary guides.

Role of statistical and probabilistic analyses

In earthquake engineering there are many situations where essential factors cannot be precisely defined because of a lack of information. For example, the physical properties of the concrete and steel are not known precisely to the engineer when he/she makes the design; nor is the quality of construction workmanship known. These could be made known to the designer by means of additional quality control, testing, and inspection, but the cost would be prohibitive, so the uncertainties are accepted. In this case it is considered to be cost efficient to accept and deal with uncertainties rather than to try to eliminate them. In the case of earthquake ground motions, it is uncertain where and when earthquakes will occur and how large they will be, and it is not known what ground motions they will produce. Again, in principle, these uncertainties could be reduced to small levels, for the problem is solvable if the states of stress and strain rate in the earth's crust were precisely known, if the failure strength of the rock were known, if all the relevant physical properties of the earth's crust through which the seismic waves travel were also known, and if ample time and money were available for computing. The difficulty and cost involved make it unlikely that the foregoing problem will ever be solved, and it is necessary, therefore, to accept a lack of knowledge and to deal with it in the best way possible. Statistical and probabilistic analyses are tools for dealing with scientific ignorance. Although factual information not already included in the basic data cannot be generated using these tools, such analyses provide subtle and inconspicuous ways of making assumptions about the basic data which can provide guidance for decision making. Because of their subtle nature, however, assumptions that are injudicious may prove misleading.

When formulating design criteria it is necessary to specify the ground motion that the structures must be designed to resist. As the characteristics of future ground motions are unknown, it is necessary to utilize other information that bears on the problem, for example statistical information on past earthquakes of various magnitudes. If the location and size of future earthquakes were known, ground shaking at the site could be estimated. This knowledge, however, is unavailable, for what is known is the location of past earthquakes. The seismic history of a region of area A conveys some information about the future, but only in the sense that the seismic record of the past N years is a data sample that is more or less similar to seismic events that will occur during the coming N years. The degree of similarity depends largely on the product, NA, there being little similarity for smaller NA and greater similarity for larger NA. It appears that for the seismicity of the 160,000 square miles of California, a value of $NA > 3 \times 10^7$ square-mile-years (a period of 200 years) represents a reasonably high degree of similarity, and $NA < 3 \times 10^6$, a very low degree. The foregoing figures indicate that for regions less seismically active than California, or smaller areas, the seismic history will not bear close similarity to the corresponding future period of years, and in order to draw conclusions it is necessary to assume the shape of the long-time frequency distribution of earthquakes in the area. This assumption permits statements to be made about probability of occurrence. A useful way of making probabilistic statements is to put them in a comparative form, for example, to compare them with the seismicity of the State of California, which is reasonably well known. The probabilistic quantities for a special region can then be compared to those for California; for example, the probability of exceeding ground shaking of a specified intensity in metropolitan Los Angeles is estimated to be p_1, which compares with p_2, calculated using the average seismicity for the State of California. Since much more is known about the occurrence of earthquakes in California than in other parts of the United States, California should be taken as a reference point, and estimates of seismicity for other regions could be better understood if comparison is made with that state.

THE MARGIN OF SAFETY
IN EARTHQUAKE-RESISTANT DESIGN

Earthquake-resistant design, like other engineering design, has the objective of achieving a functional and economical design in the face of imprecise knowledge of the forces that will act upon the structure. Because the loads and the need for a safe design are imprecisely defined, most structures have substantial margins of safety for the loads to which they are subjected in their lifetimes. They are overdesigned in the sense that lack of knowledge about earthquake forces and other loads leads to a resistive capacity that will not be fully used. This situation can be described in statistical terms in which uncertainties about the intensity of earthquake shaking and the degree of earthquake resistance are stated in terms of probabilities of occurrence and probabilities of amplitudes of response. Since it is impossible to prove that hypothetical events will not occur, it is not possible to prove that a structure will have a zero probability of failure. The probability of failure can be reduced by providing extra resistance, but it cannot be made equal to zero. In other words, engineers may be able to design earthquake-proof structures, but they cannot prove it.

The situation described above affects the way conservatism is brought into earthquake-resistant design and the way the design criteria are presented to regulatory agencies. In building codes, and in some projects, conservatism in the design is implicit in the sense that the design criteria are established by subjectively taking into account the nature and levels of forces, the types and quality of construction, the properties of materials, and the experience of structures during earthquakes. A determination of the margin of safety of each item is not attempted, and the overall margin of safety is not known.

In the case of major projects the problem should be approached explicitly; each feature of the problem should be examined separately and a decision made concerning the appropriate level. This approach has the advantage that each important aspect of the project is subjected to careful study by knowledgeable professionals, and the chance of overlooking some major point is minimized. However, a difficulty inherent in this approach, unless the project engineer keeps an overall check on the procedure, is the danger of compounding the factor of safety in the sequence of decisions that lead to the earthquake-resistant design criteria. For example, if the geologist is 1.5 times conservative on the capability of faults in the area of the site, the seismologist 1.5 times conservative on the size of the design earthquake, the earthquake engineer 1.5 times conservative on the strength of expected shaking, and the design engineer 1.5 times conservative on the allowable material responses, the final conservatism compounds to 500%. Figure 7 shows the sensitivity of the overall factor of safety to the number of sequential steps and the individual factors of safety. Without some overall assessment of the conservatism, design criteria can become excessively conservative. It is, of course, also possible to end up with deficient criteria if the consultants are all underconservative, but this seems to be a very remote possibility in the present climate.

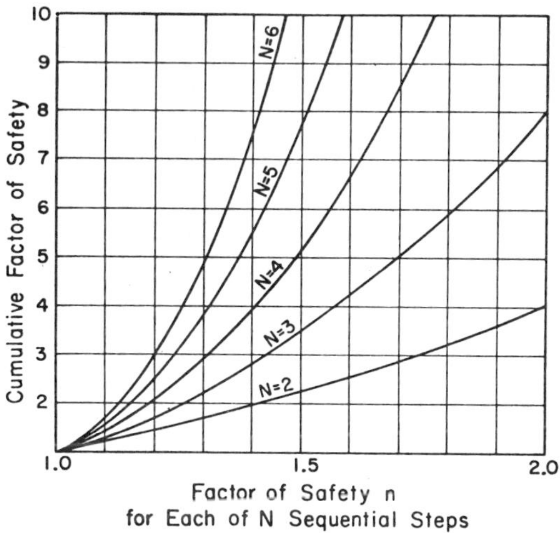

Figure 7 Cumulative factor of safety resulting from a sequence of N steps each having an individual factor of safety n.

Compounding of conservatisms can also occur when the design criteria are reviewed by regulatory agencies or political bodies. In the best circumstances, the review panels are composed of knowledgeable people with access to consultants who are expert in various aspects of the earthquake problem. It is not usual, however, for any single panel member to have an overall view comparable to that of the project engineer, so the review tends to focus on those features of the problem that lie within the experience of the panel members and their consultants, and extra conservatism is introduced at these points without consideration of the conservatism in the other parts of the design criteria. Also, the most obvious way for a reviewing agency to show that a good job is being done is to require an increase in the design criteria. Furthermore, the panel members and the consultants have their reputations at stake to a degree, but are not directly answerable for the cost of the earthquake protection, and in many cases, the hearings and deliberations of the reviewing panel are open to the public, a feature that tends to emphasize the problems over the means and costs of providing solutions to the problems. The outcome of the regulatory process therefore tends to be a very stringent set of earthquake-resistant design criteria, and the end result tends to approach an upper bound of the judgments of all the individual parties involved, rather than a compromise value.

Another factor that should be considered when dealing with regulatory agencies is the need to solve a problem in several ways. This arises because a regulatory body can demonstrate that it is doing its job by requesting information not included in the material under review. Thus, if one method has been used to determine earthquake-resistant design criteria, reviewing panels tend to ask how the result compares to that obtained by another approach and then to require that the most conservative approach be used even though it has not been correlated with other conservatisms in the design criteria. Therefore, the project engineer should consider several approaches and be prepared to explain them, even if they are not used in setting the design criteria.

It is, of course, easier to comment critically on the review of seismic design criteria by regulatory bodies than it is to suggest alternative procedures. One step that would help, however, would be to involve more knowledgeable engineers in the reviewing process. The engineering viewpoint is always well represented on the side of those applying to regulatory bodies, but it is often underrepresented on the reviewing panels themselves.

CONCLUSIONS

A project manager faced with setting earthquake design criteria should keep in mind that the criteria should specify the desired performance of structures under future conditions. Because it is not possible to prove that hypothetical events will not occur in the future, it is not possible to formulate design criteria for zero probability of failure. The project manager should thus expect the recommendations of the geological, seismological, and earthquake engineering consultants to be based on explicit probability considerations. Project managers cannot be experts in geology, seismology, and earthquake engineering, so they must rely on consultants; however, it is essential that they know the proper questions to ask. The following are examples of questions that should be asked:

1. What active faults are located within 50 miles of the site; particularly, what faults are close to the site, and in what sense are they active?
2. What significant earthquakes have occurred within 50 miles of the site? What were their characteristics?
3. What is the estimated frequency of occurrence of future earthquakes of various magnitudes in the general site vicinity?

4. What is the estimated intensity of ground shaking at the site that will be exceeded once per N years? (N may be one or more of the following: 50, 100, 200, 1000, 10,000.)

5. What accelerograms, response spectra, or average spectra are representative of the aboveground motions, in terms of intensity, duration, and frequency content?

6. What would be the consequences to the structures and facilities to be designed of various degrees of overstressing and straining beyond the elastic limits?

7. What would be an acceptable level of damage as balanced against probability of occurrence?

8. What ductility capability should the structure have, as balanced against the cost of providing it?

9. In view of the foregoing, what design spectrum should be used; what design values of damping should be used; and what allowable stresses and strains should be used?

10. What resistive capability will the use of the design spectrum, the design damping, and the allowable stresses and strains actually provide?

Being aware of these questions and their answers, project managers will be in a better position to make the necessary technical decisions and to guide projects through the regulatory processes.

APPENDIX:
Earthquake Magnitude

When a fault ruptures there is a sudden reduction of shear stress (stress drop) at the fault plane which transforms static strain energy in the rock into stress waves. As the stress waves travel away from the fault they produce shaking of the ground surface, whose intensity attenuates with increasing distance. Because of nonhomogeneities in the earth's crust, complex waves are produced, including compression waves, shear waves, Rayleigh waves, Love waves, and so on. These waves approach a site on the surface of the ground from different directions, both in azimuth and in elevation, with the predominant transport of energy being away from the fault. In general, the larger the slipped fault area, the greater the amount of strain energy released and the larger the surface area affected by strong shaking, as well as the "felt area" of ground shaking. Any measurement that characterizes the size of the area of strong shaking, or the size of the felt area, could serve as an indication of the "size of the earthquake." C. F. Richter's earthquake magnitude scale uses, as the pertinent measurement, the peak amplitude recorded by a standard Wood–Anderson seismograph, which has a natural period of 0.8 s, approximately 80% of critical damping, a magnification of 2800, and is located 100 km distant.[2] The peak amplitude, A, of Wood–Anderson seismograms varies over the surface of the ground in a manner similar to the variation of intensity of ground shaking, being small at large distances from a fault and thousands of times larger close to a fault; so for a measure, the $\log_{10}(A/A_0)$ is used, A_0 being a constant. A schematic plot of $\log(A/A_0)$ for an earthquake is shown in Fig. A.1, where it is seen that the contour

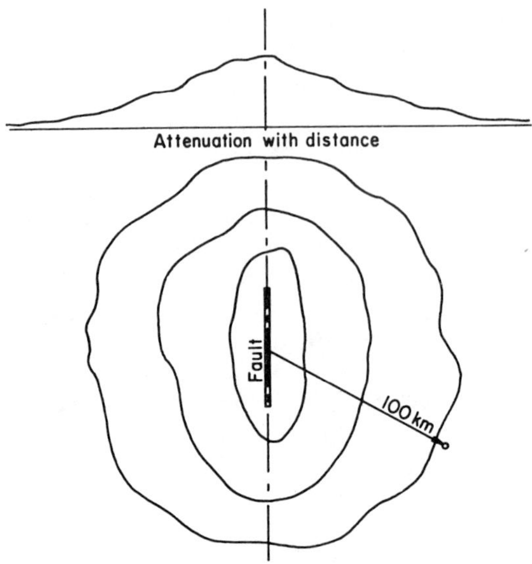

Figure A.1   Contour lines of equal values of $\log(A/A_0)$. The Richter magnitude is defined to be $M_L = \log(A^*/A_0)$. A^* is the maximum amplitude of a Wood–Anderson seismograph at 100 km. A_0 is the amplitude corresponding to $M = 0$.

lines of constant values are rather irregular oblong curves. The plot of $\log(A/A_0)$ forms a hill-shaped surface and it is clear that the volume of the hill,
$M_v = \int_{-\infty}^{+\infty} \int_{-\infty}^{+\infty} \log(A/A_0)\,dx\,dy$, although a good measure of the size of an earthquake, would be impractical to evaluate. A less precise, but more practical measure was defined by Richter:

$$M_L = \log(A^*/A_0)$$

$$A^* = \text{amplitude at 100 km}$$

$$A_0 = 10^{-3}\,\text{mm} = A^*\ \text{for}\ M = 0$$

---

[2]This instrument can be compared with the standard seismoscope, which has a natural period of 0.75 s and 10% of critical damping, and could also be used for magnitude determinations.

Actually, modern seismographs are not Wood–Andersons and are not 100 km from the center of the earthquake, but seismologists can correct for instrument characteristics, and for distance, to obtain an equivalent M. Because of the noncircular shape of the contour lines of $\log(A/A_0)$, two seismographic stations will not, in general, compute the same value of M_L, and the "official" value is usually the weighted average of several. A more stable measurement would be one based on the spectrum of the seismogram rather than on the peak amplitude, but the work involved renders this impractical. Owing to the oblong shape of the contours close to the fault, it is desirable to measure A at a distance that is large compared to the fault length; however, at different distances, different seismic waves are predominant in the seismogram. For example, within a few miles of the fault, the ground motion recorded by a strong-motion accelerograph is predominantly short-period shear waves and compression waves, from which a magnitude can be computed. At a distance of several thousand miles from a large earthquake, surface waves of 20-s duration are prominent, from which M_S can be computed. At least four different magnitudes are, in practice, computed, and they do not, in general, give the same numerical value, although there are techniques for converting from one to another. For example, a strong-motion accelerograph close to the center of a $M_S = 7$ earthquake fault (40 miles long) could be expected to record shaking of approximately the same intensity as for a $M_S = 8$ event (200 miles long), other things being equal, so in this case the magnitude based on strong-motion records could not distinguish between $M_S = 7$ and $M_S = 8$. For each type of magnitude designation, the ability to distinguish between two sizes of earthquakes is at some point in the magnitude scale lost. Unfortunately, it is not always made clear which magnitude is being used, and this can lead to confusion.

For engineering purposes, the magnitude can be taken as an approximate measure of the size of the earthquake, that is, the area affected by strong ground shaking.

REFERENCES

[1] F. L. Fox, and C. T. Spiker, "Intensity Rating of the Attica (N.Y.) Earthquake of August 12, 1929," Earthquake Notes, Eastern Section of Seismological Society, vol. 48, nos. 1 and 2, 1977.

[2] D. K. Jephcott, and D. E. Hudson, "The Performance of Public School Plants during the San Fernando Earthquake," Earthquake Engineering Research Laboratory, California Institute of Technology, 1974.

STRUCTURAL DESIGN
AND CONSTRUCTION IN BUILDINGS

*Yoshikatsu Tsuboi**

INTRODUCTION

Structural design depends largely on the way in which each structure is to be used, just as architectural design depends on the function of each building. The discussion in this paper is restricted to large-span spatial structures and to high-rise buildings.

Throughout the paper the idea of structural expression is maintained. One could say that structural expressionism is dead if by that statement one were referring to the fashionable trend that lasted until a few years ago, in which many architects employed structures to realize their subjective ideas.

The term structural expressionism as we use it here differs from that. It is what we find in the earliest modern works achieved by a few pioneering engineers and architects in the nineteenth century, and what some of our contemporaries have been seriously and continuously pursuing independent of fashionable trends.

---

*Dr. Eng., Prof. Emeritus, University of Tokyo.

STRUCTURAL DESIGN
AND ARCHITECTURAL DESIGN

The author understands structural design to be an engineering operation to foresee how rationality, safety, workability, and economy of construction will be used to satisfy particular goals. Restricting our discussion to building structures, we should be aware of the importance of architectural design. In many cases architectural design precedes structural design, which usually starts after the total image of the structure has, to some extent at least, been determined by the architect. There are many cases, however, in which structural design is on its own, encouraging creativity to establish the basic image of a building and exerting a vital influence over the building's plan and elevation.

Like the outcome of any other human endeavor, a building is a product of history. It takes form by traditional means as well as by the introduction of modern technology.

Architectural styles have gradually become international. Once it came to be generally believed that architecture could be a better form of spatial expression than sculpture could be, structural design assumed an important role in the work of most architects. In the nineteenth century, engineers were aware of this concept far in advance of architects. Knowing the opinions of those who fear that structural design may remove the humanity and all religious sentiment from architecture, the author feels that the structural designer should sometimes get ahead of architects—that, indeed, designers should play an important role in guiding and assisting architects to create rational and realistic structures. Structural designers should have a primary part in the architectural process rather than working in secret or being forced to rely on what may, as a result, be irrational engineering.

ARCHITECTURE OF STRUCTURAL EXPRESSION

The late genius of architecture, Eero Saarinen, wrote in his autobiography:

1. "Our architecture is too humble. It should be prouder, more aggressive, much richer and larger than we see it today."
2. "To express structure is not an end itself. It is only when structure can contribute to the total and the other principles that it is important."
3. "I align myself humbly with Le Corbusier and against Mies van der Rohe. . . ."

The first remarks imply a need in architecture for structural engineering and structural design, which does not necessarily mean only the development of methods of calculation. The second statement does not diminish the structural contribution to architecture; on the contrary, it cautions both architects and engineers to be more aware of the great extent to which engineering actually affects architecture, whether or not it is intended.

The implication of the third remark depends largely on one's interpretation of structural expressionism, thereby making general commentary regarding it very difficult. However, it can be appreciated as a straightforward remark by an architect who values the dynamic over the static, and whose own excellent work reflects this belief.

The author does not necessarily believe that the tenets of structural expressionism are universally understood or accepted. But whether they are or not, structural design should be carried out coolly, yet boldly. The very nature of structural design when compared, for example, to decorative design, necessitates that it reflect the scientific and engineering level of the day. In every case, structural design is far more responsible than is architectural design for the final product and for the degree of safety in its use.

The mitigation of hazards is an important feature of structural design. In Japan, much study in this area has been conducted in connection with earthquake-resistant design, which will be discussed later. The question the author wishes to raise now is how, and to what extent, this feature, which is negative in nature, is incorporated into the creative process of designing new structures and environments.

SHELL CONSTRUCTION AND SPATIAL STRUCTURES

It may be said that the architecture of structural expression began to attract attention with the theoretical and experimental development of shell structures. In the beginning of the century, engineers such as Dischinger and Finsterwalder pioneered the technology of reinforced concrete cylindrical shells. It was the late Eduardo Torroja who elevated the design of shell structures to an art. The development of large-span structures that followed presented abundant opportunities for unifying structures and their means of expression.

Concrete shell construction contributed greatly to establishing the importance of structural design. The engineer's experience with the three-dimensional behavior of concrete shells stimulated the development of steel space frames, as well as spatial structures of two-way prestressed cables, when the "hanging technology" involved in bridge design was extended to higher dimensions.

From the viewpoint of an engineer who has designed many such structures, it appears doubtful that architects and engineers have always been correct in understanding and approaching structures in the course of such development. The author's primary fear is that some designers may have adopted structural expression for capricious or fashion-conscious reasons. Second, it is feared that they may have been blinded by seemingly limitless spaces and deluded into attempting an art they were poorly trained to deal with, disregarding such realities as the limitations of materials and contemporary techniques. An even stronger fear has been expressed by Professor Leonhardt, who warns that structural engineers today are in constant danger of becoming "computing slaves" under the control of architects.

There are two ways of approaching spatial structures. One is a functional approach, in which the possibility of applying spatial structures to artificial environments is pursued. A major structure which serves free and diverse internal spaces on an urban scale is one such example. Engineers' experiences in designing long-span bridges, auditoriums, stadiums, and hangars have prepared them for this new direction. Such experiences may also have convinced engineers of the superiority of tensile structures over others.

The second approach is technological and may involve, for example, the possibility of prefabrication and assembly. When speaking of covering a certain area by means of, say, a reinforced concrete dome, or of macroscopically dividing a space into exterior and interior zones, one is taking the first approach, i.e., the structure is viewed as a continuous medium. When the image of the structure is somewhat clearer to the designer and space is to be constituted using a structural pattern, such as with steel frames, trusses, or concrete panels, the second approach is taken, i.e., the structure is regarded as the object of the uses that have shaped it.

The definition of structural design as an engineering operation whose goal is to foresee the rationality of construction requires that the engineer grasp the general behavior of not only the continuous parts of a structure, but also of their connecting details and the relation of these connections to the surrounding parts of the structure. In this sense the two approaches discussed above are very often unified in the design operation.

ORDER IN DESIGN

In discussing the relation of architectural subjects to structural design, one must consider the architectural desire for aesthetics on one hand, and its resistance to functionalism on the other. The modern architectural style of steel structures owes much of its existence to Mies van der Rohe and the school of expression he founded, which dealt in the art of steel and glass buildings. His idea that "a building and its environment should be one and continuous to each other"[1] released space from the enclosure of stones and bricks.

This philosophy was realized in his design for Farnsworth House, a design that was later to evolve into the style of his skyscraper buildings. It was not surprising that controversy attended this radical new art of steel and glass construction. When the author visited the construction site of Wayne University in Detroit in 1959, he watched as a reinforced concrete wall panel was affixed to a steel frame. It was an example of the way in which the architect, M. Yamasaki, dealt with curtain walls of a type opposite to that used by van der Rohe. Structural expression has been approached in a wide variety of ways—not only in the treatment of curtain walls but in the structural system itself—in combination with functionalism and not a few times with commercialism, all of which may have stimulated improvement in structural technology.

The John Hancock Building in Chicago and the Federal Reserve Bank in Minneapolis are examples of a straightforward expression of structural principles. The elegant exterior and rich interior of the I.D.S. Center, designed by Philip Johnson, are unified by means of a deformed space frame; the structure can be regarded as an attempt to develop the functional prototype created by Mies van der Rohe into an art form of its own.

Professor Torroja once made a philosphical inquiry into the structural nature of columns, in which he characterized their expression by assigning to them the role of endurance. In reality, however, it is rather difficult to conceive of columns in line with Torroja's statement. Wide flanges or box sections (most popular in Japan) of columns are extremely exaggerated in size by coverings that provide for fire resistance and finishing, and they are hardly given opportunities to "express themselves."

When we think of structural expression, we must be aware of those factors that are not controlled by the principles of structural mechanics. Modular coordination is one such factor. Others besides Le Corbusier, who presented his own design module, have long since tried to establish a criterion for "beauty." Some of the conditions proposed are the golden number, whirling squares, and the Fibonacci series, which is known as an approximation of the golden ratio.

A module that dictates the locations of columns, for example, not only produces a controlled rhythm of expression, which is considered essential to satisfy a condition of beauty, but, in obtaining repetitions of structural

---

[1]As L. Hilberseimer has said: "To form this contained or enclosed space, as well as the shell which enclosed it, to relate it to outside space by placing it in unformed, limitless space, is one of the basic architectural problems which Mies van der Rohe has solved in an unprecedented manner."

patterns, makes realization of the building rational, easy, and economical. In this way a good module contributes greatly to both order in expression and rationality in construction.

While designs for many skyscrapers have adopted well-controlled modules to produce a style of modern buildings, P. L. Nervi created, on the basis of highly realistic structural design, an order that he successfully applied to many of his buildings. Yamasaki succeeded in expressing his ideas of structure when he repeated hyperbolic paraboloidal shells in Dahran Airport to create an orderly and moderate building on the desert by means of modern technology and yet in line with Muslim style.

VARIATION IN DESIGN

Although order and repetition are useful in the design process, they sometimes produce useless internal spaces and a monotonous external atmosphere. To prevent this, variation in design is often attempted to make the spaces more dynamic, less static. In modern architecture, which excludes decoration, this is sometimes achieved only by means of structural variation.

When structural and architectural designs fuse and enrich each other, the resulting form of well-controlled repetition with vivid contract inspires us with a feeling of "calmness of dammed up force flow and its rushing power where it is released."

Excellent works of this kind were seen in the developing stage of shell construction. Spherical and cylindrical shells are inherently stable and create a static atmosphere. When stress disturbances around shell connections to the boundary structure result in bending moments and transverse shears, the use of such disturbances may be welcomed as lending dynamic movement to the shell forms, even though membrane states are structurally essential in major parts of thin shells. Spherical and cylindrical shells are sometimes cut or supported along lines that differ from those of the surface curvature to provide suitable cover. Stress disturbances along such lines can also be a useful means of making simple geometric forms pleasing to the human eye.

Saarinen's Kresge Auditorium at MIT, which has a spherical shell cut along great circles, and Yamasaki's St. Louis Air Terminal, composed of intersecting cylindrical shells, are examples of how, despite some analytical complexity, excellent boundary treatment can invigorate plain geometric structures. Using sound visual judgment, Torroja altered the curvature of the geometric hyperbolic paraboloidal roof surface of the grandstands at the Madrid horseracing track and in so doing, converted a static hyperbolic paraboloidal geometric form into a dynamic architectural surface. This suggests that at times, the optimum creation exists as a deviation from mechanical perfection.

Jørn Utzon's finished drawings for the design of the Sydney Opera House show, for the roof surface, a cluster of two-dimensional Riemann spaces (parts of a sphere) expressed as two kinds of orthogonal curvilinear coordinates. This design, in which the oblique coordinates complicate the metric tensor, is of a type that does not permit analytical treatment. In shell theories, tensor equations are normally simplified in the coordinates of the lines of curvature. This fact notwithstanding, Utzon's technique, which uses oblique coordinates to transform a simple sphere into a cluster of beautiful lotus flowers, should be evaluated as a model for designing buildings of this kind.

About 1950, when shell construction was at its most prolific, anticlastic surfaces were applied to two-way prestressed network roofs, and hyperbolic paraboloidal surfaces were applied to shell structures.

The Olympic Indoor Stadium in Tokyo (see Appendix 1), designed by Tange and the author, is covered by a suspension roof. Unlike suspension bridges, however, the two main cables of the roof had to be "opened" in the central span to allow space for a skylight. This gave us the opportunity to create a very special device, called "rotating saddles," on top of the main columns, to facilitate opening the main cables during construction.

Another feature of the stadium was that it provided the possibility to design "semirigid" structures. The plan of this stadium consists of a circular main area for two swimming pools, a set of grandstands, and two access areas formed by two tangential lines to the central circle, antisymmetric to each other. It was readily seen that neither of the twin roof surfaces designed between one main cable and the circumference of the plan above could be realized effectively by a cable net of negative Gaussian curvature. Because the main cables were convex downward and the boundary upward, the hanging members, demanding a downward curve, would produce a surface that possessed part of the zero Gaussian curvature and on which bracing cables could not exert their stiffening effect. It was also found that a surface thus obtained was neither visually exciting nor aesthetically appealing. The solution to these problems was to provide the hanging members with bending stiffness. These members were built-up steel sections having depth-to-span ratios of about 1:100, and were three-hinged to absorb deformation during construction.

The purpose of the semirigid structures thus obtained was three fold: consistent formation of a surface that partially lacked the condition of negative Gaussian curvature for prestressing; ample flexibility in architectural form; and effective rigidity against partial deformation due to wind to which cable networks are not very resistant unless highly prestressed. This example indicates that variation in boundary structures demands a number of spatial constituents and structural technologies.

IDEALIZATION OF STRUCTURES
AND PRINCIPLES OF FRAMING

Structural design functions to develop insight regarding the substance of a structure. It is thus the responsibility of structural design to recognize the outline of stress flow around the connections and corresponding structural mechanisms, as well as the required dimensions of the continuous parts of the structure. This recognition can be regarded as preparation for the architectural expression of the structure.

The following two examples are illustrations of this principle.

Spherical steel dome of Tokyo International Exhibition Hall (1959)

Modeling and simplification of a structure are two of the most important steps in the structural design process. In this example (see Appendix 2), substitution of a continuum for a latticed spatial structure and simplification of the boundary conditions are shown to be sufficient for estimating stresses in the truss members.

The plan of circumference and the opening were represented by means of bipolar coordinates, by which the location of the truss members were also determined. The spatial pattern thus obtained was then modified to reduce variation in the length of members. For the analysis, the structure was represented by a continuum based on the bipolar coordinates. This continuum was then analyzed for a rectangular plan referred to x, y coordinates by means of the finite difference method. The solution was obtained by manual calculation because computers were not available at that time. A detailed analysis was made afterward.

In the original manual calculation for meshes of 3×4, the numerical errors totaled less than 10% compared to recent results for finer meshes of 6×8, and was sufficient to estimate the stresses in the steel shell of a double-layered grid truss.

Structural plan for hotel in Iran (construction in progress)

In the design of a multistory building, it is important to minimize torsional deformation due to horizontal earthquake forces. This is because torsional deformation due to concentrated shear forces at some parts (particularly at the corner) of a building is attended by a loss of uniformity in column sections and increased economic liability.

Shear walls can be placed so as to adjust the sections of the remaining parts of the frame, to control lateral deformation of the building and to regulate the lateral stiffness of the structure under seismic loads. If shear walls are properly arranged, lateral deformation caused by earthquake forces in a rigid frame structure can be adjusted in such a way that the building's torsional component is sufficiently small for design. In this kind of structure, the rational arrangement of shear walls according to the building's functional requirements determines the structural soundness of an earthquake-resistant design.

In this example (see Appendix 3), shear walls were arranged as shown so that the centers of rigidity and gravity would be located as close to each other as possible in the sense stated above. In a three-dimensional frame analysis, the 800 joints and 2800 members were subjected to static lateral loads equivalent[2] to 100 gals, which gave a lateral displacement of 35.0 cm at the top floor in the direction of the applied loads and a corresponding maximum displacement, caused by torsional deformation of the structure, of 9.3 cm in the transverse direction. If the shear walls had been rearranged, the torsional deformation could have been made smaller still, but the result discussed above sufficed for the first stage of design.

Dynamic analyses were also carried out on the structure for the following simulations:

Simulated earthquake:	N–S: El Centro
	E–W: Taft
	Pacoima Dam
	Parkfield
Input acceleration:	100 gals for elastic response
	(flexure-shear model)
	300 gals for elastic–plastic response
	(equivalent shear model)

PRINCIPLES OF EARTHQUAKE-RESISTANT DESIGN

Rigid structures

Between 1955 and 1976, the Japan Housing Corporation built a number of four- to eight-story reinforced concrete apartment houses, comprising some 1 million dwelling units. The structure of most of these houses was designed using the wall-structure system, a method that had been recommended by

---

[2]A lumped mass model with lateral eccentricities was subjected to simulated earthquake motion of 100 gals and the resulting lateral loads were statically applied to the three-dimensional model.

the author and his group for the design of fire-resistant dwellings. The system which excludes the use of columns, consists solely of cast-in-situ reinforced concrete walls and floors. More recently, however, some attempt has been made at prefabrication.

Following are some fundamental factors used in the design of the system:

Wall thickness	15–18 cm
Wall ratio	225–270 cm²/m²
(wall length)	15–20 cm/m²)
Weight (including superimposed loads)	1.0–1.2t/m²
Base shear coefficient	0.2
Mean shearing stress	3.1–3.9 kg/cm²

Full-scale as well as reduced model experiments were carried out to establish the standard specification of the system. Through experiments the standard house based on this system proved capable of withstanding a lateral force corresponding to a static base shear coefficient of more than 0.5. Destructive earthquakes experienced by this type of house occurred at Niigata in 1964 (photo, Appendix 4) and off Tokachi in 1968.

In the latter, no damage was observed in wall-structure buildings, but many reinforced concrete framed structures suffered irreparable damage. During the Niigata earthquake, some houses of the latter type inclined or even overturned, owing to liquefaction of the supporting sandy ground, but the structures themselves suffered no damage and there was no loss of life.

For this type of very rigid structure, the natural periods of the super-structures are not very important because deformation and vibration of the ground dominate and a rocking vibration prevails. The formulas[3] for standard natural periods and base shear coefficients are

$$T_s = (0.06 \sim 0.10)N \quad \text{(0.10 for steel, 0.06 for others;}$$
$$N \text{ is the number of stories)}$$

$$C_B = \frac{0.15}{T} \sim \frac{0.30}{T} \quad \text{(moderate} \sim \text{destructive)}$$

These formulas are clearly restricted to very rigid structures having a small number of stories.

Response spectra for low- and middle-rise reinforced concrete structures generally show undesirable features when used with a damping factor of 0.03, the factor most often used for the first-mode vibration of reinforced

[3]Recommendation for Highrise Building Design, A.I.J.

concrete structures.[4] The principle of structural design for this kind of structure is as follows:

> Reinforced concrete members should be designed to withstand the greatest possible ultimate deformation, so that the whole structure is sufficiently ductile to absorb as much energy as possible during earthquake vibration. This is especially important for column design. Shear walls with strong frames along their peripheries also contribute substantially to this absorbing effect in their cracked states.

The concepts that the dimensions of frame members should be smaller so as to lengthen the natural period of the frame to reduce its response to earthquakes, and that shear walls are harmful because they are too stiff, are not sound and should be dismissed in the design of reinforced concrete structures.

Such "tricks" in seismic technology do not work in comparatively low reinforced concrete structures of considerable rigidity.

High-rise buildings

The 1977 recommended draft for earthquake-resistant design of high-rise buildings (30 to 75 m above ground) issued by the Building Center of Japan states:

1. The plan of a structure shall be close to rectangular and such that torsion will be low under seismic effects.
2. As a rule, the upper stories should be constituted by repetition of the same configuration of plan.

The intent of these statements is to minimize torsional vibration.

The first statement recommends that the center of gravity (mass) and the center of rigidity (similar to a shear center) coincide with each other on each floor, while the latter controls "eccentricity" (as it is called in vibration theory), which occurs in a building with setback. It is not difficult to estimate

[4] 1. Project for Housing Corporation under investigation at the Tsuboi Laboratory. See Appendix 5.

2. In 1976, when an 18-story building with a two-story basement experienced a moderate earthquake, the installed SMAC's recorded maximum accelerations of 61 gals at the B-2 floor and a value as high as 297 gals on the top of the building.

$$\begin{array}{lll} (elastic) & & (elastic\text{–}plastic) \\ T_1 = 0.87 \text{ s} & T_2 = 0.27 \longrightarrow & 1.030, 0.496 \ (h = 0.07) \\ T_1 = 0.78 & T_2 = 0.28 \longrightarrow & 0.964, 0.454 \ (h = 0.06) \\ T = 0.77 \text{ s (torsion)} & & \end{array}$$

Because of the abundance of shear walls, T is small (rigid) in comparison with $T_s = 0.06 \times 18 = 1.08$ s.

statically the effect of torsion, but it should be noted that static analysis underestimates torsional angles. Calculations for a single-story structure suggest that static eccentricity should be increased by at least 50% for a suitable dynamic equivalent (Appendix 7). Moreover, the response stresses vary with the direction of the ground motion up to 1.2 times the stresses due to the motion in the principal directions of the building.

Recommendations drafted by the Ministry of Construction facilitate the application to high-rise buildings of simplified design methods (Appendix 6) which are similar to those of the SEAOC. The base shear coefficients adopted for 27 buildings in Japan designed between 1963 and 1967 are 0.06 to 0.24, with natural periods of 0.71 to 5.10 s.

Simplified static design methods can hardly be applied to buildings that deform with torsion, especially those buildings whose upper stories do not repeat the configuration of lower stories. In such cases, three-dimensional lumped mass models must be used, making it difficult to discriminate between the periods of shearing and torsional vibrations.

However, the difficulty of understanding torsional vibration should not restrict the engineer's freedom in planning. On the contrary, engineers should be interested in this problem for the sake of producing flexibility in architectural design. The author's own design experience shows the adequacy of structures evaluated using response analysis and vibration experiments (see Appendix 8).

WIND- AND EARTHQUAKE-RESISTANT DESIGNS: CREATION, INVENTION, AND IMPROVEMENT

Although structural design is usually based on findings from previously designed analogous structures or on codes or recommendations, true structural design should be creative and inventive and work to improve conventional technology. The following examples may serve to clarify this idea.

Oil dampers for wind-resistant design

In the design of hanging roofs, which are generally light and flexible, the aerodynamic effects of wind are extremely important. The natural periods of the roof structure for the Olympic Swimming Pool Stadium in Tokyo were obtained from $\frac{1}{30}$-scale model experiments:

1. 0.89 s (0.42).
2. 0.86 s (0.67).
3. 0.52 s (0.18).
4. 0.48 s (0.24).

(Figures in parentheses are for the actual building and are shorter because of the shell action of the roof.)

A wind tunnel test on an idealized one-way model suggested that the structure was aerodynamically stable, but, in view of the closeness of the natural periods in the first and second modes, it was felt that provisions should be made against unanticipated wind effects. It was therefore decided to design a damping system (see Appendix 1) of 12 oil dampers, each with a stroke of 250 mm. The damping factor of the system thus obtained was 0.44. Six of the dampers were mounted near the top of each main column and were visible from the outside, thereby participating in the architectural expression of the whole building.

Invention of slit shear walls and braces with variable stiffness

In earthquake-resistant design, the nature of a structure should generally be thought to vary for moderate and destructive loads, that is, loads of over 200 gals, with 400 gals being the most destructive. Against moderate earthquakes a structure should be rigid for small displacements, and under destructive earthquakes it should be soft enough so that its natural period will be longer than the characteristic site period. This idea has been incorporated into the design of reinforced concrete shear walls, which become ductile even after cracking takes place, thus withstanding a large deformation along with the remaining parts of the frame.

SLIT SHEAR WALL. The slit shear wall (Appendix 9) is also a reinforced concrete shear wall, but differs from conventional walls in being slit. A conventional shear wall yields in shear with cracks caused by diagonal tensioning, but a slit shear wall will yield in both bending and shear.

The latter type is superior to the former in deformability or ductility, and has comparatively less rigidity in both elastic and elastic–plastic regions. The collapse of ordinary shear walls is brittle and such walls lose their capacity for lateral resistance. The slit shear wall is thus based on the idea that a shear wall should not be overly stiff.

This concept of slit shear walls was conceived and developed by K. Muto, who has devoted himself to the establishment of earthquake engineering in Japan using observation records of destructive American earthquakes such as El Centro and Taft, and has also been involved in the development of vibration engineering.

BRACING SYSTEM WITH VARIABLE STIFFNESS. Braced frames (Appendix 10) are similar to conventional shear walls: their rigidity is extremely high compared with that of other parts of a framed structure, and their deformability is difficult to preserve unless buckling and breaking of the bracing members are prevented.

Steel or concrete devices have been contrived by which braced frames are modified. Presented here is a system developed by the author that utilizes steel because of its excellent ductile properties in shear and bending. Desirable stiffness for the system can easily be developed through proper choice of the length l. This technique as well as that of the slotted shear wall, enables the design of a framed structure that will remain elastic even during a destructive earthquake, and that can be easily repaired after such an earthquake. The rigidity of both the slit shear wall and the bracing system described above can change according to the intensity of an earthquake.

Structural planning, that is, the location of columns and shear walls, is one of the most important stages of structural design for determining the static and dynamic characteristics of the structure.

Shear walls that play a part in controlling the character of a building and in providing it with structural integrity demand positive yet careful treatment in structural design, however free its configuration in plan may be.

Innovations in connections of trusses and frames

For structural expression as well as for the structure, connecting points, or joints, of steel structures are never less important than their connected members. This is especially true in a space frame, where numerous members come together at a joint, which plays an important visual role as the "collecting center" of forces, even if the three-dimensional stress states in the joint are intuitively difficult to understand.

Neither visual nor productional judgment allows excessive variation in the types of joints. For a number of reasons, joints generally possess higher safety factors against yielding and fracture than do the connected linear members. A well-designed joint is thus adequately uniform in shape and size, without giving the impression of being overdimensioned. It can be regarded as a crystallized concentration of resisting forces acting against external agencies. A series of joints for space frames developed many years ago by K. Wachsmann recalls this.

New designs of connections or joints constantly appear. Many joints with special features have been exhibited at the two conferences on Spatial Structures organized by Z. X. Makowski. Even more possibilities have been attained with developments in welding technology.

From the standpoint of structural expression, joints are constituents as essential to a spatial structure as the connected members, and should be designed to be aesthetically pleasing, or at least to appear not unnatural in form.

JOINTS FOR TRUSSES. This example is limited to the soon-to-be-dismantled grand roof structure designed by the author and M. Kawaguchi for the Festival Plaza of Expo '70. Today, a structure of this scale can be realized in many ways; the author took the opportunity of designing the exposition structure to investigate the following problems.

1. Can a large structure be better realized by large members and large joints?
2. Can such a structure produce an image that will prefigure future megastructures?
3. Can a very simple double-layer grid frame be fused into nature?
4. To what extent can such a structure be industrialized?
5. Can the structure be designed to be lifted and lowered (for disassembly) as a whole?
6. Can the design accommodate provisions against winds or earthquakes during construction?

Cast steel ball joints invented for the structure resolved most of these problems. The joints have a fully mechanical connecting system that can be operated from the outside and can absorb various types of errors anticipated during assembly.

Purity of truss mechanism was pursued in the design of the whole structure, allowing loads to be applied only on the joints. This principle gave the structure clarity of expression, which contrasted well and was in harmony with the environment. This principle of clarity of design was further extended to the absorption of thermal deformation of the roofing substructure (expandable beam) as well as to the lifting system (mechanical equalizer). Transparent pneumatic structures of plastic skin were realized on a large scale for the first time as a roofing system for the space frame.

RIGID CONNECTIONS IN FRAMES. In the course of researching and developing wall-type reinforced concrete housing, the author experimented on panel zones under shearing forces by means of photoelasticity. With the increasing number of designs for high-rise structures, research such as that done by L. Beedle on shear lag in panel zones of steel framed structures has drawn attention to the importance of panel zones in earthquake-resistant design. The author has long maintained that connections of steel frames can be better achieved by means of cast steel.

The following summarizes the problems encountered in experiments on present joints and emphasizes the advantages of using cast steel joints (and see Appendix 12).

Problems:

1. Welding is difficult between rolled members and cast steel joints which contain higher percentages of carbon.
2. Quality control of cast steel is difficult to assess.
3. Such joints cannot easily be mass produced.

Advantages:

1. Important elements, such as diaphragms, can be integrated without difficulty.
2. Cast steel joints are free of brittleness caused by heat, unlike built-up joints; distinct plastic flow is observed in panel zones.
3. Such joints are also free of potential problems caused by lamination in the connection of a penetrating column.
4. The construction period can be reduced, as was true in the present case.

The preceding study and development was undertaken with the co-operation of S. Aoki and Y. Suenaga.

CONCLUSIONS

In this report the author has described structural design based on his experiences. In doing so, he has restricted his discussion to large-span spatial structures and to high-rise buildings. The object of structural design is to create an artificial environment that is safe, sound, and rich in every sense. Its achievement requires sensitivity on a large urban scale. Of techniques, there are many: order and variation in space, repetition and contrast in expression, feeling of forces, and so on. Of course, this entails having strong supporting data and the fundamental knowledge necessary to utilize them. The most powerful data can often be obtained after the design of a structure is completed. In any particular design, it is common for some kind of approximation to be adopted in analysis. Yet, a postdesign study in which attempts are made to refine the design procedures and to understand more fully the behavior of the designed structure very often provides supporting information that lends insight into the nature of structures to be designed in the future.

The most important requirement, however, is a progressive attitude toward creation and the unyielding pursuit of relevant problems through continued research, all of which is in proportion to aesthetic sensibility.

APPENDIX 1: Structure of Swimming Hall
for Tokyo Olympics (1964)
(by K. Tange and Y. Tsuboi)

A system similar to that of a suspension bridge is employed in the roof structure of the swimming hall. Two main cables span 126 m between the two main columns and 65 m (visually, 44 m) outside them in the longitudinal direction, with the sag of the cables in the center span being 9.653 m. The cables are parallel for the side span at an interval of 2.580 m, but the interval is gradually widened to 16.800 m for the center span to provide a space for skylights and artificial lights. Each of the main cables, 330 mm in external diameter, consists of 31 ropes of 52 mm diameter and 6 ropes of 34.5 mm diameter. The design strength of the steel wires constituting the ropes is 150 kg/mm² (actually, 165 to 170 kg/mm²), and the apparent Young's moduli of the ropes are 16,000 to 17,000 kg/mm².

A series of hanging members spans the area between the main cables and the boundaries surrounding the seats. The space between adjacent hanging members is 4.500 m. A series of bracing cables that runs approximately along the geodesic lines of the roof surface penetrates these hanging members. After the roofing was in place, the bracing cables were stretched to a tension of 20 tons per cable. Then all the members of the roof structure were stressed, which considerably raised the rigidity of the structure. After stressing, the tensile force in each of the main cables was 1350 tons and the stress in the component wires was 25.1 kg/mm².

Therefore, the safety factor for the main cables is about 5. Safety factors for the cables of suspension bridges are usually taken to be about 3, but as the object of this design is a building, a larger section was preferred in order to obtain higher rigidity. The tensile forces in the main cables produce about 2000 tons of compression in each of the main columns and descend through the backstay cables down to the anchor block to produce an uplift force of 1270 tons in the anchor block and a compressive force of 2380 tons in the struts. The weight of each anchor block is about 2800 tons. If the weight of the soil just above the base slab of the anchor block is considered, the total weight, about 4900 tons, is sufficient to resist the uplift force. There are two concrete struts with a cross section of 1.50 m × 2.00 m in the side spans, but in the center span the observation tunnels running along the sides of the swimming pools serve as compression struts.

In the first stage of design, it was proposed to use ropes not only for the main and the bracing cables, but also for the hanging members, with stiffening steel sections attached along them. As described later, however, it was concluded that the sharp roof surface desired by the architects would prove uneconomical if this shape were attempted only by adjusting the tensioning in the cable network. Therefore, it was decided to forego the use of ropes and to design the stiffening steel members to take the additional part of the hanging members. Steel members designed in this way combine three roles: they are essentially hanging members; they have enough bending rigidity to maintain their shapes that differ considerably from catenaries; and they act as stiffeners to prevent deformation due to partial loading. These I-shaped steel members, or hanging girders, have flanges of 22 mm × 190 mm and webs 12 mm thick, vary in depth from 500 mm to 1000 mm, and are set at 4.500 m, center to center.

The bracing cables that run through the hanging girders are ropes 44 mm in diameter. Their same mechanical properties are identical to those of the main cables. They are arranged along the approximate geodesic lines of the roof surface at intervals of 1.500 to 3.000 m and are clamped to the hanging girders at every intersection after stressing.

Roofing steel plates 4.5 mm thick are bolted to the light-gauge steel rafters fastened to the hanging girders at intervals of about 1.5 m. Owing to the importance of the building, the main cables are provided with oil dampers in order to absorb vibrational energy due to unexpected violent winds.

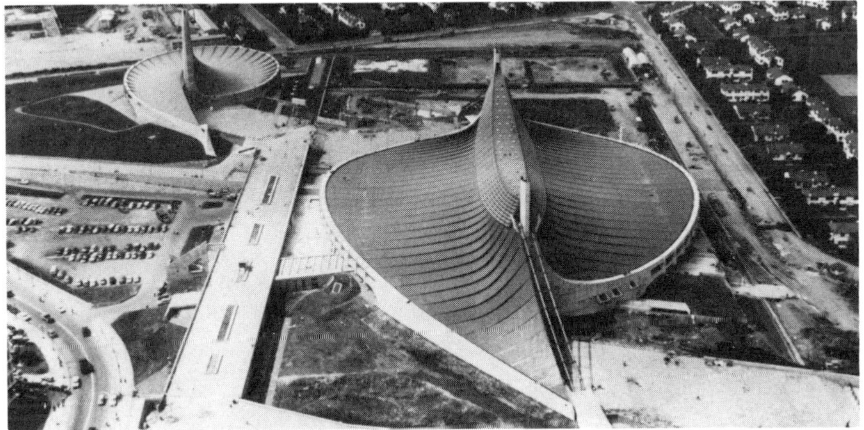

Photo 1.1   Aerial view of stadium.

Photo 1.2   Exterior view of stadium.

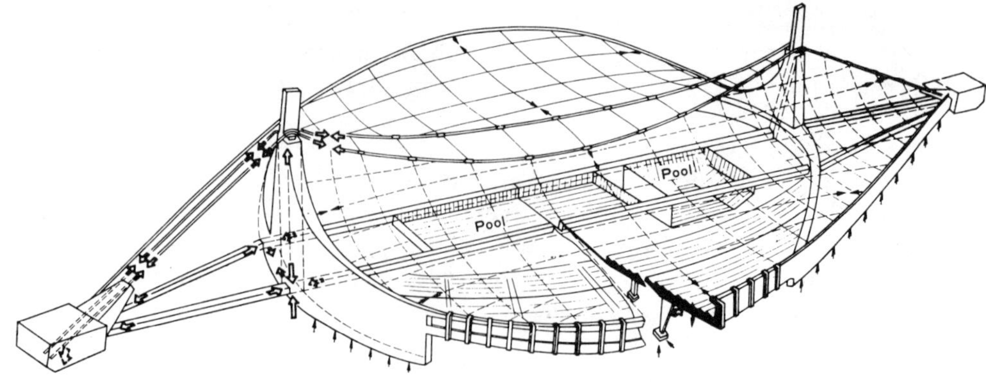

Figure 1.1   Schematic structural system.

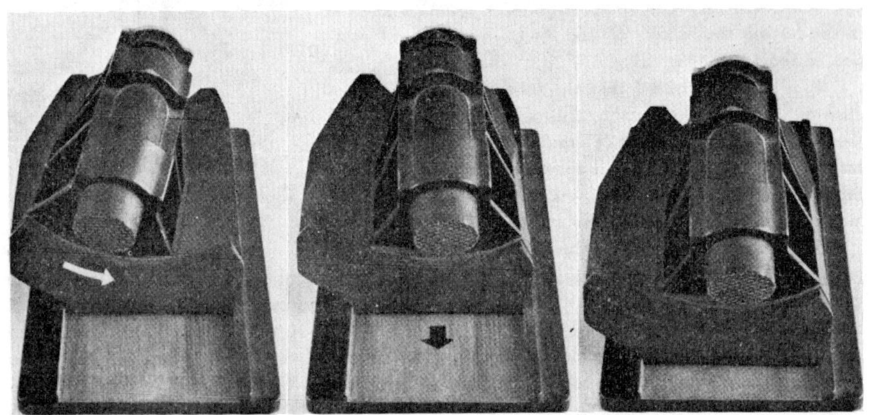

Figure 1.2   Operation of saddle.

Mode Shapes	Natural Periods (sec)		T_{real}/T_{exp}
	T_{exp}	T_{real}	
No. 1	0.89	0.42	0.47
No. 2	0.86	0.67	0.78
No. 3	0.52	0.18	0.35
No. 4	0.48	0.24	0.50

Figure 1.3   Natural periods of roof structure.

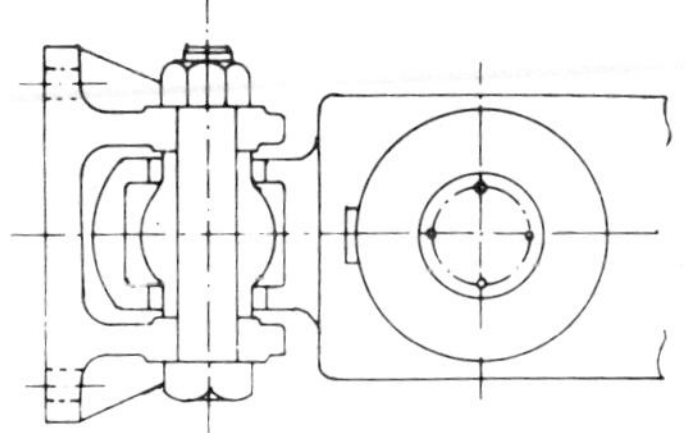

1	PISTON ROD	8	SUCTION VALVE (B)
2	PISTON	9	BEARING
3	INNER CYLINDER	10	OIL SEAL
4	OUTER CYLINDER	11	RING KEY
5	RETURN SPRING	12	FLANGE
6	SUCTION VALVE (A)	13	OIL GAGE
7	RELIEF VALVE	14	DAMPER OIL

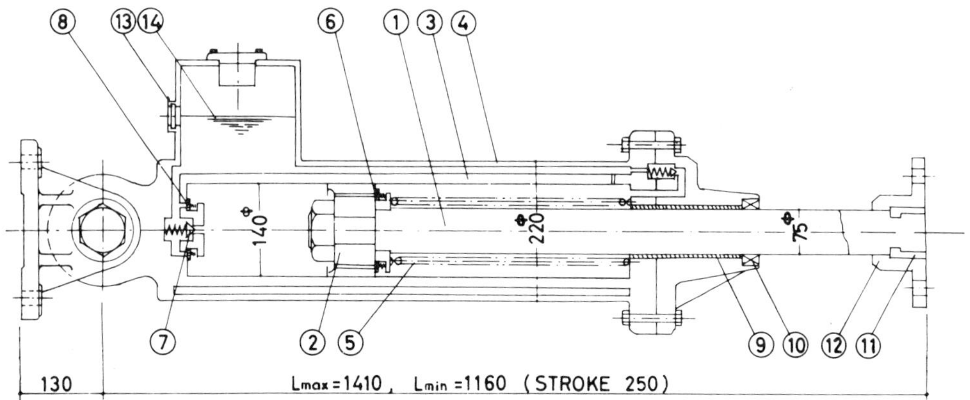

Figure 1.4 Damping mechanism.

References

[1] Y. Tsuboi, and M. Kawaguchi: "Design Problems of a Suspension Roof Structure," *Report of the Institute of Industrial Science*, University of Tokyo, Vol. 15, No. 2, 1965.

[2] Y. Tsuboi, and M. Kawaguchi: "Probleme beim Entwurf einer Hängedachkonstruktion anhand des Beispiels des Schwimmhalle für die Olympischen Spiele 1964 in Tokio," *Der Stahlbau*, Heft 3, Berlin, 1966.

APPENDIX 2: Tokyo International Exhibition Hall
(by Y. Tsuboi and R. Nasukawa)

Reference

[1] Y. Tsuboi: Die Stahlrippenkuppel auf dem Messegelande in Tokio, *Der Stahlbau*, 1962, Heft. 10.

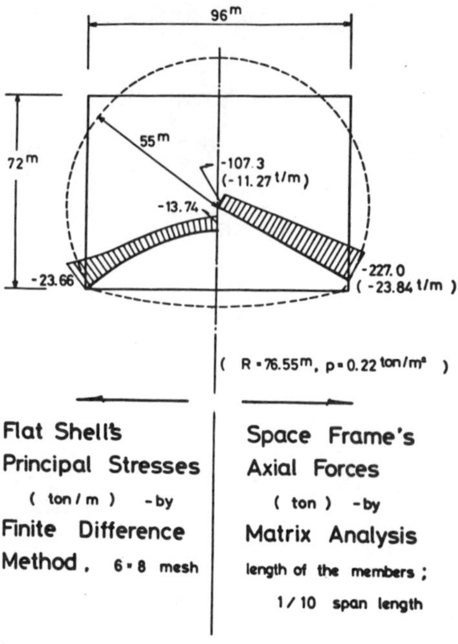

Figure 2.1

Photo 2.1   Interior view of hall.

336

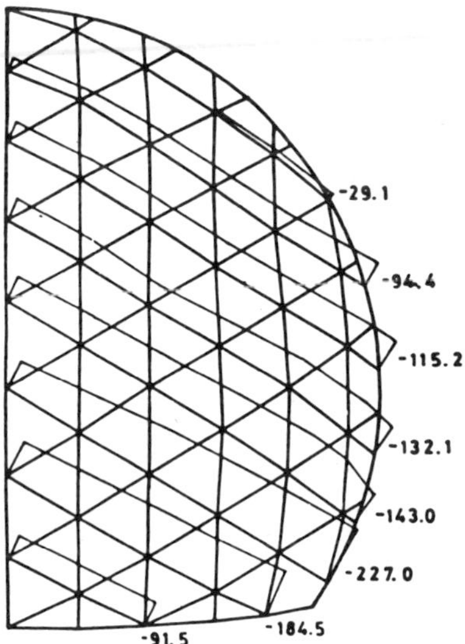

Figure 2.2 Axial force in members (tons).

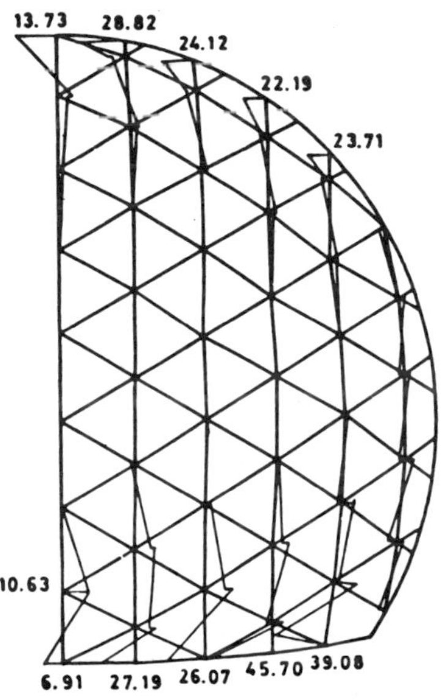

Figure 2.3 Bending moments in members (ton-m).

337

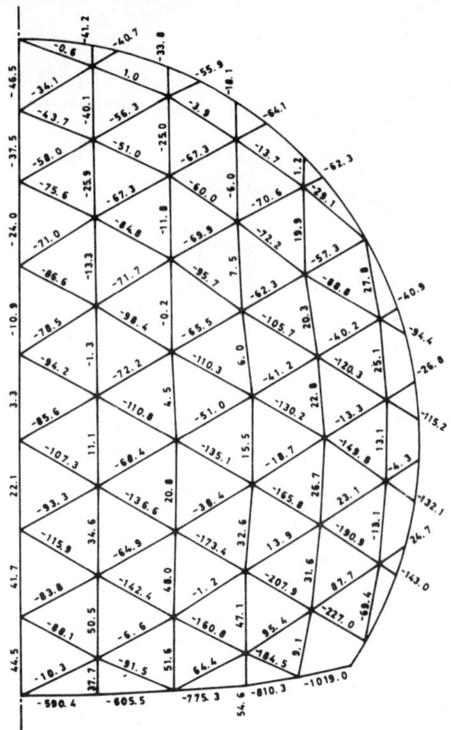

Figure 2.4 Axial force in members (tons).

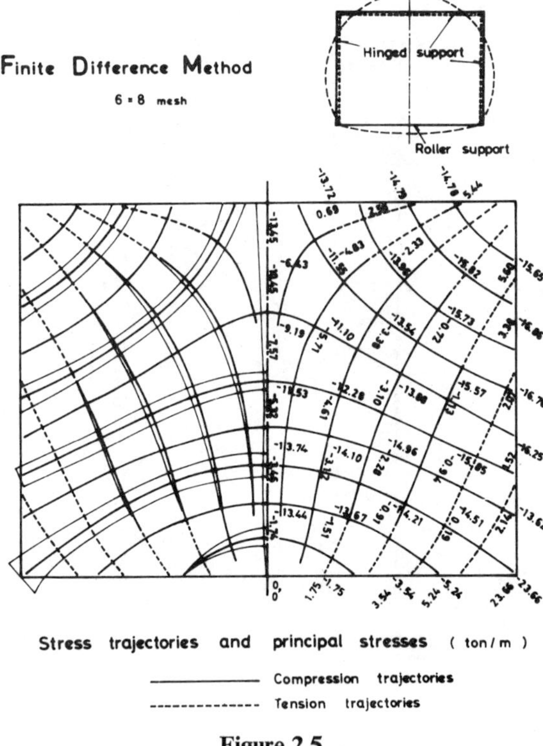

Stress trajectories and principal stresses (ton / m)

———————— Compression trajectories

------------- Tension trajectories

Figure 2.5

APPENDIX 3: Structural Plan for Hotal in Iran—
Computer Analysis
(with the cooperation of M. Yoshimura)

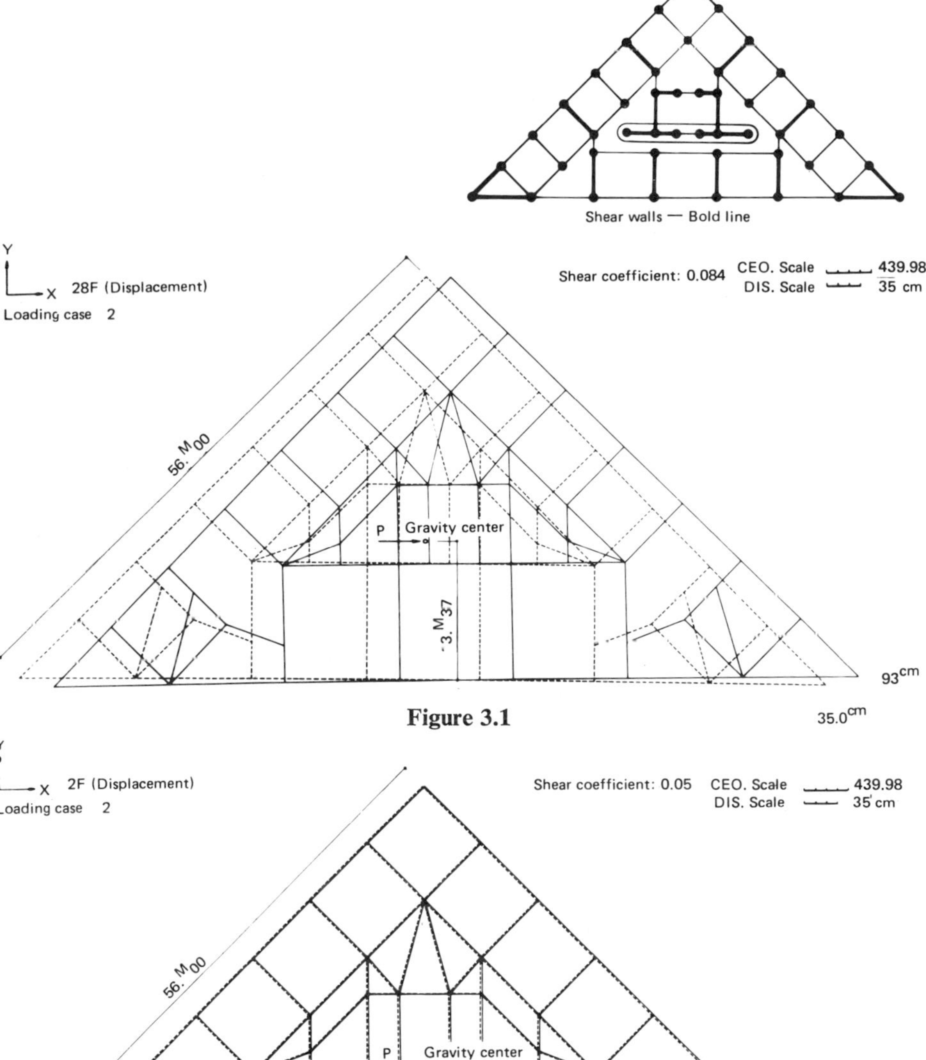

Figure 3.1

Figure 3.2

339

Photo 4.1   Aerial view of overturned buildings.

Photo 4.2   Overturned building.

APPENDIX 5: Project for Housing Corporation—Mechanism of Shear Transmission in Checkered Shear Walls Subjected to Lateral Force
(with the cooperation of M. Takeda)

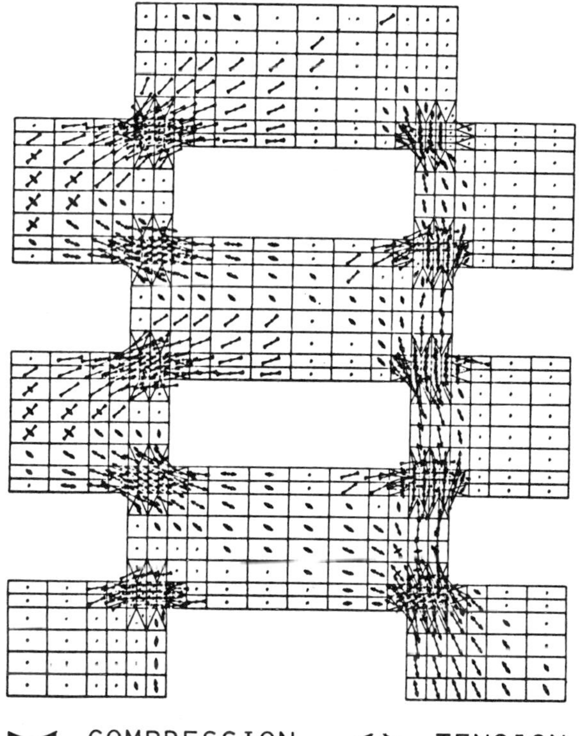

Figure 5.1 Principal stress in walls (F.E.M.).

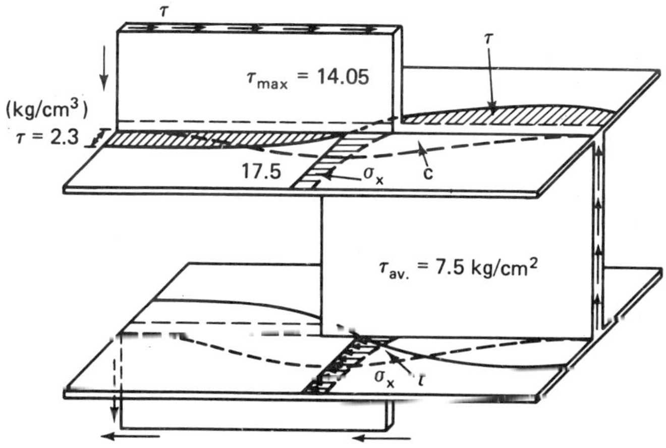

Figure 5.2 Stress propagation of slabs from walls.

APPENDIX 6: Draft for Earthquake-Resistant Design Methods (1977), Ministry of Construction

The principal design procedure for earthquake-resistant structures has been suggested by the Japanese Ministry of Construction in "Draft for Earthquake-Resistant Design Method (1977)." The procedure consists of three major categories or six minor categories, most of them comprising two stages: elastic design and check of ultimate capacity. The following procedure corresponds to the first stage of the second category.

1. Estimation of fundamental period of the structure:

$$T = (0.06 \sim 0.09)N \text{ s, where } N = \text{number of stories}$$

2. Base-shear coefficient[5]

$$C_B = Z \cdot G(T) \cdot C_0$$

where Z = seismicity coefficient
$G(T)$ = design spectra considering site condition $G_0(I)$ or $[G_1(T)]$ should be properly taken at each stage
C_0 = basic base shear coefficient (set at 0.2)

A structural factor K is introduced in the second stage; an importance factor I may be considered during any stage.

3. Lateral force at level i:

$$F_i^6 = K_i \cdot W_i$$
$$K_i = [1 + \ell(\xi_{i+1}^2 + \xi_{i+1} \cdot \xi_i + \xi_i^2 - 1)]LC_B$$
$$\ell = 1 - \frac{2}{1 + N}, \qquad L = 0.5 + \frac{1}{0.1(N - 1) + 2}$$
$$\xi_i = \frac{\sum_{j=1}^{i-1} W_j}{W}$$

where W_i = weight of level i
K_i = lateral force coefficient at level i

According to the U.B.C., lateral force distribution is considered triangular, with a concentrated force on the top.

4. Limitation for deflection

---

[5]$G(T) \times C_0$ corresponds to $C \times S$ in the Uniform Building Code (U.B.C.), but $G(T)$ represents both the effects of C and S, because C_0 is set constant at 0.2 to fit the current code.

[6]Derived from story shear distribution in the form of a third-degree function of ξ_i and transformed into a similar term K_i to conform to the current code.

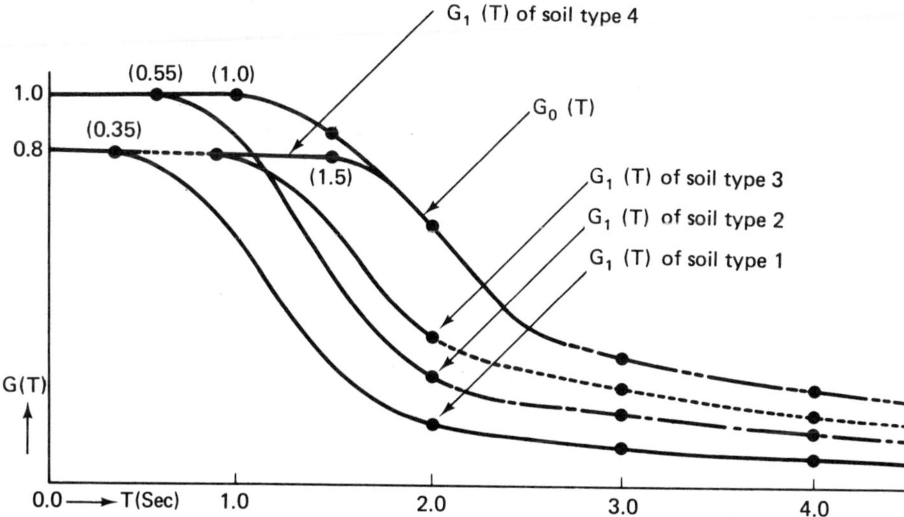

Figure 6.1 Design spectra considering soil condition.

APPENDIX 7: Three-Dimensional Aseismic Design of Building Systems (by S. Hirashima)

A desired aseismic design should be achieved by the exact prediction of building response, considering the characteristics of earthquakes and the dynamic properties of buildings. The response of non-eccentric buildings under one horizontal earthquake component has been predicted with sufficient accuracy for the purposes of structural design. However, three-dimensional response forecasts of structures appear to be insufficient. The following data should supply information useful for the three-dimensional aseismic design of structures.

Changes of maximum dynamic stresses in members of a non-eccentric building (Fig. 7.1) under two horizontal earthquake components are shown in Figs. 7.2 through 7.4. These figures indicate that the ratios of maximum dynamic stress to standard response ($\theta = 0$) are less than 1.3.

In Table 7.1 the results of response analyses on a simple eccentric model are summarized (Fig. 7.5). The dynamic eccentric distance is shown to be about 1.5 times as great as the static eccentric distance.

Results from more detailed studies of these analytical data are shown in Table 7.2. This table has been arranged according to relations between maximum dynamic stresses due to various combinations of earthquake components. The square root of the sum of the squares (SRSS) is conspicuous, resulting in fairly nonconservative estimates; therefore, the mean value of

Table 7.1

DYNAMIC ECCENTRICITY/STATIC ECCENTRICITY

Model	Wave	A e_{D1}/es	A e_{D2}/es	B e_{D1}/es	B e_{D2}/es	C e_{D1}/es	C e_{D2}/es
S3	EL	2.73	1.36	1.77	1.57	1.99	0.89
	TA	1.36	1.36	1.14	0.71	0.54	0.49
	TO	2.06	1.39	1.91	1.35	1.75	1.13
S2	EL	1.53	0.94	3.10	2.38	10.31	4.15
	TA	1.75	1.75	0.53	0.38	2.58	1.71
	TO	2.11	2.11	0.61	0.46	5.18	0.69
S1	EL	1.32	1.02	1.87	1.38	7.41	1.30
	TA	−0.44	7.38	0.29	0.25	1.28	0.99
	TO	2.25	2.25	0.28	0.12	1.31	−1.07
T1	EL	0.51	0.20	−0.45	−0.81	−3.66	−7.74
	TA	0.82	0.16	0.05	0.04	−0.27	−0.57
	TO	1.45	1.29	−0.29	−0.36	−1.83	−5.37
T2	EL	−0.43	−0.97	−1.45	−1.81	−3.89	−5.68
	TA	−0.23	−0.86	0.04	0.04	−0.68	−0.89
	TO	0.80	0.38	−0.27	−0.33	−1.49	−4.32
T3	EL	0.71	−1.55	−1.16	−1.66	−1.01	−1.62
	TA	0.61	0.07	0.10	−0.49	0.03	−2.27
	TO	0.56	−2.44	−0.49	−0.95	−1.18	−1.40

where $e_{D1} = (Q_1/_HQ_1 - 1)/l$

$\qquad e_{D2} = (1 - Q_2/_HQ_2)/l$

$$es = \begin{cases} 3:0.25l \\ 2:0.05l \\ 1:0.03l \end{cases}$$

Table 7.2

COMBINATION OF MAXIMUM RESPONSE STRESSES

	combining method	accuracy 0.7 0.8 0.9 1.0 1.1 1.2 1.3 1.4 1.5
combining of horizontal two components	sum of absolute	
	SRSS method	
	mean of the upper methods	
combining of horizontal and vertical	sum of absolute	
	SRSS method	
	mean of the upper methods	
combining of horizontal and torsional	sum of absolute	
	SRSS method	
	mean of the upper methods	

SRSS and the sum of the absolute should be adopted in structural design. Thus:

$$S_{DX} = \alpha \cdot S_X, \qquad S_{DY} = \beta \cdot S_Y$$

$$S_{DXY} = \tfrac{1}{2}(|S_{DX}| + |S_{DY}| + \sqrt{S_{DX}^2 + S_{DY}^2})$$

$$S_{DXYZ} = \tfrac{1}{2}(|S_{DXY}| + |S_Z| + \sqrt{S_{DXY}^2 + S_Z^2}) \qquad (1)$$

$$S_{TXY} = \tfrac{1}{2}(|S_{TX}| + |S_{TY}| + \sqrt{S_{TX}^2 + S_{TY}^2})$$

$$S_{DXYZT} = \tfrac{1}{2}(|S_{DXYZ}| + |S_{TXY}| + \sqrt{S_{DXYZ}^2 + S_{TXY}^2})$$

where S_X, S_Y, S_Z = maximum response stresses to three spatial components calculated separately

α, β = increase factor due to earthquake direction changes (1.0–1.3)

S_{DXY} = combined stress of horizontal two components

S_{DXYZ} = combined stress of three components

S_{TX}, S_{TY} = torsional stress due to horizontal two components

S_{TXY} = combined torsional stress of horizontal two components

S_{DXYZT} = combined stress of three components and torsion

Figure 7.6 is a flowchart of the effect of torsion on both static and dynamic aseismic design and considers both the results of Table 7.2 and the expressions of eq. (1). A practical model of an eccentric building for dynamic analysis is shown in Fig. 7.7.

earthq. \ direc.	S − N	E − W	U − D
EL. CENTRO, 1940	341.7 gal	210.1	206.3
TAFT , 1952	152.7 gal	175.9	102.9

INPUTTED EARTHQUAKE WAVES

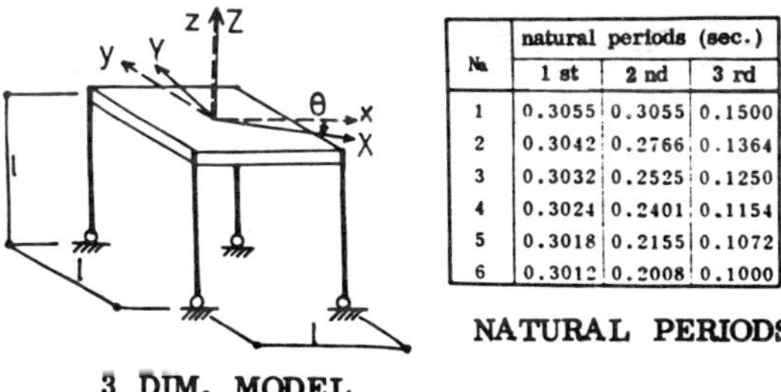

No.	natural periods (sec.)		
	1 st	2 nd	3 rd
1	0.3055	0.3055	0.1500
2	0.3042	0.2766	0.1364
3	0.3032	0.2525	0.1250
4	0.3024	0.2401	0.1154
5	0.3018	0.2155	0.1072
6	0.3012	0.2008	0.1000

NATURAL PERIODS

Figure 7.1

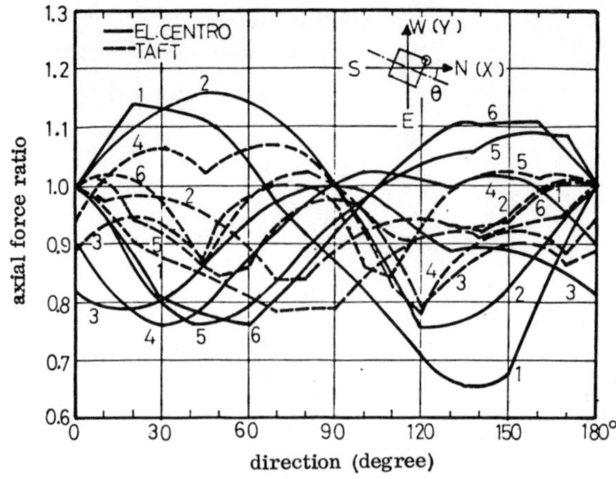

Figure 7.2 Earthquake direction and change in column shear force.

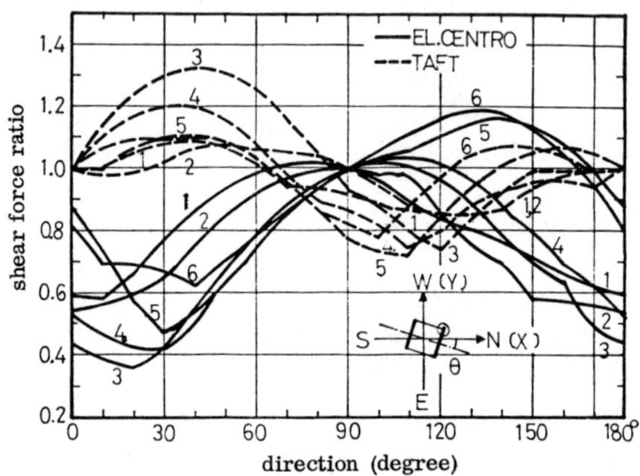

Figure 7.3 Earthquake direction and change in column shear force (shorter natural period direction).

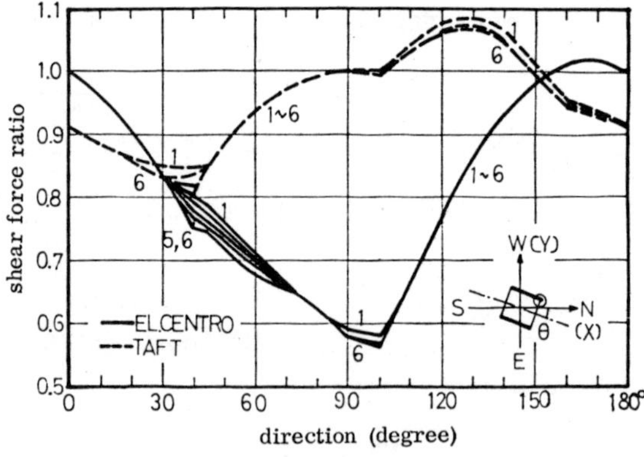

Figure 7.4 Earthquake direction and change in column shear force (longer period direction).

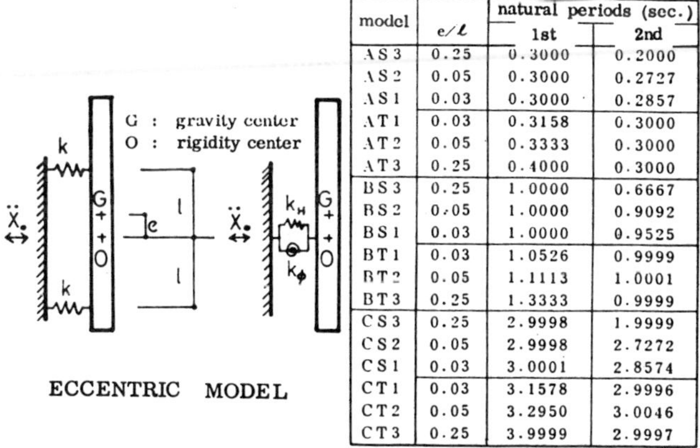

model	e/ℓ	natural periods (sec.)	
		1st	2nd
A S 3	0.25	0.3000	0.2000
A S 2	0.05	0.3000	0.2727
A S 1	0.03	0.3000	0.2857
A T 1	0.03	0.3158	0.3000
A T 2	0.05	0.3333	0.3000
A T 3	0.25	0.4000	0.3000
B S 3	0.25	1.0000	0.6667
B S 2	0.05	1.0000	0.9092
B S 1	0.03	1.0000	0.9525
B T 1	0.03	1.0526	0.9999
B T 2	0.05	1.1113	1.0001
B T 3	0.25	1.3333	0.9999
C S 3	0.25	2.9998	1.9999
C S 2	0.05	2.9998	2.7272
C S 1	0.03	3.0001	2.8574
C T 1	0.03	3.1578	2.9996
C T 2	0.05	3.2950	3.0046
C T 3	0.25	3.9999	2.9997

Figure 7.5   Eccentric model.

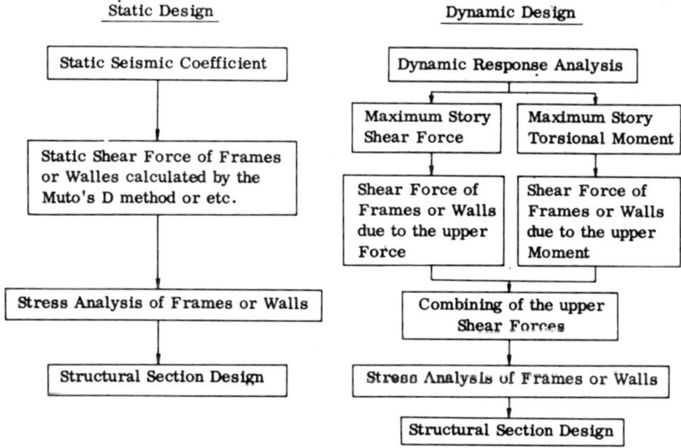

Figure 7.6   Flowchart of torsional effect in aseismic design.

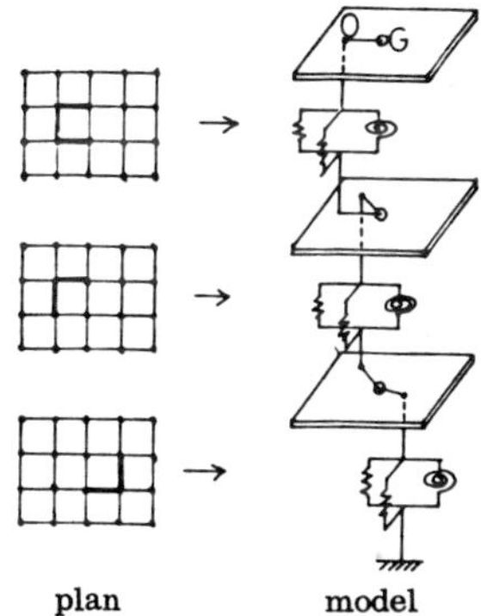

Figure 7.7   Practical model of building.

APPENDIX 8: Mode Shapes Determined from Participation Factors of an 18-Story Building
(with the cooperation of H. Tajimi and T. Ishimaru)

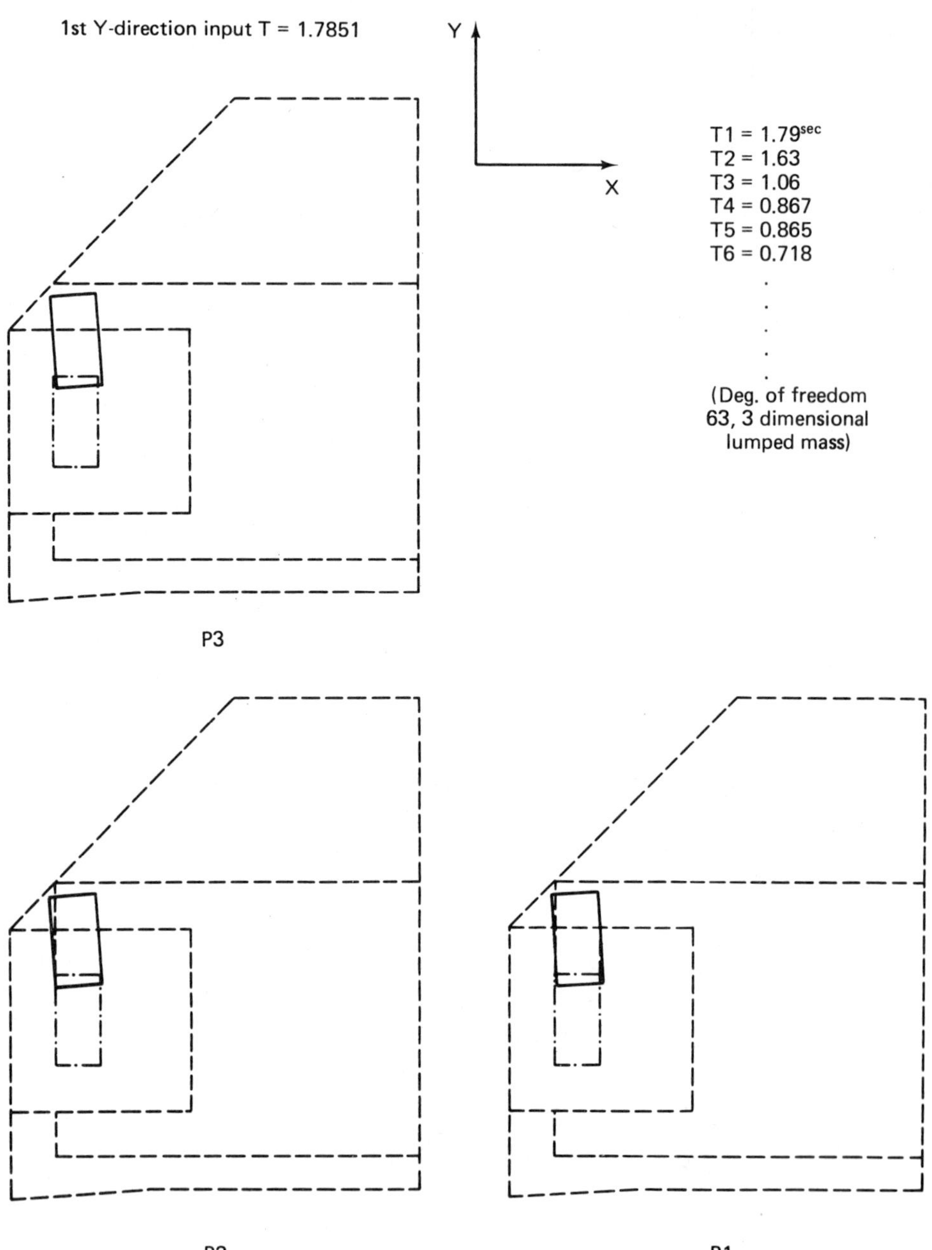

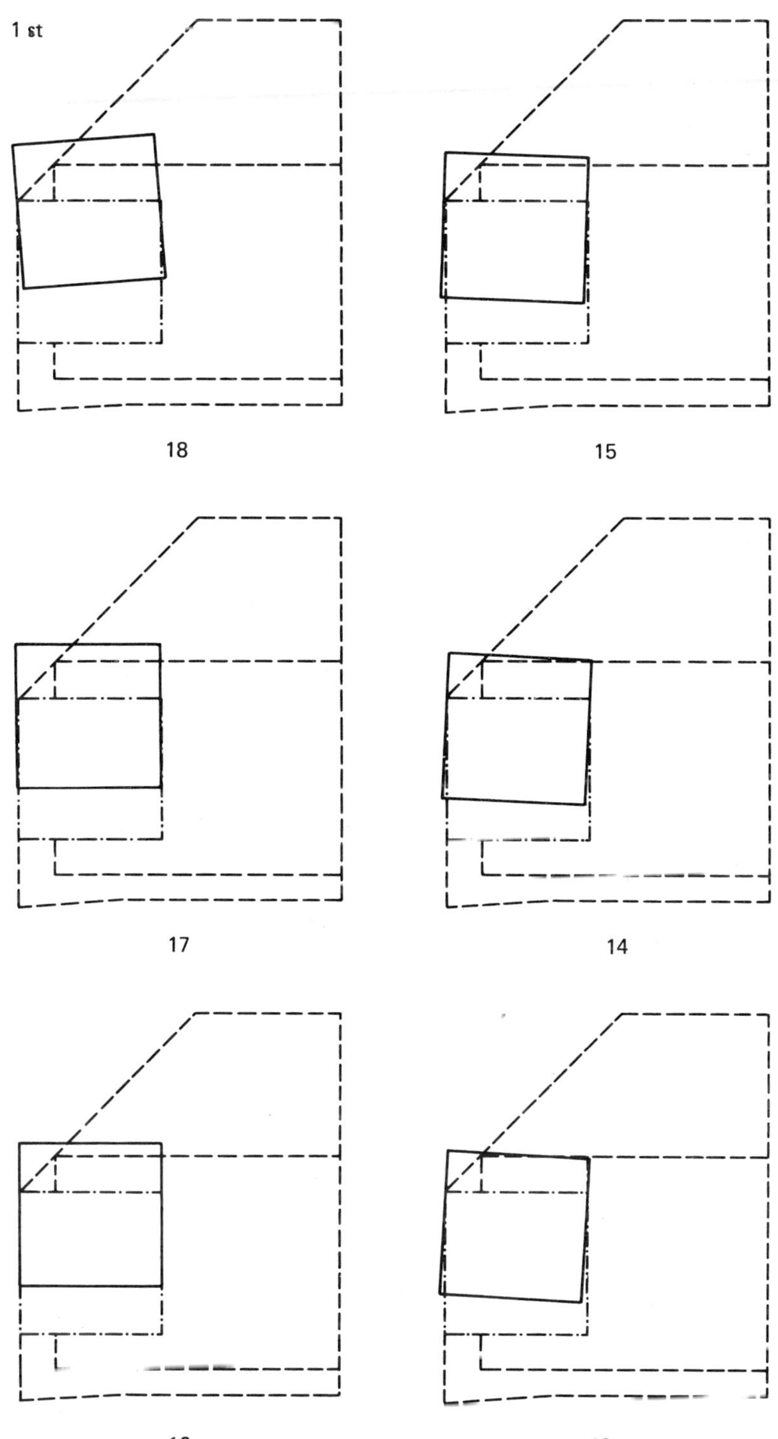

1 st
18
15
17
14
16
13

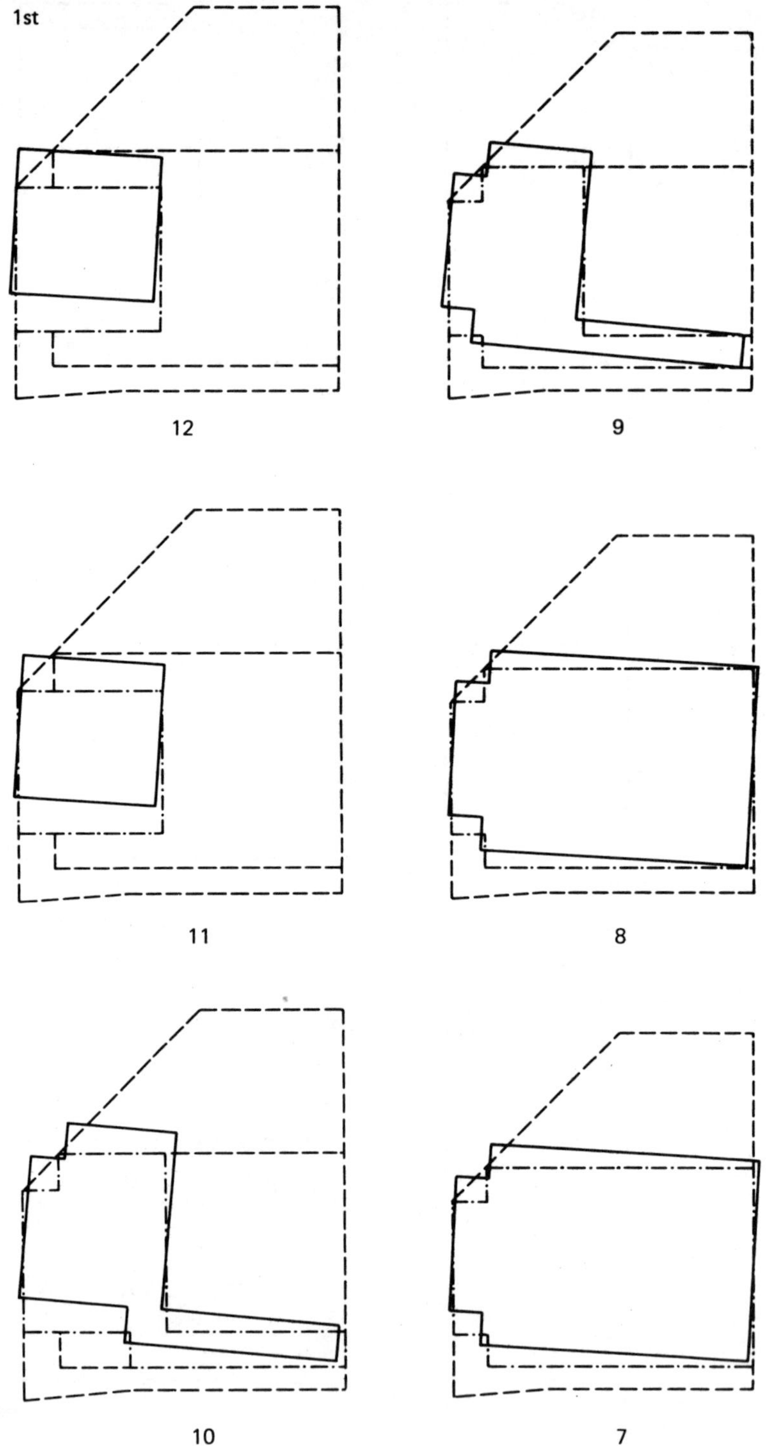

1st
12
9
11
8
10
7

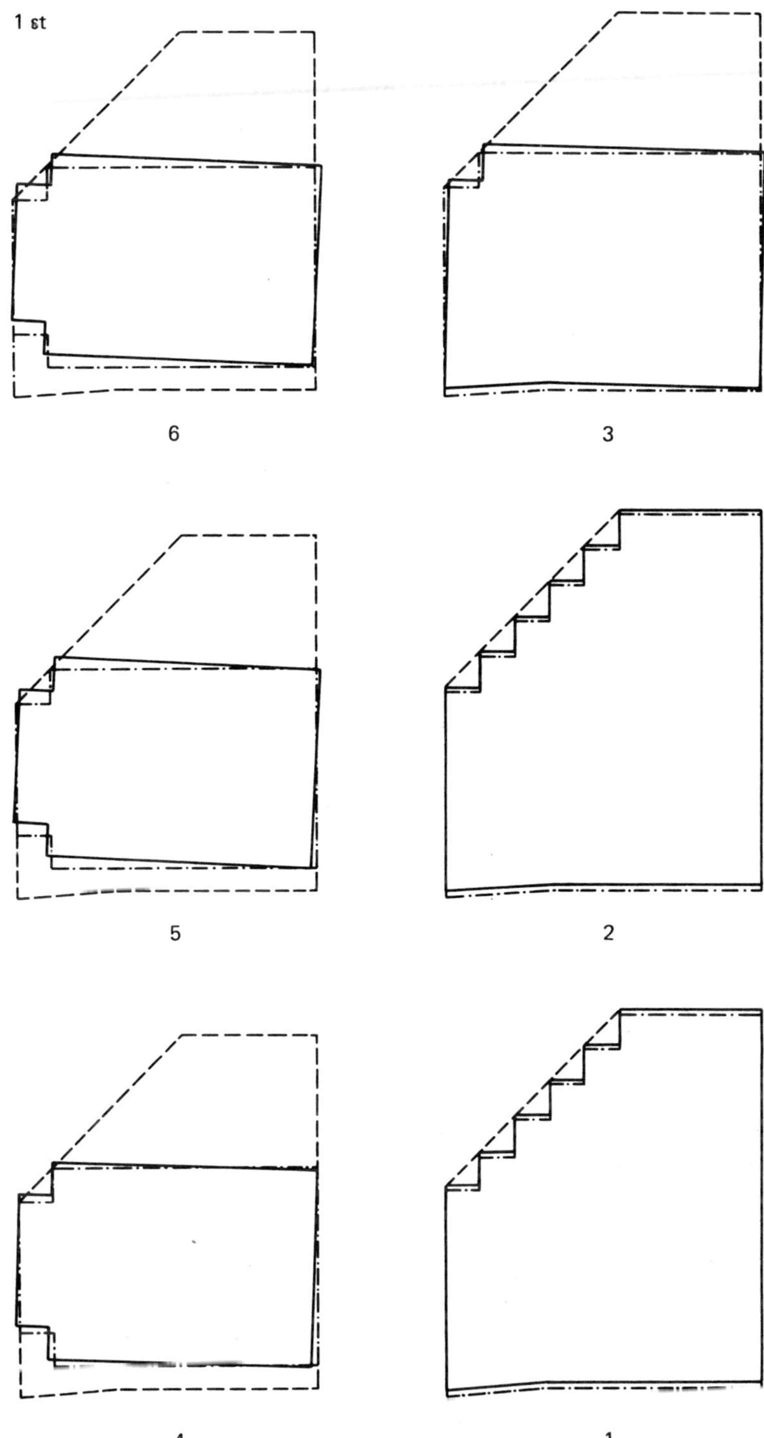

1 st
6
3
5
2
4
1

APPENDIX 9: Reinforced Concrete Slit Wall
for Tall Building Structure
(by K. Muto)

Tall building structures consisting solely of steel frames are particularly susceptible to high winds, which can induce large, swaying movements that create discomfort and insecurity among building occupants. Reinforced concrete shear walls are useful in the design of ordinary buildings, which is concerned primarily with providing high rigidity and the capacity to resist lateral force. However, when an ordinary, monolithic, reinforced concrete shear wall is combined with tall steel frames, its rigidity becomes too high, resulting in brittle shear failure at even low levels of deformation.

To eliminate such defects in ordinary shear walls, a new type of ductile wall, a slit wall, has been developed. The vertical slits in the reinforced concrete wall and the arrangement of the unique steel bar reinforcements cause this wall to behave as a series of vertical wall columns separated at the slits. It is consequently possible to reduce the initial rigidity of the wall and to distribute the burden of any horizontal forces more favorably between wall and steel frame. Furthermore, even when the wall is in the plastic range, ductility maintains the force-resisting capacity of the structure and capability to withstand large deformations along its steel frames (Photos 9.1 and 9.2).

The structural characteristics of this type of wall have been confirmed by numerous experiments and analyses. Such walls have been used in a number of tall buildings as an important design device not only for structural safety, but for economy—their use reduces the number of costly steel frames—and they function as fire-protective diaphragm walls surrounding the elevator shaft.

Figure 9.3 shows the design of the tallest building in Japan, the 60-story, 226-m-tall Ikebukuro Subcenter, being constructed in Tokyo. In this building the ratio of the resistance of the slit walls to seismic force is about a uniform 30% along the height of the building.

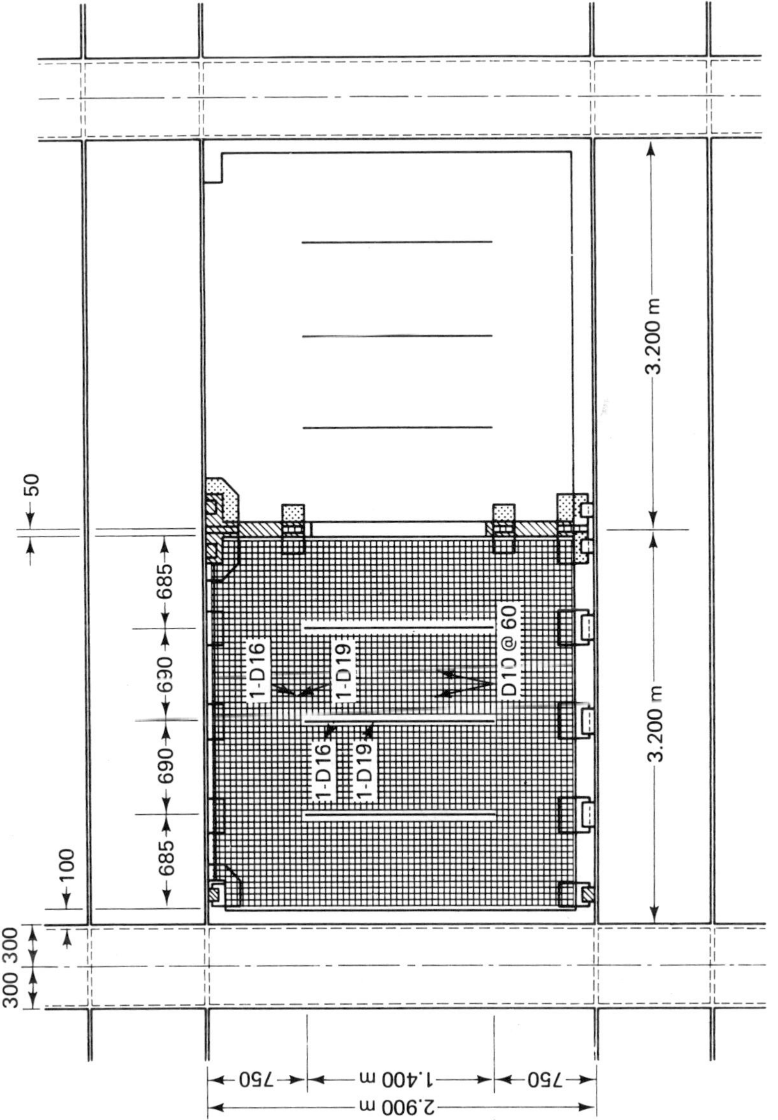

Figure 9.1 Example of slit layout and reinforcement for slit wall.

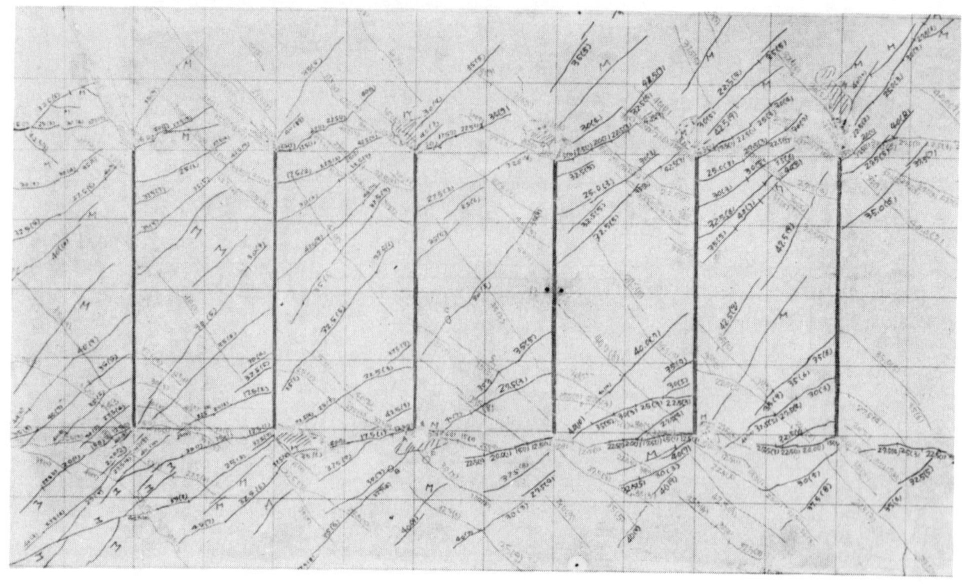

Photo 9.1   Crack pattern in a slit test wall; when the applied shear force is 45.8 kg/cm², the cracks are distributed finely, with no major diagonal cracks.

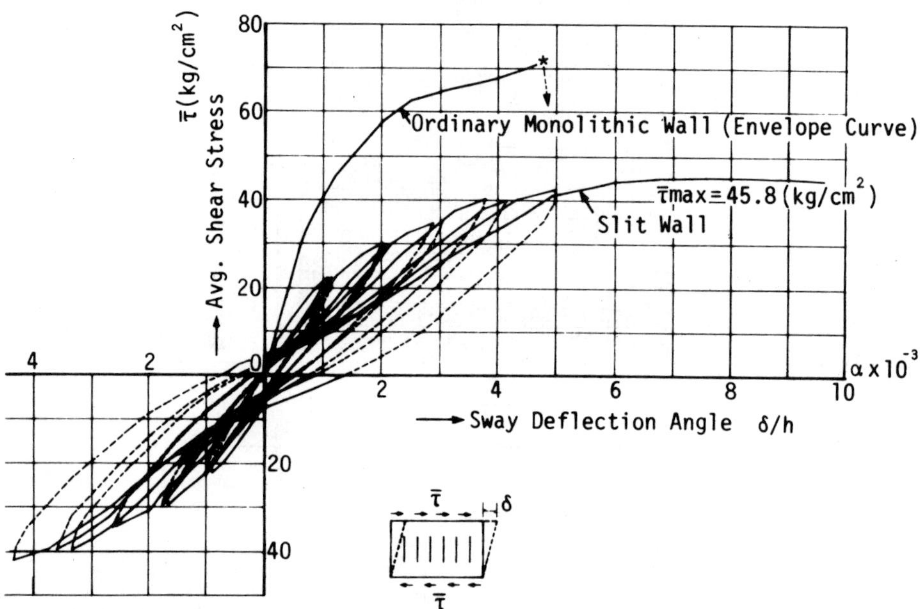

Figure 9.2   Shear stress versus sway deflection angle; the deformability of the slit wall is more than twice that of an ordinary monolithic wall.

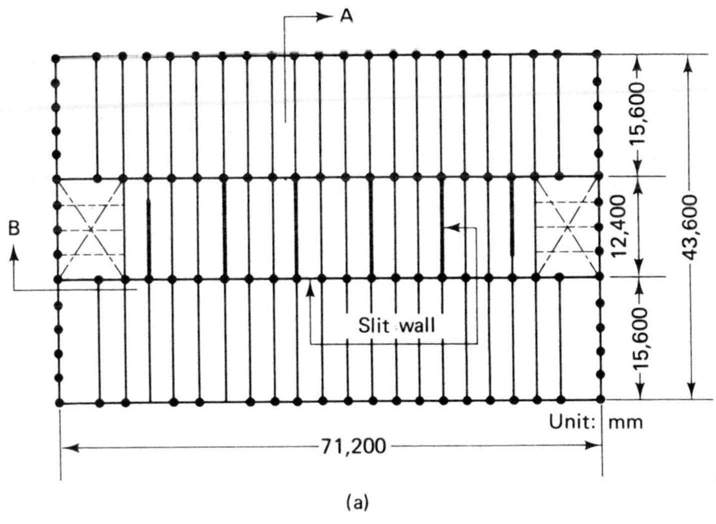

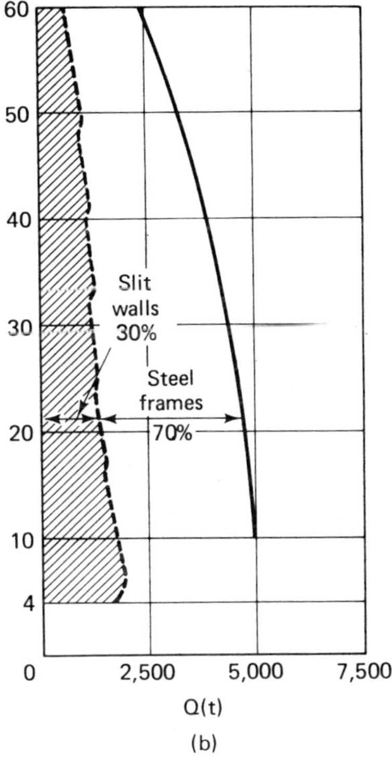

Figure 9.3 Ikebukuro Subcenter, Tokyo: (a) framing plan, (b) resisting ratio to seismic force (transverse direction).

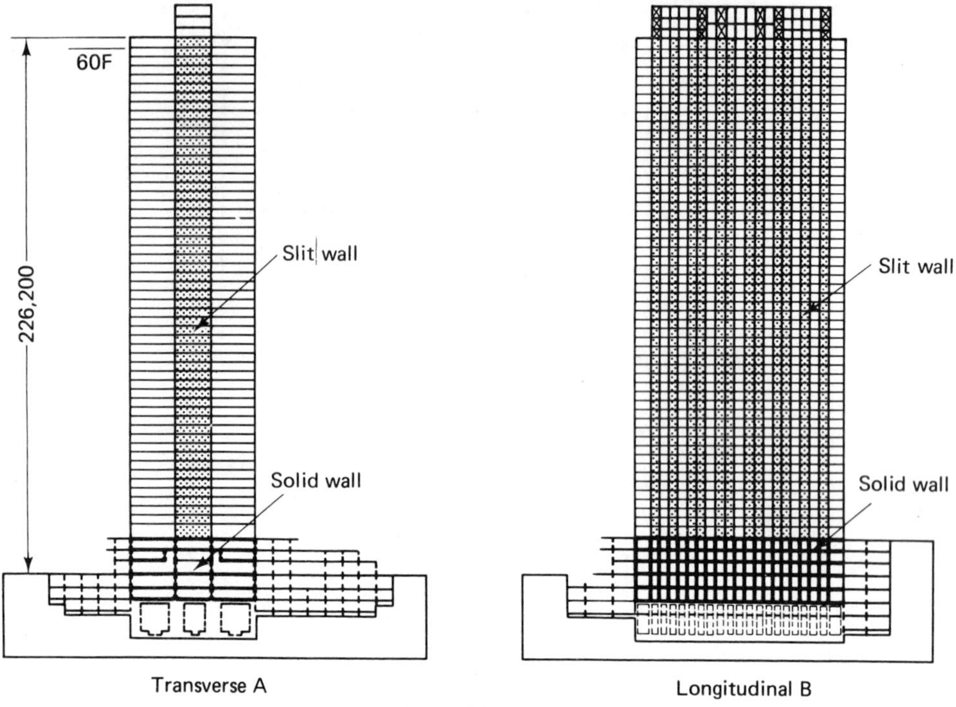

Figure 9.3 (Continued) (c) framing section.

Photo 9.2 "Slit wall" installed in steel framing.

APPENDIX 10: Ductile Bracing for Steel Structure
(by Y. Tsuboi, K. Saida, and Y. Suenaga)

The typical X-shape bracing is too rigid and results in brittle failures of steel structures under seismic load. To eliminate such defects of ordinary bracing, ductile bracing was developed for steel structures, considering recent developed concepts in seismic dynamics, by adding a horizontal member between diagonal members (Fig. 10.1). This modification provides better ductility and greater flexibility (Table 10.1). The former is achieved by having the web of the horizontal member yield in shear before bending or axial yield of all members occurs (Figs. 10.2 and 10.3). The flexibility is increased by varying the length and depth of the horizontal member.

The structural characteristics of this modified bracing method have been confirmed by numerous experiments and analyses, and the method has been incorporated into the structural design of many steel frame buildings.

In Figs. 10.2, 10.4, and 10.5, the load-displacement relations for three diagonal members are illustrated. In Fig. 10.6 compares the stiffness of static loading with that of dynamic loading. In this figure, D. L. indicates dynamic loading and the small numbers show the loading times.

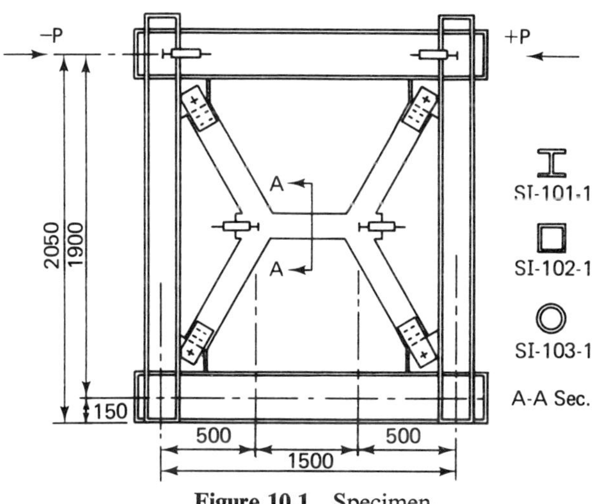

Figure 10.1 Specimen.

Table 10.1

MECHANICAL PROPERTIES OF MEMBERS

		σ_y	I x (cm^4)	A_p (cm^2)	Z_x (cm^3)
Column	II-200 × 200 × 8 × 12	30.3	4720	63.53	472
Beam	H-300 × 150 × 6.5 × 9	29.6	7210	40.78	481
Bracing	H 126 × 126 × 6.5 × 9	27.2	847	30.31	136
Bracing	□ P-136 × 136 × 6	31.2	808	25.16	97.8
Bracing	○-165.^{2} × 5	36	860	30.96	99.9

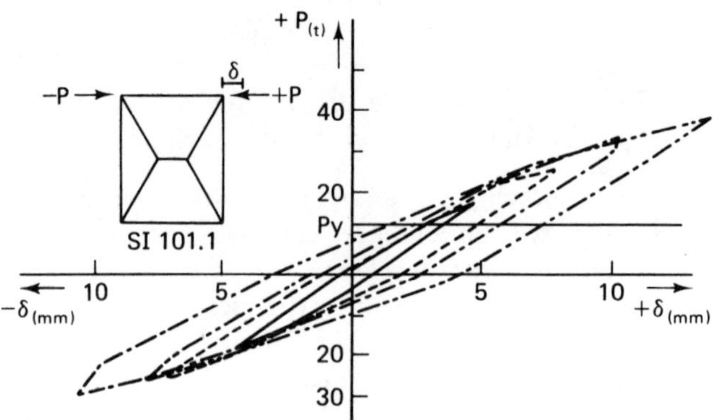

Figure 10.2 Relation of load displacement.

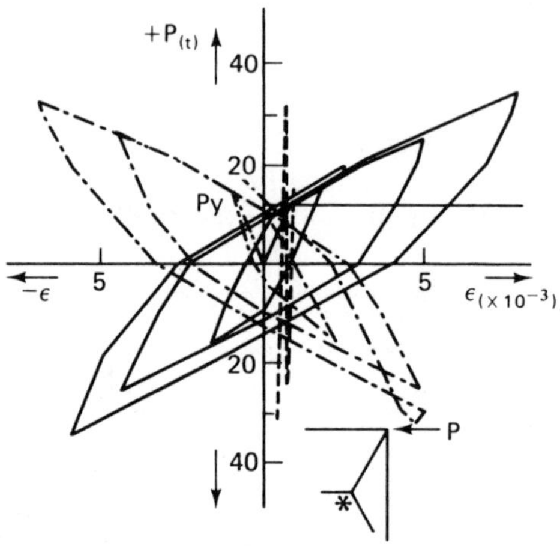

Figure 10.3 Relation of load-shearing strain of horizontal member.

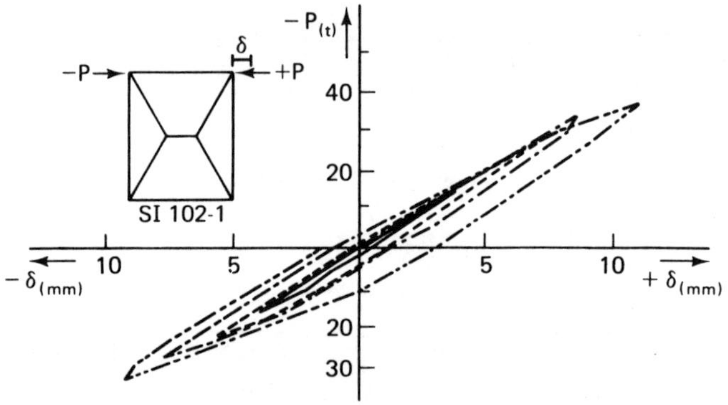

Figure 10.4 Relation of load displacement.

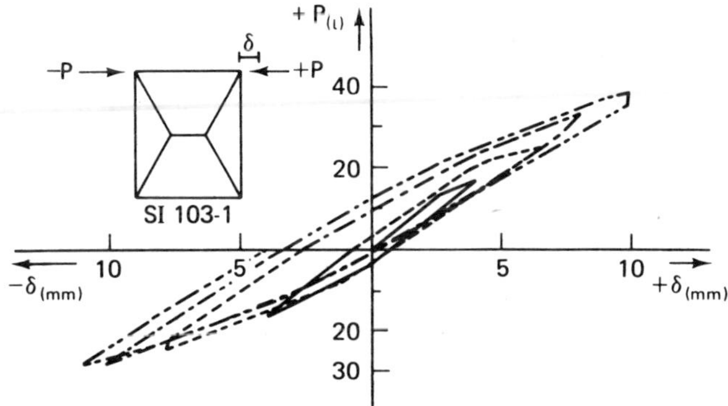

Figure 10.5   Relation of load displacement.

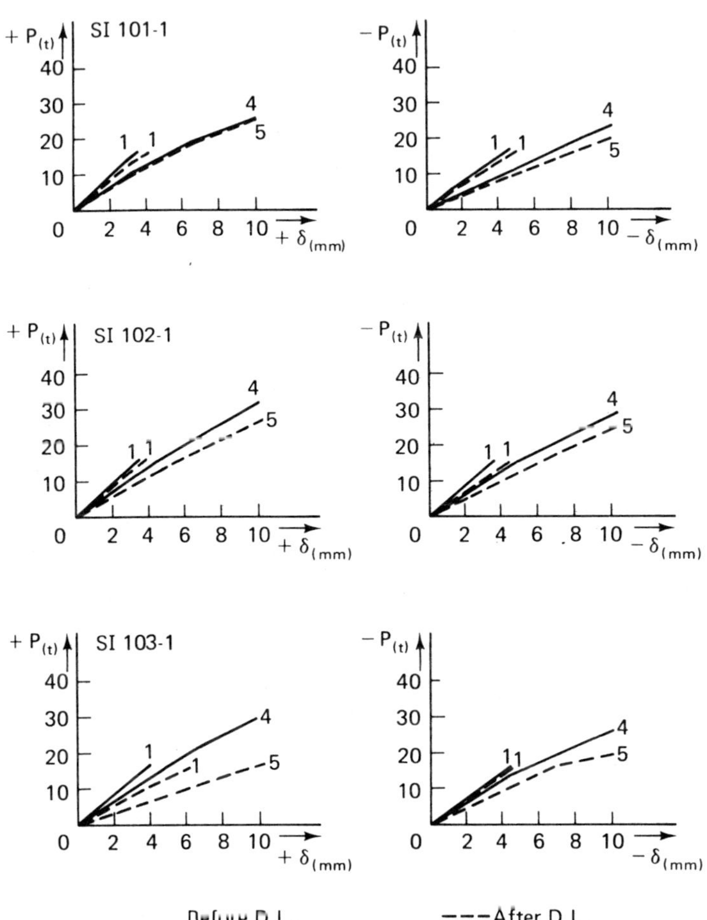

Figure 10.6   Comparison of stiffness.

Photo 10.1 Experimental setup.

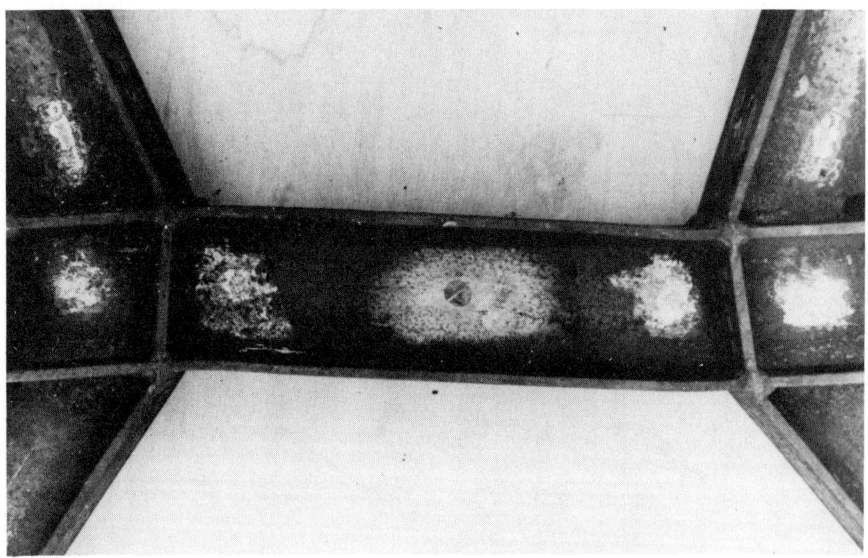

Photo 10.2 Deformation of the specimen.

APPENDIX 11: Space Frame for the Symbol Zone of Expo '70 (by Y. Tsuboi and M. Kawaguchi)

The transparent roof covering the Symbol Zone of Expo '70 is a huge space frame structure supported by six columns. Linear members of the space frame are connected solely by mechanical ball joints that can compensate for errors in angle and length. The functions of a connecting system for space frames are discussed and possible errors in the present system are evaluated. The fixing and error-absorbing mechanism of the connecting system is reported. Results of full-size tests made on the ball joints are also presented. The roof is assembled on the ground level and lifted by pneumatic jacks along the central posts of the columns. To facilitate the lift of the roof and its connection to the columns, a mechanical capital joint is designed.

Reference

[1] Proc. 1971 IASS Pacific Symposium Part II on TENSION STRUCTURES and SPACE FRAMES, Tokyo and Kyoto, pp. 893–904, Paper 11-1, 1972, Architectural Institute of Japan.

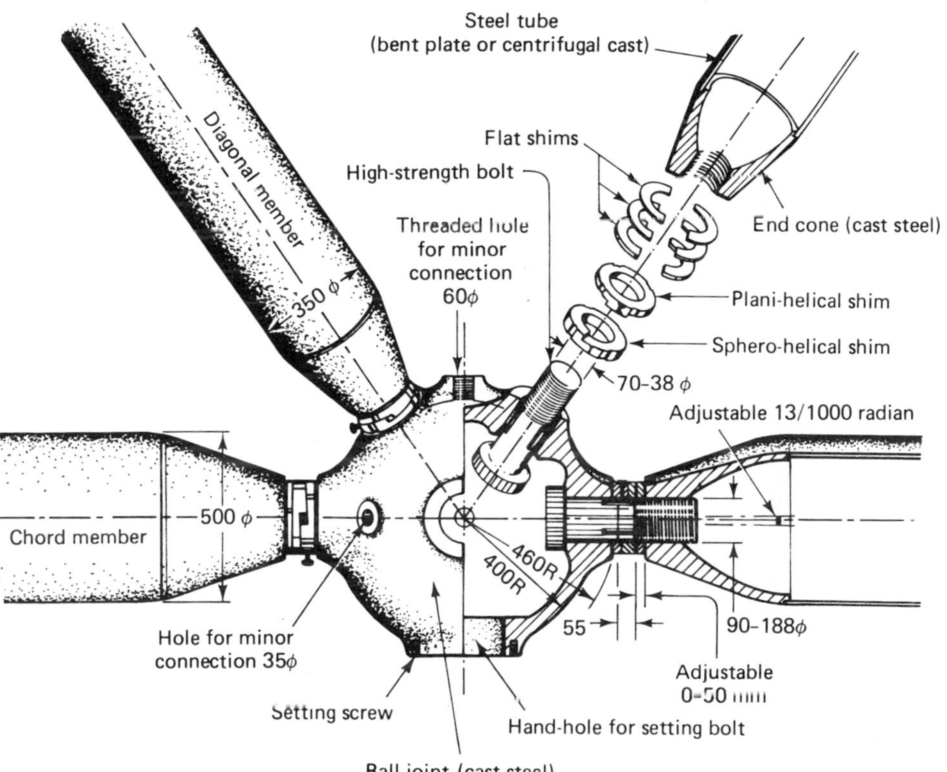

Figure 11.1 Mechanism of the ball joint.

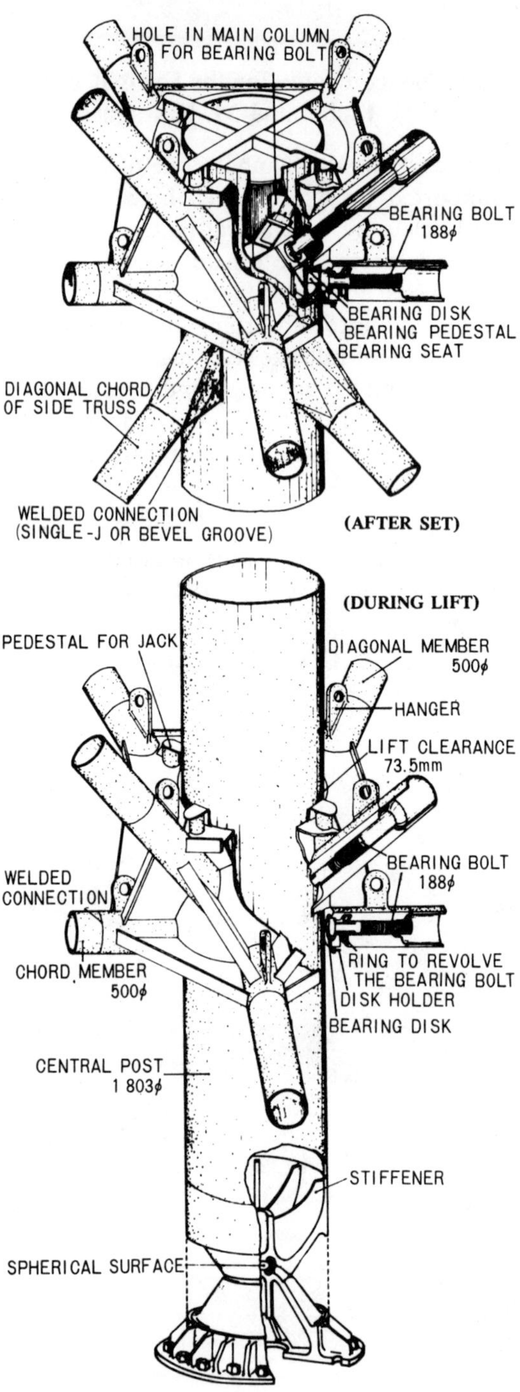

Figure 11.2   Mechanism of the capital joint.

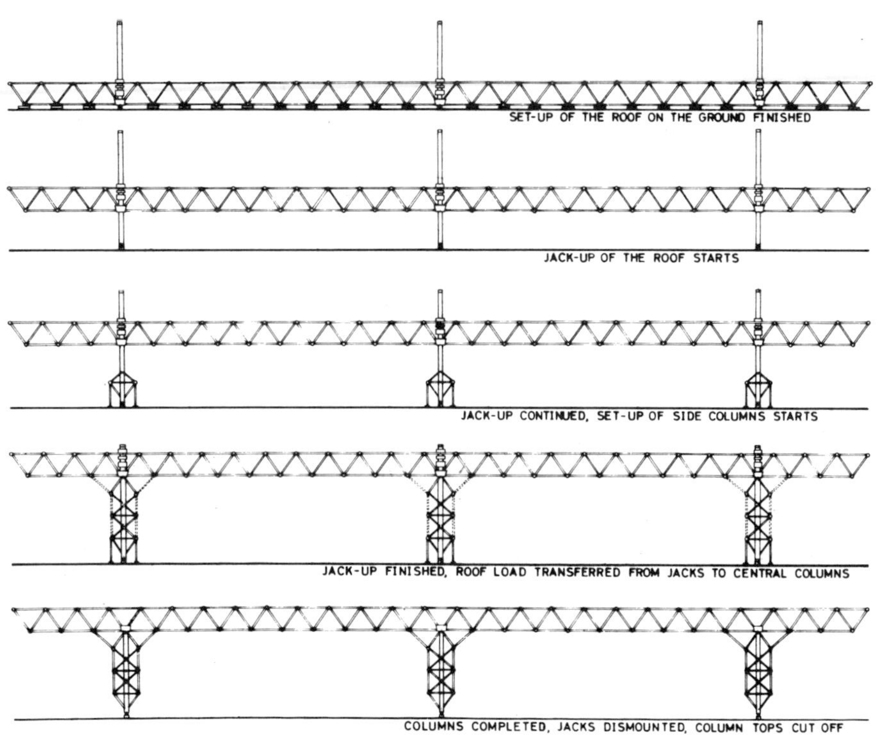

Figure 11.3 Lift processes.

Photo 11.1 Aerial view of the structure.

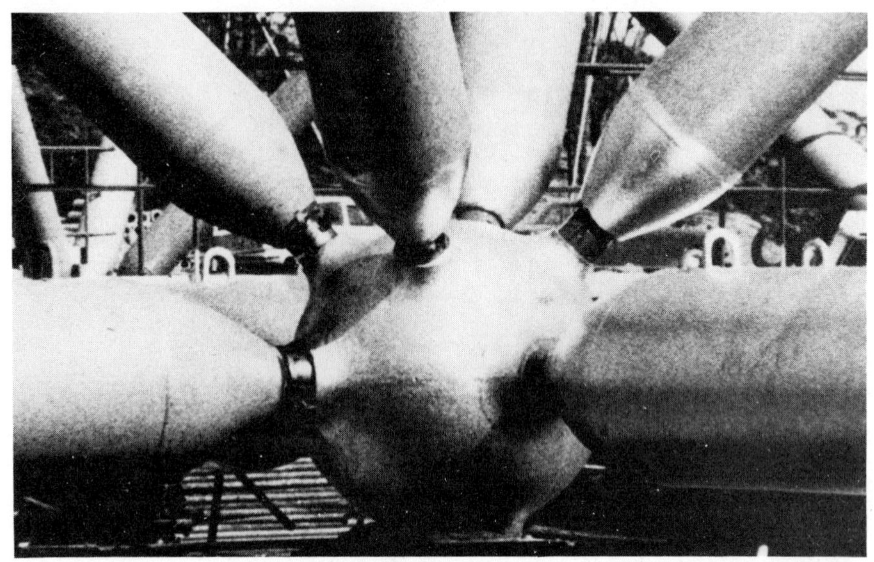

Photo 11.2   A ball joint.

Photo 11.3   The structure being lifted.

Photo 11.4 Overall view of structure being lifted.

APPENDIX 12: Rigid Connections in Frames
(with the cooperation of Y. Suenaga and S. Aoki)

An experimental study on the static behavior of newly developed cast-steel joints to be employed between box-section columns and H-section beams was conducted to improve the connections of rigid frames.

The behavior of cast-steel box joints was experimentally compared with that of conventional, built-up connections, mainly with regard to their shearing deformation. In Fig. 12.1 the shearing stress–strain curves of the panel zones of several specimens are shown. In this figure, SJPBC-7 represents the typical joint.

The results of the experimental study can be summarized as follows:

1. Observed initial shearing rigidity of the cast-steel box joint is higher than that of built-up joints.
2. Cast-steel box joints exhibit a clear yield point with little strain hardening after yielding.
3. In the case of built-up box joints, the stress–strain curve after yielding increases gradually as a result of the residual stress existing in the joint, which in turn is a consequence of the thermal influence of welding during fabrication.

Cast-steel joints are relatively free of residual stress during either the manufacturing or fabrication; consequently, the mechanical property of the material is well preserved and is reflected in the behavior of the panel zone.

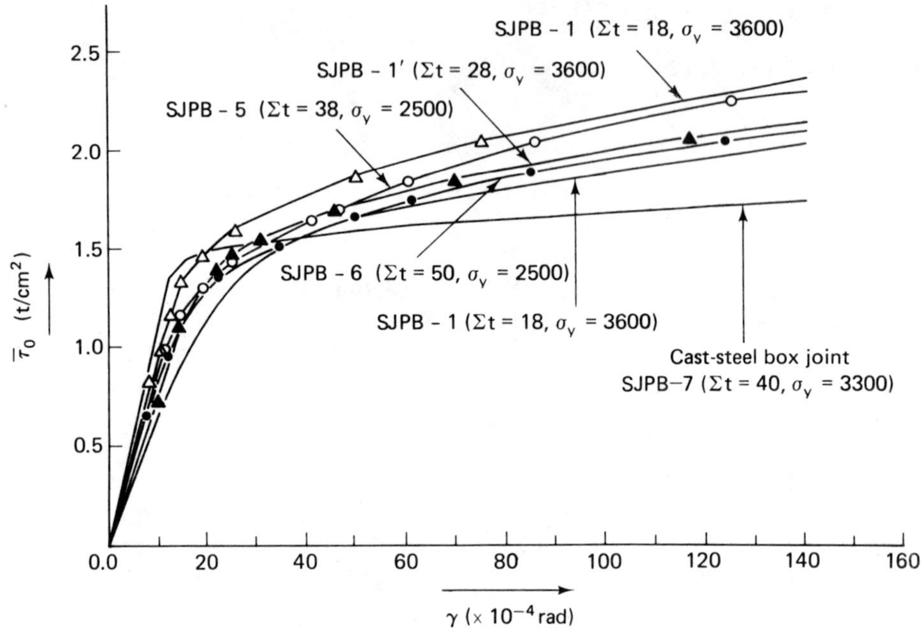

σ_y = yield point, normalized to yo = 2.46 t/cm^2
$\bar{\tau}_0$ = mean value of shearing stress
γ = shearing strain
Σt = total thickness of panels

Figure 12.1   Results of two-way connection test.

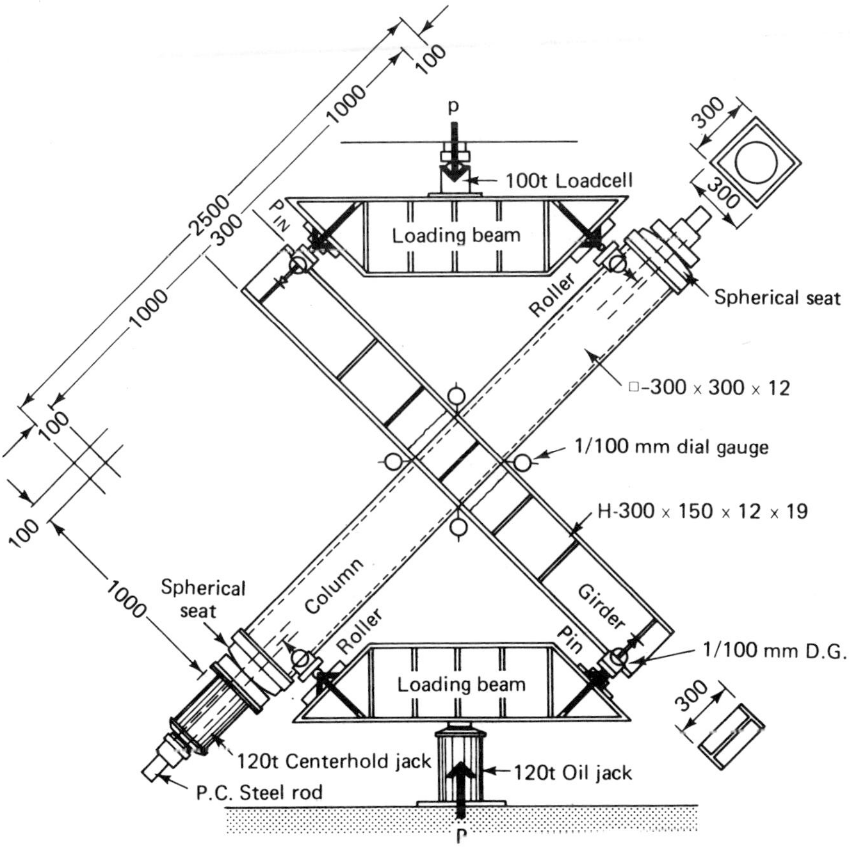

Figure 12.2 Specimen and load-measuring mechanism.

Photo 12.1 Cast-steel box joint.

367

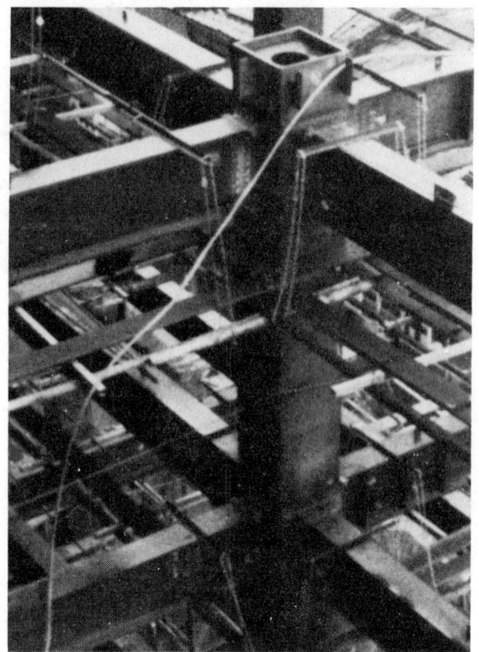

Photo 12.2 Steel frames with cast-steel box joints.

OFFSHORE CONCRETE STRUCTURES:
WAVE LOADS ON GRAVITY PLATFORMS

*Per K. Larsen**

INTRODUCTION

During the last few years governmental agencies in a number of countries have issued regulations for the design, construction, and certification of offshore structures. The design regulations for fixed offshore structures issued by the Norwegian Petroleum Directorate give specific rules regarding load determination, stress analysis, and foundation testing and design [15]. The main points regarding the environmental loadings can be summarized as follows.

Environmental loads are caused by waves, currents, wind, ice, and snow. Hydrodynamic loads, such as wave and current loads, must be specified by a design wave, a design spectrum, or long-term statistical models. The loads must be determined from model tests or by computational methods provided that their accuracy is verified by model tests or full-scale measurements. Local loads caused by slamming or shock pressures are to be assessed.

These regulations went into effect as of May 1977 and reflect to a large

*Associate Professor of Civil Engineering, The Norwegian Institute of Technology.

extent the state-of-the-art concerning wave-load determination for gravity platforms. The purpose of this paper is to elaborate on various aspects of wave-load determination in the context of the experience gained from work performed on CONDEEP, TRIPOD, and other gravity structures. It should be pointed out that this discussion will exclude the dynamic response caused by hydrodynamic loads.

DESCRIPTION OF SEA STATES

The most common models for describing the sea state are:

> Deterministic design waves with given probability levels of occurrence and specified wave heights and periods; specified current profiles due to storm surge and tidal action.

> Short-time stochastic models, which describe the waves as zero mean ergodic Gaussian processes in terms of one- or two-dimensional wave spectral densities.

> Models based on the joint probability density of the significant wave height and average wave period.

Independent of the choice of description, wave-load models must be established to determine the loads caused by a harmonic wave with given period and frequency, either as part of the deterministic analysis or as part of the determination of hydrodynamic transfer functions.

The deterministic description is conceptually the simplest, and results in a design wave with a 100-year return period having wave heights of $H = 28$ to 31 m and periods $T = 14$ to 17 s for North Sea sites. These harmonic design waves bear little resemblance to the observed sea states, and the more realistic statistical models are commonly preferred. For short-time stochastic models, one-dimensional wave spectra have been used almost exclusively, with the JONSWAP, Pierson–Moskowitz, and Darbyshire–Scott spectra being the most common. These spectra differ mainly in the energy distribution over the frequency range, with the JONSWAP spectrum being the most peaked. Figure 1 shows three such spectra, all having the same significant wave height, $H_{1/3} = 15.0$ m.

While knowledge regarding the wave conditions in the North Sea was scant when the first structures were being planned, extensive programs for acquisition of environmental data are now under way. The original data generally tended to underestimate severity of the wave climate, as illustrated by the fact the wave height of the 100-year wave at Ekofisk originally was set at a height $H = 24$ m, which has since been increased to $H = 27$ m. There is still considerable uncertainty regarding the spectral representation of the sea state, but most interest is presently centered around the JONSWAP

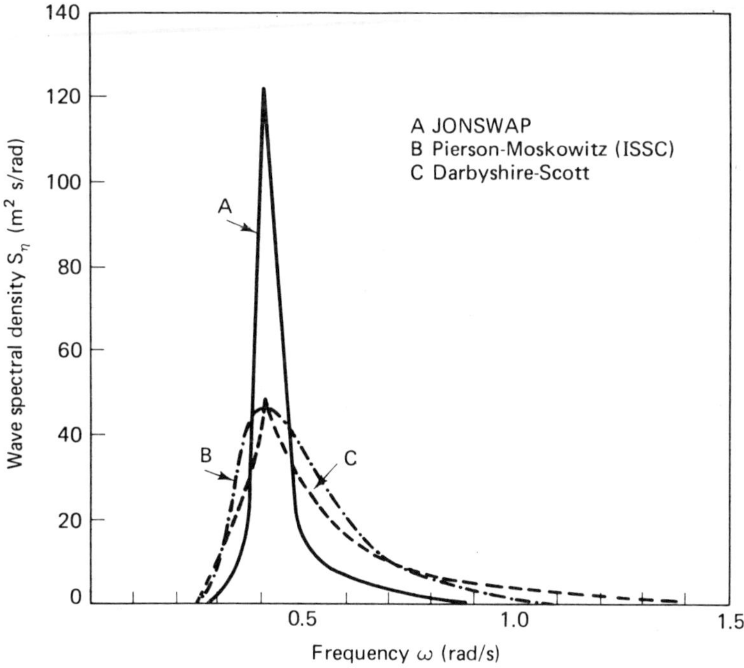

Figure 1 Typical wave spectra representing a sea state with $H_{1/3} = 15.0$ m.

spectrum. It should be pointed out, however, that energy content in the higher-frequency range of the Darbyshire–Scott spectrum may be of particular importance with respect to the fatigue design of the deck structure.

DETERMINATION OF HYDRODYNAMIC LOADS

The fluid-induced loads that act on an offshore structure are of the hydrostatic and hydrodynamic types. The hydrostatic loads are primarily determined by the construction procedure, that is, construction and floating out of bottom section, casting of caisson and towers, submergence for placement of deck structure, and so on. Hydrodynamic loads are caused by wave and current action, and are in general more difficult to determine.

The wave loads on structures are commonly classified as loads due to drag, inertia, and diffraction. Drag loads result from flow separation induced by the relative velocity of the fluid; inertial loads are caused by the pressure gradients caused by the relative acceleration; and diffraction forces appear due to the scattering of waves as the incipient wave interacts with the structure.

These forces may be denoted as global loads, as they can be determined from wave kinematics based on some appropriate wave theory. Local wave loads, such as slamming forces and shock pressures from breaking waves, are more difficult to determine, as they involve a much more complicated physical process. Furthermore, nonstructural aspects of the wave loads, such as scouring, must be considered in the design of the structure.

The following sections briefly review the most common methods for determination of wave and current loads due to drag, inertia, and diffraction loads.

MODEL TESTS

The overall design of the first North Sea concrete structure, the Ekofisk tank, was to a large extent based on wave loads established by model tests. This was primarily because of a lack of confidence in the reliability of the available computational methods when applied to the complicated perforated wall design. The objectives of these model tests were to determine the total wave loads needed as input to the foundation analysis, the loads on exterior and interior walls, and the wave heights inside the wave-breaking wall. For both the Ekofisk tank and the later concrete gravity structures, the inertial loads predominated, and model tests could be carried out without serious scale effects. For most of these structures the model scale varied between the range of 1:80 to 1:120.

The objectives of the later model tests for CONDEEP, TRIPOD, and so on, were primarily concerned with the feasibility of the design with respect to foundation loads and forces on structural components such as towers. A typical test program was concerned with the following aspects:

1. Total horizontal and vertical forces and overturning moment at the mudline.
2. Total forces and moments on individual towers.
3. Wave-induced pressures on caisson walls and top shells.
4. Increase in wave-crest elevation, column run-up, and local shock pressures due to wave–caisson interaction.
5. Determination of allowable sea-state parameters for dowel penetration during installation on the field (hydroelastic tests).

Both regular and irregular waves with and without superimposed wind were used; the main input parameters were wave height and period, direction of the incident wave, and type of wave spectrum.

For gravity structures the reliability of such tests is quite good. Since the drag force constitutes only 5 to 10% of the total force on the structure, the scale effects related to differences in the Reynolds number in model and prototype can be neglected. For jacket-type structures these scale effects are a major problem, because it is impossible to obtain similitude in Reynolds number in the available model tanks. A jacket member with a diameter of 1.0 m subjected to a wave with $T = 14$ s and $h = 30$ m will have a Reynolds number of 7×10^6, while a Froude model in a 1:15 scale with a wave height of 2.0 m has a Reynolds number of only 1.2×10^5. This means that even for such large scales the model will lie in the subcritical region, while the prototype is in the supercritical range (Fig. 2).

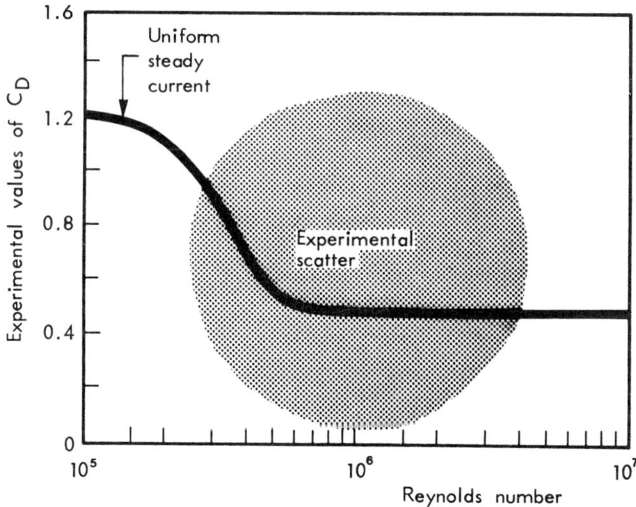

Figure 2 Drag coefficient C_D as function of the Reynolds number.

For the preliminary design of large submerged bodies, wave-load calculations based on the diffraction theory or the finite-element method may be unfeasible because both methods require access to a large computer. For this purpose simple design formulas have been derived for the inertial loads acting on such structures [9,17]. The pressure distribution based on the undisturbed incipient wave is here integrated over the body surface and then modified by experimentally determined force coefficients to yield total forces and moments acting on the body. Extensive model tests have been carried out to determine such coefficients for bodies of various geometrical shapes and wave height/water depth ratios.

Model tests on gravity structures have almost exclusively been carried out with rigid models; the dynamics of the fluid–structure interaction have been neglected. For the structural components in the caisson structure (top shells, cylindrical panels, etc.), the natural frequencies are much higher than the frequencies of the wave-force excitation, and the dynamic effects are small. The overall dynamic properties of the gravity structures are governed largely by the stiffness and damping properties of the foundation, which are frequency-dependent and hence difficult to represent in a hydroelastic model. Furthermore, the natural frequencies of the structure will in a Froude model, quite often fall outside the frequency domain of the wave-generating equipment. However, hydroelastic model tests have been carried out in order to verify some of the concepts used in the dynamic analysis.

COMPUTATIONAL METHODS

Morison's equation

Morison's equation was originally formulated on the basis of measured wave loads on piles, and gives the wave load per unit length of one-dimensional bodies. The equation is based on the assumption that the presence of the structure does not significantly alter the flow field, that is, that the cross-sectional dimensions of the body are small compared to the wave length.

For a slender cylindrical body perpendicular to the flow field, the total hydrodynamic force per unit length is

$$F_t = \tfrac{1}{2}\rho C_D D \,|\dot{u} - \dot{r}|(\dot{u} - \dot{r}) + \rho A C_M \ddot{u} + \rho A C_M'' \ddot{r} \qquad (1)$$

where ρ = water density
D, A = cross-sectional diameter and area
C_D = drag force coefficient
$C_M = C_M' + C_M''$ = virtual mass coefficient
C_M' = Froude–Kryloff force coefficient
C_M'' = inertial force coefficient
$\dot{u}, \ddot{u}$ = water-particle velocity and acceleration
$\dot{r}, \ddot{r}$ = structural velocity and acceleration

There exists considerable uncertainty regarding the validity of Morison's equation for oscillatory flow around cylinders. This pertains particularly to the drag term, where experiments give C_D values with large scattering (Fig. 2). Depending on the Reynolds number and surface roughness, C_D is normally taken in the range $C_D = 1.2$ to 0.6, even though values as high as $C_D = 1.6$ have been measured for oscillatory rectilinear flow [16]. The inertial coefficients C_M' and C_M'' are most commonly taken as unity. However, for values of the ratio of cylinder diameter to wave length $D/L > 0.2$, diffraction effects become important, and the inertial coefficient should be varied according to diffraction theory [13] (Fig. 3).

For structures with members not perpendicular to the flow field the velocities and accelerations are commonly decomposed vectorially into normal and tangential components, and eq. (1) is applied to the normal component. Similarly, water-particle velocities due to currents are superimposed vectorially on the wave flow field.

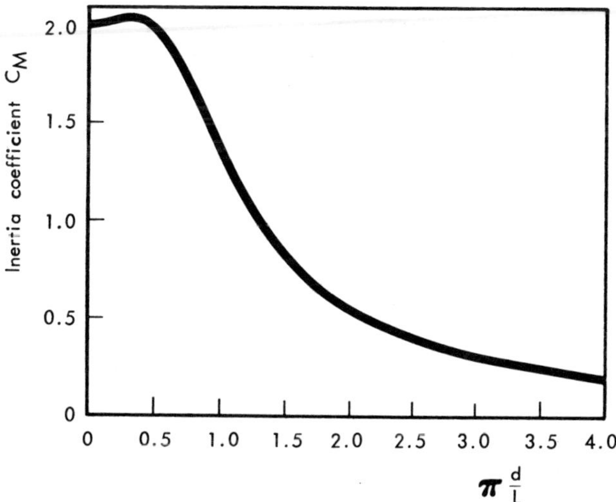

Figure 3 Inertial coefficient C_M based on diffraction theory.

For most gravity platforms the structural velocity $\dot{r}$ is small ($\dot{r}_{max} <$ 0.2 m/s for $H = 30$ m and $T = 15$ s) and is commonly neglected, unlike the water-particle velocity, when computing the wave loads. The fluid–structure interaction described by the drag term in eq. (1) is represented by the introduction of a damping term in the dynamic equilibrium equations of the structure.

Definition of the diffraction problem

When the characteristic dimension D is not negligible compared with the wave length L (i.e., $D/L > 0.2$), the interaction between structure and incident wave must be considered. Viscous effects are neglected, and the governing equation for an incompressive inviscid fluid is given by the Laplace equation,

$$\nabla^2\phi = \frac{\partial^2\phi}{\partial x^2} + \frac{\partial^2\phi}{\partial y^2} + \frac{\partial^2\phi}{\partial z^2} = 0 \tag{2}$$

where ϕ is the velocity potential. Within the context of a first-order (linear) wave theory, the linearized boundary conditions for a submerged body interacting with the fluid are (Fig. 4)

$$v = \phi_{,y} = 0 \qquad \text{for } y = -d$$

$$\phi_{,y} - \frac{\omega^2}{g} = 0 \qquad \text{for } y = 0 \tag{3}$$

$$\phi_{,n} = 0 \qquad \text{on all structural boundaries}$$

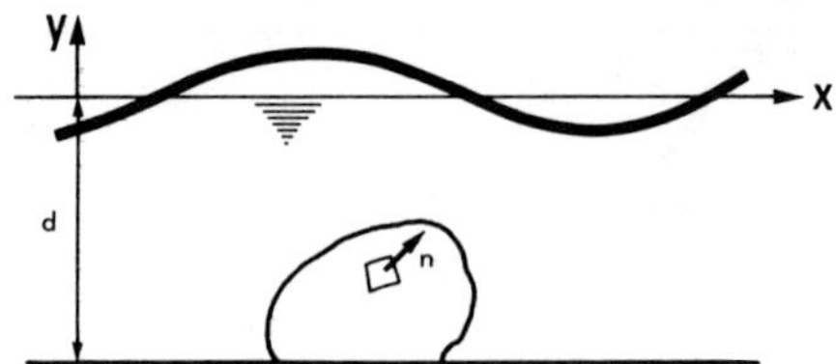

Figure 4 Definition of fluid–structure interaction for submerged body.

Alternatively, if the structure has some normal velocity v_n^s, the latter boundary condition is replaced by

$$\phi_{,n} = v_n^s \tag{4}$$

For physical reasons it is further required that the interaction effects tend to zero as the distance from the structure goes to infinity. This radiation boundary condition is commonly given as [20]

$$\lim_{r \to \infty} \sqrt{r}\,(\phi_{,n} - ik\phi) = 0 \tag{5}$$

where $k = 2\pi/L$ is the wave number and r is the distance from the structure.

The solution of the given boundary-value problem can be obtained by interior (Rayleigh–Ritz, FEM) or exterior, that is, boundary (Treffz), methods, or by a combination of the two types [20]. The former method is based on an expansion of the field variable ϕ in terms of coordinate functions (shape functions) that satisfy exactly neither the field equation nor all the boundary conditions. The generalized coordinates (nodal parameters) are determined from some variational principle such that the governing equations are satisfied approximately. In the boundary-solution method the fied variable ϕ is expanded in terms of coordinate functions that satisfy the field equation exactly. The unknown generalized coordinates are now determined according to averaged boundary conditions.

For boundary-value problems where the domain has a complicated geometrical form the FEM has been the most powerful method because of the difficulty involved in constructing global coordinate functions that satisfy the field equation. However, for problems concerning half-spaces or domains that tend to infinity in one direction, the boundary methods are quite attractive, as is shown in the case of the diffraction method.

Boundary integral methods

In the context of offshore structures the most common boundary integral method is the diffraction, or source-sink, method [2,3,5]. (The name is misleading, since diffraction problems can also be solved by, for example, the FEM.) Here the velocity potential is given as

$$\phi(x, y, z) = \phi_i(x, y, z) + \phi_s(x, y, z) \tag{6}$$

where ϕ_i represents the potential of the incident wave and ϕ_s is the scattering potential caused by the fluid-structure interaction. The value of ϕ_i is given by Airy's first-order wave theory.

The scattering potential at a given point (x, y, z) in the fluid is expressed in terms of sources distributed over the surface of the body,

$$\phi_s = \frac{1}{4\pi} \iint_S f(\xi, \eta, \zeta) G(x, y, z; \xi, \eta, \zeta)\, dS \tag{7}$$

where (ξ, η, ζ) are the coordinates of a source on the surface, $G(x, y, z; \xi, \eta, \zeta)$ is the source potential (Green's function) at the point (ξ, η, ζ), and $f(\xi, \eta, \zeta)$ is the unknown source distribution. As is typical for boundary integral methods, G must satisfy the differential equation and all boundary conditions except the one related to the normal velocity of the body, eq. (4). An extensive discussion of the choice of source functions and the computational aspects of the method is given by Garrison [6].

For a fixed, rigid body the kinematic boundary condition on the body results in the equation

$$-f(x, y, z) + \frac{1}{2\pi} \iint f(\xi, \eta, \zeta) \frac{\partial}{\partial n} G(x, y, z; \xi, \eta, \zeta)\, dS + 2\frac{\partial}{\partial n}\phi_i(x, y, z) = 0 \tag{8}$$

Generally, this integral equation should be satisfied at all points on the body surface, but in practice eq. (8) is satisfied by a collocation method or by some averaging procedure using weighting functions. The result of any such method is a set of linear algebraic equations with discrete values of $f(\xi, \eta, \zeta)$ as unknowns. Garrison uses a collocation method, where the source strength is assumed constant over an area element associated with each collocation point [6]. With this procedure bodies of any geometric shape can be treated, subject only to possible numerical sensitivities.

With the discrete values of the source function $f(\xi, \eta, \zeta)$ determined, the total potential $\phi = \phi_i + \phi_s$ is known, and wave heights, velocities, accelerations, and wave pressures can be determined in the caisson neighborhood. Total forces and moments acting on the body are obtained by integrating the wave pressures, and hydrodynamic added mass and damping can similarly be computed by exciting the body in the desired modes.

While the use of simple sources for the Green's function is the most common method, Garrison also indicates the possibility of applying doublet distributions over the surface of bodies with both exterior and interior fluid domains [6]. In this case the normal velocity is continuous over the wall thickness, while the potential ϕ undergoes a jump. By the introduction of a relationship between normal flow velocity and potential differential over the wall thickness, this type of doublet distribution should allow the modeling of pervious boundaries, such as the wave-breaking walls of the Ekofisk tank.

Finite-element methods

Although the finite-element method (FEM) has been extensively used for fluid-flow problems, its application to wave-load calculations has been limited. For some structure–fluid interaction problems (i.e., dam–reservoir problems) a structure has commonly been modeled by elements, while closed-form solutions have been sought for the fluid domain [11]. When elements were used to model the fluid proper, the radiation boundary condition was often neglected by discontinuing the element mesh at some arbitrary distance from the structure. The accuracy of some of these methods may be sufficient for a fluid–structure system subjected to high-frequency earthquake loading, but is in general unacceptable for diffraction problems where the scattering of waves is significant.

The problem of the radiation boundary condition is also encountered in soil mechanics problems related to the dynamics of footings on elastic or viscoelastic half-spaces. Here, the structure and a subdomain of the foundation are modeled by elements, while the radiating boundary of the element mesh is represented by discrete dashpots or by closed-form solutions for the wave equation in the infinite soil [12].

A systematic discussion of the use of FEM in diffraction problems, posed in a variational form, is given by Zienkiewicz and coworkers [20]. The structure and the neighboring fluid are here represented by a finite-element model, and four methods are proposed for the representation of the infinite fluid:

1. Boundary dampers derived from the functional equivalent of the radiation boundary condition.
2. An exterior element based on global coordinate functions that satisfy the two-dimensional wave equation and the radiation boundary condition.
3. An exterior element based on source distribution on the radiating element boundary.
4. Infinite elements with shape functions that display both periodicity and exponential decay in the radial direction (i.e., normal to the surface of the structure).

The boundary damping elements include the possibility of partially reflecting boundaries. Other elements that represent pervious boundaries by allowing potential discontinuities can be used for structures with pervious boundaries similar to the Ekofisk tank.

Zienkiewicz and coworkers apply these methods to two-dimensional diffraction problems for both submerged bodies and bodies that penetrate the water surface [20]. For the latter case the results are excellent, but the accuracy for the wave forces on the submerged body are less accurate. The latter deficiency may be resolved by using three-dimensional elements in the domain next to the body.

Whereas some of these may be classified as pure finite element formulations, others may be considered as the coupling of an element model representation of the structure with a boundary integral solution for the fluid. The relative merits of these formulations, when compared to the diffraction theory of the preceding section, are difficult to assess. Such comparisons can only be made on the basis of extensive computer studies concerning computing time, accuracy, and numerical sensitivity. At the present time, only limited experience with these methods has been gained, while the diffraction method has been widely used. However, the FEM may be particularly promising for cases where nonlinear effects are important (i.e., structures in shoaling waters).

Hydrodynamic impact loads

Wave impact loadings may be classified as either entry-type loads or shock pressures. The former type is primarily associated with steel jackets, where horizontal truss members in the free surface zone are subjected to repeated slamming loads. These loads are caused by the sudden immersion of the member as the wave passes and data regarding the load intensities are available from both experiments and theory [4]. Shock pressure is caused by breaking waves hitting a structural member protruding above the water line, and is well known from breakwater design. No prototype measurements of shock pressure are available for cylindrical shafts of a concrete gravity platform, but model tests in wind–wave flumes have indicated the possibility of such loads.

Since most data regarding shock pressures from breaking waves are based on measurements of breakwaters in shallow water and on planar surfaces, such loads on gravity platforms must be estimated from theoretical considerations. Such considerations are based on estimates of the wave height and period for the maximum breaking deep-water wave with a given probability of occurrence, and for the water jet velocities of this wave. A model describing the interaction between the water jet and the circular body, and the resulting pressure distribution, must also be obtained.

Based on work by Nath and Ramsey, a sea state described by a JONSWAP spectrum with $\gamma = 7.0$, $T_p - 15.5$ s, and $H_{1/3} = 14.3$ m will result in a maximum breaking wave of $H = 21.2$ m and $T_b = 8.9$, and a return period of 100 years [19]. By scaling experimental jet velocities, which are described in [14], the corresponding prototype jet velocity is estimated to be $\dot{u} = 16.5$ m/s.

Two alternative models are used to determine the impact pressure. One is based on the occurrence of a hammer-type shock (i.e., a water jet hitting a plane structure), where the maximum pressure can be determined as

$$P_{\max} = \rho \cdot c_e \cdot \dot{u} \tag{9}$$

where ρ is the water density, $\dot{u}$ the water velocity, and c_e the velocity of sound in the mixture of water and air in the jet. The value of c_e, and hence the maximum pressure, strongly depend on the percentage of air in the jet as depicted in Fig. 5. For the breaking wave height given above, eq. (9) leads to maximum pressures of 23.5 and 0.33 MPa, respectively, for 0% and 30% entrained air. In the case of a breaking deep-water wave hitting a circular surface, the latter value is considered the most likely and has been used for design.

Alternatively, the shock pressure can be estimated by the use of the formula governing the slamming force on cylinders hitting a water surface,

$$p = \tfrac{1}{2}\rho C_s \dot{u}^2 \tag{10}$$

where C_s is a slamming coefficient. Depending on the cylinder shape, surface roughness, and so on, experiments have given values on the order of $C_s =$

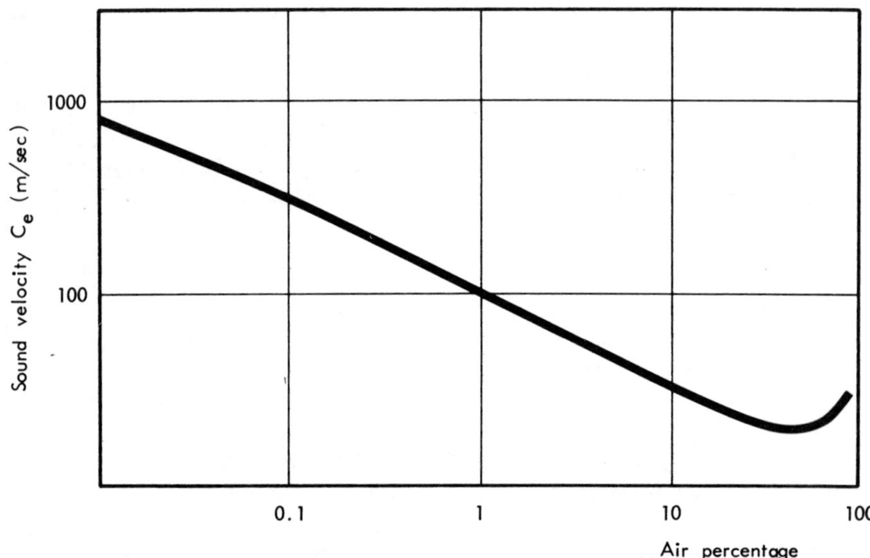

Figure 5   Velocity of sound in water–air mixture.

0.6 to 3.0. With the previous wave-particle velocity, this gives a slamming pressure of $p \simeq 0.4$ MPa. However, as the mechanics of a breaking wave hitting a vertical cylinder differ somewhat from those of a cylinder hitting a plane water surface, eq. (10) should be used with care.

For CONDEEP towers, the total impact load from breaking waves is estimated to be $P = 2$ MN, or 5% of the total wave load acting on an individual tower. This load necessitates a 4% reinforcement in the tower wall in the splash zone.

APPLICATIONS TO CONDEEP-TYPE STRUCTURES

For a typical gravity platform the wave loads are computed according to the Morison equation for the towers, and from the diffraction theory or a semi-analytical method [9] for the caisson. The justification for this approach is illustrated in Fig. 6, where the regions of predominant diffraction and viscous effects are indicated, and where the positions for caisson and towers are given for a wave period of $T = 15$ s. Both the undisturbed and the diffracted flow fields based on first- or fifth-order wave theory have been used in the Morison equation, and the accuracy of the various approaches has previously been compared with the results of model tests [7,8]. Table 1 gives the total forces at the mudline, based on three alternative computational procedures for caisson/tower calculations:

Case 1 : Diffraction/diffracted first-order incident wave.
Case 2 : Diffraction/diffracted fifth-order incident wave.
Case 3 : Semianalytical/undisturbed fifth-order incident wave.

All three analytical models show excellent agreement with the model tests, the maximum deviation being 3 to 5%. As the accuracy of the model test is approximately $\pm 2.5\%$, the deviations are clearly negligible. Because of its computational simplicity, the semianalytical method has been used in

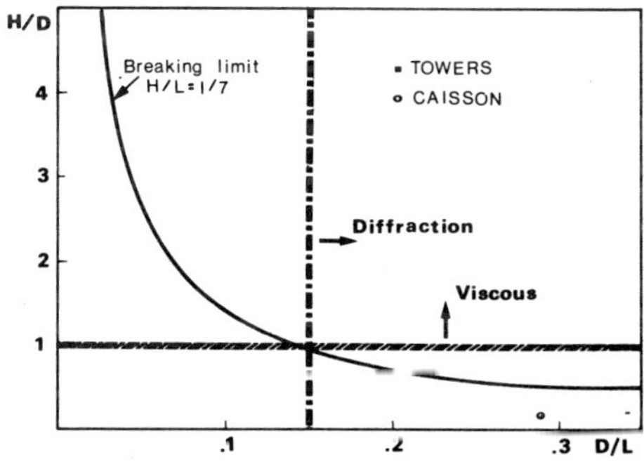

Figure 6 Classification of viscous and diffraction domains.

Table 1

WAVE FORCES ON A CONDEEP-TYPE STRUCTURE
($T = 15.5$ s, $H = 29.2$ m)

Case	Horizontal (MN)	Vertical (MN)		Moment (MN·m)
		Up	Down	
1	447	380	340	1.35×10^4
2	453	373	332	1.30×10^4
3	470	388	357	1.36×10^4
Model tests	455	380	340	1.30×10^4

the global, stochastic dynamic analysis of the soil–structure water system [1], while diffraction analysis has been used primarily for design wave calculations. However, for caisson structures that deviate from simple circular, rectangular, or hexagonal shapes, diffraction analysis must be used exclusively.

The effect of wave–structure interaction on the flow field is given in Fig. 7, where the normalized velocity and acceleration profiles over the caisson are depicted for a wave period of $T = 15$ s [7]. Whereas diffraction effects are negligible near the free surface, increases of approximatly 30% in velocity and acceleration are observed near the caisson. The use of the undisturbed flow field will hence underestimate the tower forces in the region near

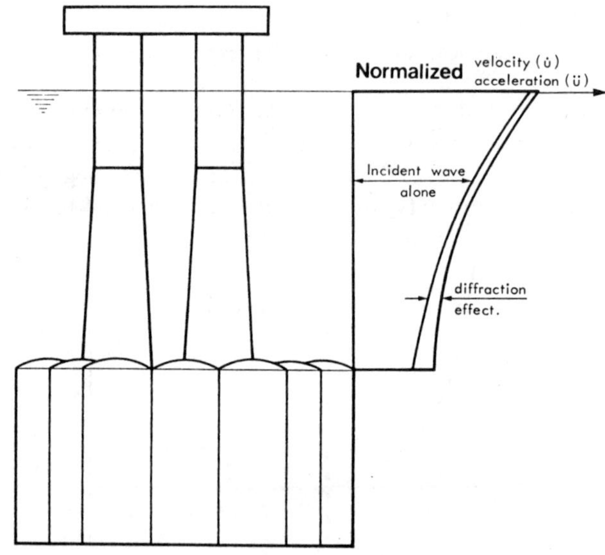

Figure 7   Normalized velocity and acceleration field over caisson.

the caisson, while the use of the increased water surface elevation as givne by fifth-order wave theory better represents the forces near the top of the towers. The net result is, as seen from Table 1, quite satisfactory. The drag-force contribution can be estimated from Fig. 8, where both the drag and the inertial component of the wave forces are given. Because the two force components are 90° out of phase, the net contribution of the drag term is clearly small [7].

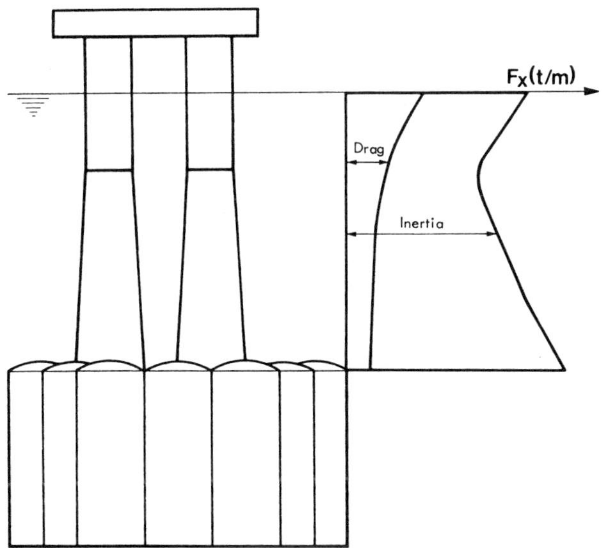

Figure 8 Distribution of drag and inertial loads on columns.

Owing to the simple geometry of the CONDEEP caisson, diffraction analysis can be carried out using only a few points or elements, as illustrated by the two meshes in Fig. 9. The relative errors in the force components as compared to a 36-point mesh are given in Fig. 10. Whereas the horizontal force can be determined quite well with a rather crude mesh, the vertical force and overturning moment are more sensitive to the discretization of the caisson surface. This indicates the importance of the pressure distribution around the perimeter of the top surface of the caisson.

By exciting the caisson as a rigid body, the hydrodynamic added mass and the hydrodynamic diffraction damping in translation and rotation are also determined by diffraction analysis. The frequency-dependent damping coefficients for a typical CONDEEP caisson are depicted in Fig. 11. Note that the damping is negligible in the range of the resonant frequencies of a typical platform ($f = 0.25$ to 0.6 Hz). It should also be pointed out that the hydrodynamic radiation damping gives approximately 0.2% of the critical damping in the fundamental mode.

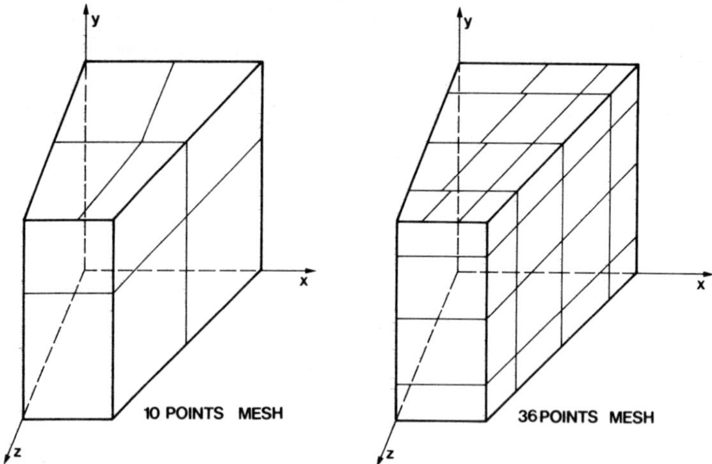

Figure 9   Typical models for analysis of CONDEEP caisson.

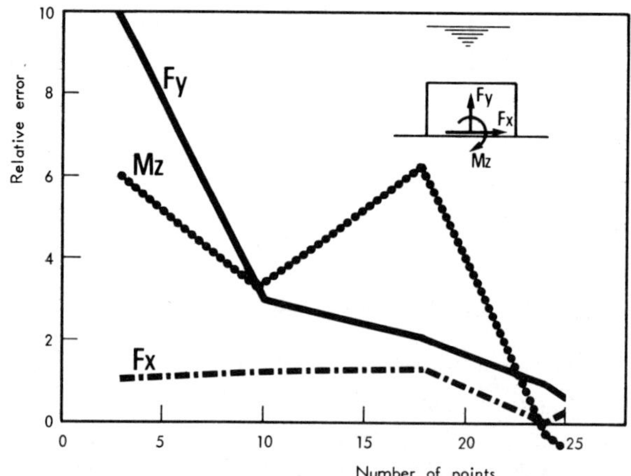

Figure 10   Relative errors of diffraction forces (36-point mesh taken as basis).

For the present generation of gravity platforms the caissons have generally been deeply submerged, and linear diffraction analysis gives satisfactory results. For most CONDEEP platforms the ratio between caisson height and water depth is approximately 0.4, and for SEATANK and ANDOC, even less. The only nonlinearity that is commonly included in the diffraction analysis of these structures is that of the velocity-squared terms in the Bernoulli equation, which give rise to the difference between the upward and downward vertical forces in Table 1. However, for gravity structures in shallow water,

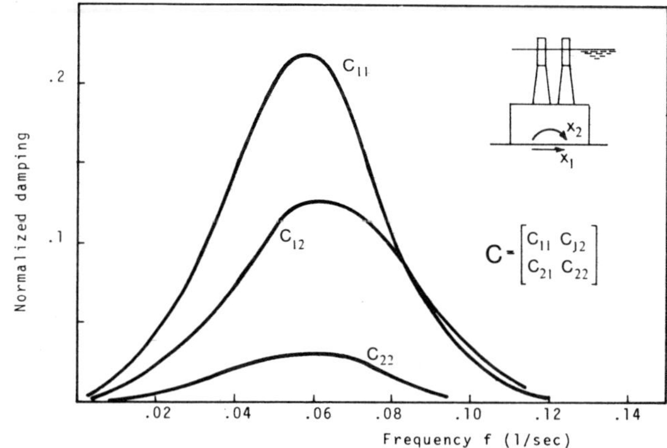

Figure 11 Hydrodynamic damping coefficients due to diffraction.

nonlinear effects may be important and a linear incident wave may thus not be valid.

For a deterministic analysis the nonlinearities in the Morison equation are easily handled. The use of a nonlinear wave theory or the vectorial superposition of current velocities is straightforward. However, in a stochastic analysis the nonlinearities caused by the distorted free surface, the drag term, and the wave–current interaction are more complicated. The drag component is commonly treated by using a linearized wave-force spectrum based on the assumption that water-particle velocities and accelerations are two-component Gaussian processes. The effect of the current cannot be incorporated directly into the hydrodynamic transfer functions, as this requires some knowledge of the correlation of wave heights and current velocities. Instead, the interaction can be treated by using a current-modified wave spectrum such as that given by Huang and coworkers [10]. It should be noted, however, that this procedure gives forces and moments that are significantly larger that those obtained by a deterministic analysis based on vectorial superposition and hence should be used cautiously [1]. The nonlinearities introduced by the use of the correct sea surface elevation cannot be treated by the linearized wave-load spectrum. Approximate hydrodynamic transfer functions are thus determined according to fifth-order wave theory and a frequency-dependent wave height. This variation in height is chosen so as to reflect the frequency variation of the wave spectrum, and the transfer functions are then obtained by scaling down the forces by the appropriate wave height. Even though this procedure gives the correct sea surface elevation and hence the correct load distribution for a single harmonic wave, there are considerable uncertainties regarding the net effect of this procedure when applied to a wave train consisting of many superimposed components. The same uncertainties apply to the use of the frequency-dependent inertia coefficient C_M, as given in Fig. 3. This coefficient is derived from the flow field of a single harmonic incident wave. However, it is not clear that the flow field of a small-amplitude, high-frequency wave superimposed on a large-amplitude, low-frequency wave will yield the same wave forces as are obtained by summing the individual components obtained from their respective C_M values.

SUMMARY AND CONCLUSIONS

The wave loads acting on large submerged bodies can be determined with sufficient accuracy by linear diffraction (source-sink) analysis. However, more research is needed to develop the FEM as an efficient tool for such calculations, and to assess the relative efficiency and accuracy of the two methods. The validity of the Morison equation is still being questioned, and extensive research is in progress to develop a more rational and accurate method of wave-load prediction for jacket-type structures. A main point of the latter effort is to develop a method to describe the oscillatory flow field around structural members due to an irregular sea state.

The recent increase in construction of offshore facilities in exposed areas has put new requirements on the efficiency and safety of offshore operations. Model testing of structures and equipment used in these operations has necessitated that existing laboratory facilities be expanded so that tests can be carried out on larger-scale models. To satisfy these requirements a new wave basin with dimensions $80 \times 50 \times 10$ m is being constructed at the Norwegian Institute of Technology. The basin will be capable of reproducing two-dimensional sea states with superimposed wind and currents and with a maximum wave height of 0.9 m.

The mechanism of local impact loads on curved surfaces caused by breaking waves is still insufficiently understood, and reliable mathematical models are unavailable. Because scale effects are important, model tests cannot yield conclusive answers to the problem and prototype measurements are needed.

While the wave loading on gravity structures can be determined within an accuracy of 5% by existing analytical methods, the uncertainties regarding the determination of wave heights are of the order of 10 to 15%. More research is needed before adequate sea-state descriptions can be obtained. Special emphasis should be given here to the determination of reliable two-dimensional wave spectra and long-term models. For a number of structures on the Norwegian continental shelf, an extensive program to acquire environmental and performance data (E and P data) has been initiated. The E data (i.e., data regarding wind, waves, and current climate) will be available shortly, and the P data, regarding structural response, will be released after a two-year period.

REFERENCES

[1] BELL, K., HANSTEEN, O. E., LARSEN, P. K., AND SMITH, E., "Analysis of a Wave–Soil–Structure System. Case Study of a Gravity Structure," BOSS'76, The Norwegian Institute of Technology, Aug. 1976.

[2] BOREEL, L. J., "Wave Action on Large Off-Shore Structures," Proceedings Off-Shore Structures, ICE, London, 1975.

[3] FALTINSEN, O., AND MICHELSEN, F., "Motion of Large Structures in Waves at Zero Froude Number," Proceedings of the International Symposium of the Dynamics of Marine Vehicles and Structures in Waves, University College, London, 1974.

[4] FALTINSEN, O., KJAERLAND, O., NØTTVEIT, A., AND VINJE, T., "Water Impact Loads and Dynamic Response of Horizontal Circular Cylinders in Offshore Structures," Paper 2741, OTC 1977.

[5] GARRISON, C. J., AND CHOW, P. Y., "Wave Forces on Submerged Bodies," *Proceedings, ASCE*, Vol. 98, No. WW3, 1972.

[6] GARRISON, C. J., "Hydrodynamic Loading of Large Offshore Structures," International Symposium on Numerical Methods in Offshore Engineering, Swansea, Jan. 1977.

[7] GARRISON, C. J., TØRUM, A., IVERSEN, C., LEIVSETH, S., AND EBBESMEYER, C. C., "Wave Forces on Large Volume Structures—A Comparison Between Theory and Model Tests," Paper 2137, OTC, 1974.

[8] GARRISON, C. J., AND STACY, R., "Wave Loads on North Sea Gravity Platforms—A Comparison of Theory and Experiments," Paper 2794, OTC, 1977.

[9] HAFSKJOLD, P. S., TØRUM, A., AND EIE, J., "Submerged Offshore Concrete Tanks," XXIII International Navigational Congress, PIANC, Ottawa, 1973.

[10] HUANG, N. E., CHEN, D. T., TUNG, C. C., AND SMITH, J. R., "Interaction Between Steady Non-uniform Currents and Gravity Waves with Applications for Current Measurements," *Journal of Physical Oceanography*, Vol. 2, Oct. 1972.

[11] LIAW, C. Y., AND CHOPRA, A. K., "Earthquake Analysis of Axisymmetric Towers Partially Submerged in Water," *International Journal of Earthquake Engineering and Structural Dynamics*, Vol. 3, 1975.

[12] LYSMER, J., AND WAAS, G., "Shear Waves in Plane Infinite Structures," *Proceedings, ASCE*, Vol. 98, No. EM1, Febr. 1972.

[13] MACCAMY, R. S., AND FUCHS, R. A., "Wave Forces on Piles: A Diffraction Theory," Beach Erosion Board, Tech. Memo 69, 1954

[14] NATH, J. H., AND RAMSEY, F. L., "Probability Distributions for Breaking Wave Heights," *Proceedings of the International Symposium on Ocean Wave Measurements Analysis*, New Orleans, Sept. 1974.

[15] "Regulations for the structural design of fixed structures on the Norwegian Continental Shelf," Norwegian Petroleum Directorate, Apr. 1977.

[16] SARPKAYA, T., "Vortex Shedding and Rough Circular Cylinders," BOSS'76, Norwegian Institute of Technology, Aug. 1976.

[17] TØRUM, A., LARSEN, P. K., AND HAFSKJOLD, P. S., "Offshore Concrete Structures—Hydraulic Aspects," Paper 1947, OTC, 1974.

[18] TØRUM, A., AND LARSEN, P. K., "Some Considerations on Shock Pressures from Breaking Deep Water Waves Against Circular Fixed Columns," Report STF60 A75086, River and Harbour Laboratory, The Norwegian Institute of Technology, Aug. 1975.

[19] VAN DORN, W. G., "Wave Breaking in Deep Water. Part II of Technical Progress Report on Stable Floating Platforms," Scripps Institution of Oceanography, University of California, San Diego, June 1974.

[20] ZIENKIEWICZ, O. C., BETTESS, P., AND KELLY, D. W., "The Finite Element Method for Determining Fluid Loadings on Rigid Structures," International Symposium on Numerical Methods in Offshore Engineering, Swansea, Jan. 1977.

OFFSHORE CONCRETE STRUCTURES: STRUCTURAL ANALYSIS AND BEHAVIOR OF CELLULAR STRUCTURES

*Ivar Holand**

INTRODUCTION

A common feature of many of the offshore structures discussed by Apeland is a cellular raft or caisson [2]. CONDEEP caissons have thus far been composed of 19 cylindrical cells (Fig. 1). SEATANK structures have rectangular cells, except for the external walls, which are circular cylindrical. Even the Tripod 300 is provided with cellular caisson foundations. The following discussion refers primarily to CONDEEP structures.

Rafts are exposed to a variety of loads as described by Larsen [13]. External hydrostatic pressure under deep submerging during deck installation largely governs the design of the main dimensions of the cell walls. Hence, only this load will be considered here.

FINITE ELEMENT ANALYSES

Extensive finite element analyses have been run to verify the structural behavior of CONDEEP rafts. Among the programming systems used are NASTRAN, ASKA, and SESAM 69. The analyses are often carried out on

---

*Professor, dr. techn., The Norwegian Institute of Technology.

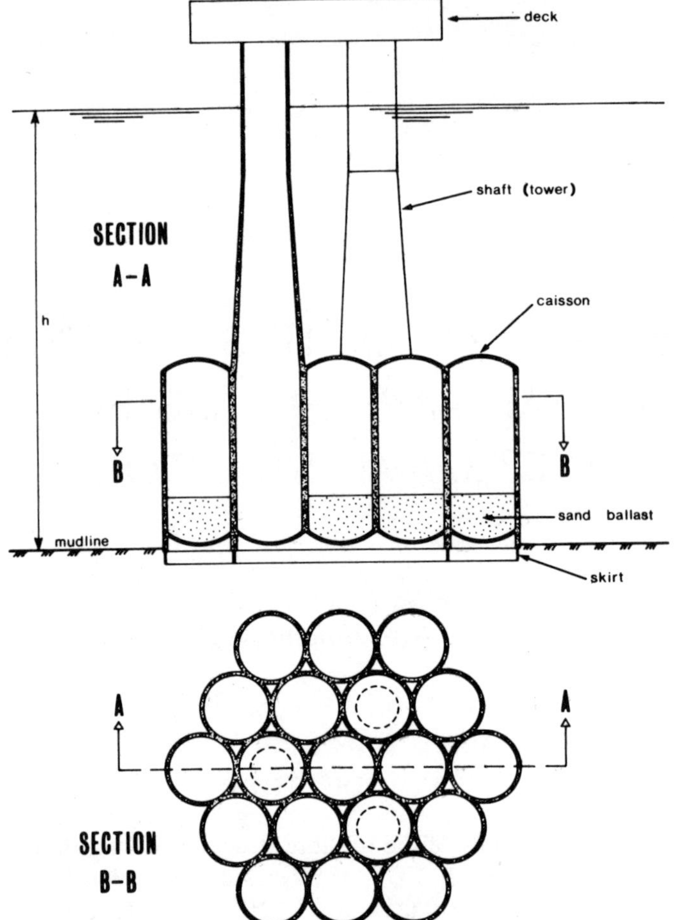

Figure 1   CONDEEP platform.

various levels, starting, for instance, with single cell analyses and ending with a global analysis of the complete structure.

A global analysis of a CONDEEP caisson using SESAM 69 has been described by Lindvik [14]. The SESAM 69 system has been described by Araldsen [3]. From a modest start by the system SESAM [6]; developed at the Norwegian Institute of Technology, the SESAM 69 system has been extended to handle large static and dynamic analyses through extensive use of a multilevel super-element technique [6].

The finite element CONDEEP model described by Lindvik comprises 6000 elements assembled through eight levels of superelements. The number of elements is reduced on the first level to less than 300 superelements of fourteen different types. The elements are superparametric thick shell and

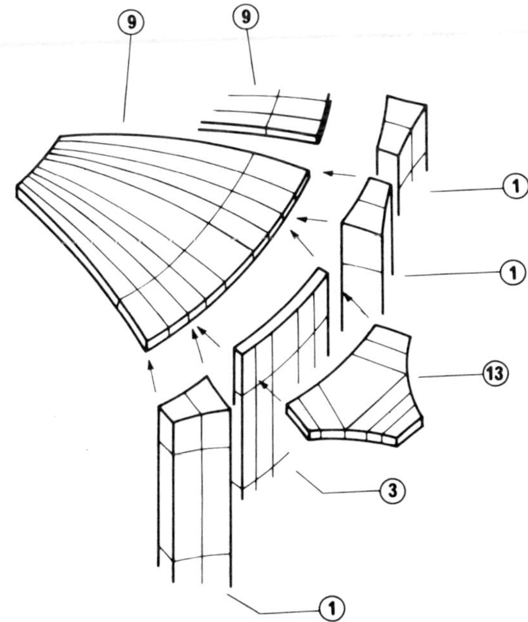

Figure 2 CONDEEP. Basic elements and four different superelements.

isoparametric solid. The combination of basic elements to first-level super-elements, and their assemblage, are indicated in Fig. 2. Eleven global load cases were considered. They were described by 180 different cases on the first level to account for the different locations of each superelement type.

The total CPU times on a UNIVAC 1108 were:

Data checks	2 hours
Assembly and reduction	16 hours
Stress calculation	2 hours
Load combination	1 hour
	21 hours

The large amounts of input and output are particularly important. It is vital that the program system include a data generator and facilitate data checking by element mesh plot facilities. The strength of the reinforced shell section must be checked at a large number of points. This is not a straightforward task because the state of stress is expressed by six resultants:

Membrane forces: N_x, N_s, N_{xs}

Bending moments: M_x, M_s, M_{xs}

Computer programs that check the concrete capacity and determine the necessary reinforcement by nonlinear equilibrium iterations have been developed [4]. To save computer time and output, such analyses should be post-processed. Thus, the main program should store the basic results in a file so that the designer can select successively the necessary data and dimension checks, preferably in an interactive operation.

CYLINDRICAL SHELL THEORY ANALYSES

Although the finite element method is widely used in such computations, a warning should be raised against the tendency to consider it as the only method for complex structural analyses. In particular, faster methods by which the main features of structural action can be described are very useful in parametric studies to establish principal dimensions. A final finite-element analysis can be run to account for structural performance not considered by the more rapid, but perhaps less accurate, methods.

A modern interpretation of traditional cylindrical shell theory opens the possibility of a fast and efficient analysis of cellular structures. Such an approach is outlined below.

The traditional theory of cylindrical shells is based on a set of linear first-order differential equations in the two surface coordinates. The equations express equilibrium of stress resultants, constitutive laws relating stresses to strains, and kinematic relations between strains and displacements. Traditionally, these equations were reduced to higher-order differential equations. If secondary terms were ignored, simpler equations could be obtained, the neatest and most systematic of which have been derived by Donnell [8]. The equations could be integrated by expanding them into single trigonometric series of the Levy type. The present solution uses trigonometric functions in the direction of the generator imposing the conditions of simple support at the shell ends.

The traditional shell theory is also utilized in the approach described here, the only difference being that the equations in the circumference coordinate for individual shell panels are integrated by the transfer matrix method. This method uses the set of first-order equations directly, and can easily include secondary terms. Thus, the accuracy of the shell theory used is on the same level as that of the improved Donnell theory [9]. The advantage of this accuracy is that the limiting case of pure arch action is described correctly. Details of the approach are omitted here but may be found in [10] and [12].

The transfer matrix method is used only to establish stiffness matrices for the individual shell panels. The remaining statically indeterminate problem of connecting shell parts, walls, and edge beams is handled by the matrix displacement method.

The system analysis is carried out for each term in the series, with shell element matrices referring to a single term. The longitudinal distribution of each term is determined according to either a sine or a cosine function.

The shell elements may be connected to beam elements along an arbitrary generator. The beam displacement is described by four displacement quantities, as indicated in Fig. 3. These displacements and the corresponding stress resultants are expanded in trigonometric series in the same manner as the shell quantities. A stiffness relation is established for each term by using conventional bending theory and St. Venant's theory of torsion. The centroid is assumed to coincide with the shear center. If the beam is of an open, thin-walled type, the beam must be subdivided into plate elements, or beams or shells, in a manner which accurately satisfies the assumptions.

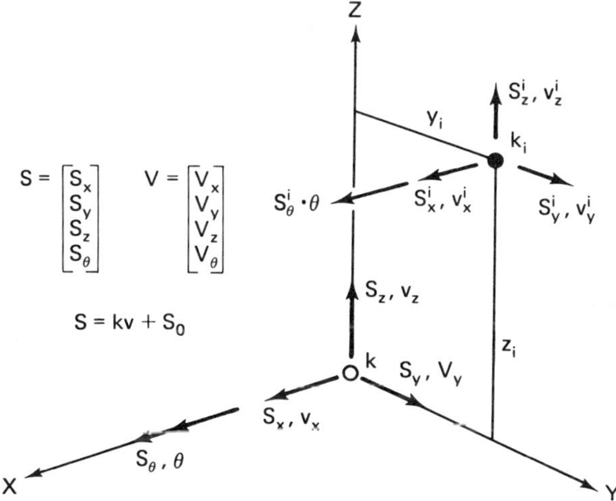

Figure 3   Forces and displacements at nodal point (k) and slave point (k_i).

The stiffness matrix of a beam element is 4×4, and the displacement components and stress resultants entering this matrix are referred to the centroid k (see Fig. 3). The term S_x is the shear force applied at the center line.

When technical bending and St. Venant torsion are discussed, the four displacement components of the center line of the beam define uniquely the corresponding displacement components of an arbitrary generator of the beam. Thus, the centroid may be considered to be the master node and all other points in the cross section to be slaves that follow the master. Figure 3 shows stress and displacement components at a node k and a slave point k_i.

Although shell elements are generally connected to slave points, all quantities in the stiffness and load assembly must be referred to the nodes. Thus, the shell elements are considered to be eccentrically connected to the nodes, and their stiffness matrices are transformed accordingly.

The approach of introducing only the beam centers as nodes and transforming shell element matrices to these node parameters is one of the decisive features of the method. It minimizes the number of unknowns, permits simple and systematic programming, and accounts for all linear relationships between the displacements of different points in a beam *ab initio*.

Because the analysis is based on a Fourier expansion in the longitudinal direction and on a numerical integration along the circumference, all loads can be handled in principle. In the program, however, the longitudinal distribution is restricted to a piecewise linear variation, which is obtained by the superposition of linear cases extending over part of the structure.

The load on the shell elements is expected to be constant in the circumferential direction within each segment. A rapid variation in the *s*-direction must therefore be reflected by narrow segments. All beam loads are line loads, including line moments about the *x*-axis.

NUMERICAL RESULTS; STRUCTURAL BEHAVIOR

As a numerical example using the method outlined, consider the analysis of a cellular platform raft of the type shown in Fig. 1. A radial external load with a slight variation over the height is assumed as shown in Fig. 4. Owing to symmetry only $\frac{1}{12}$ of the total section need be analyzed (see Figs. 5 and 6).

The analysis assumes a beam element to be associated with each nodal

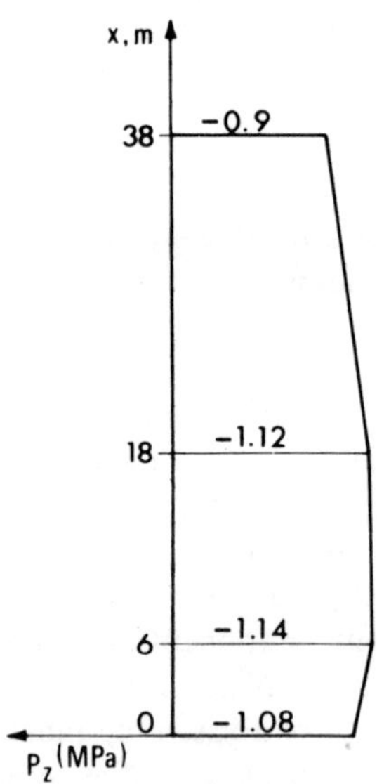

Figure 4 Distribution of radial load on cell walls.

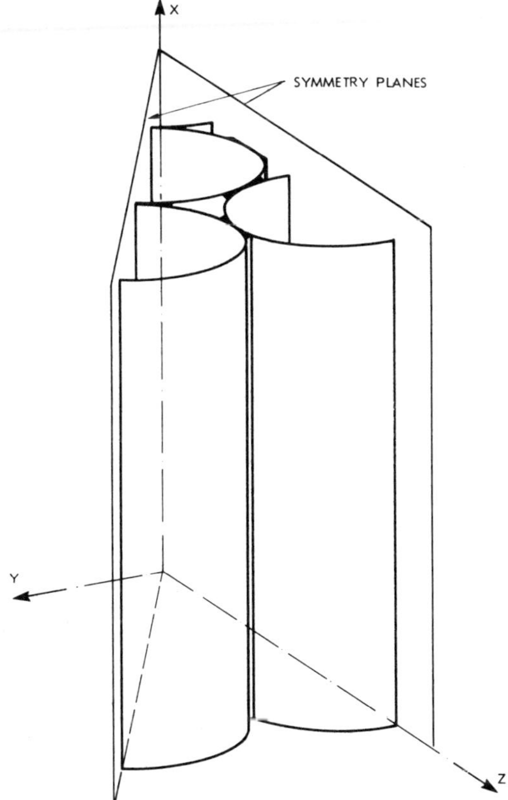

Figure 5 Example structure.

point. At node 9, where only two shell elements meet, the beam must be assigned zero stiffness. This node has been introduced to reduce the number of element types.

A more refined model was used in the numerical results included. In this model fictitious shell elements were introduced between the shell and beam elements to simulate the flexibility of the connection between shell and beam elements. Characteristic data for the two models are given in Table 1. The example shows that the number of degrees of freedom is kept small, even for a very complex structure. This is also reflected in the computing time, which was 30 s CPU in an analysis with 15 Fourier terms for the refined model.

Typical results from the analysis are shown in Figs. 7, 8, and 9, which indicate direct forces, moments, and shear forces in a ring section at mid-height, 19 m from the bottom. The simpler model of Fig. 6 showed an increase of peak moments by about 5%, whereas the differences in hoop forces were negligible.

A typical feature of the results is the pronounced deviation from mem-

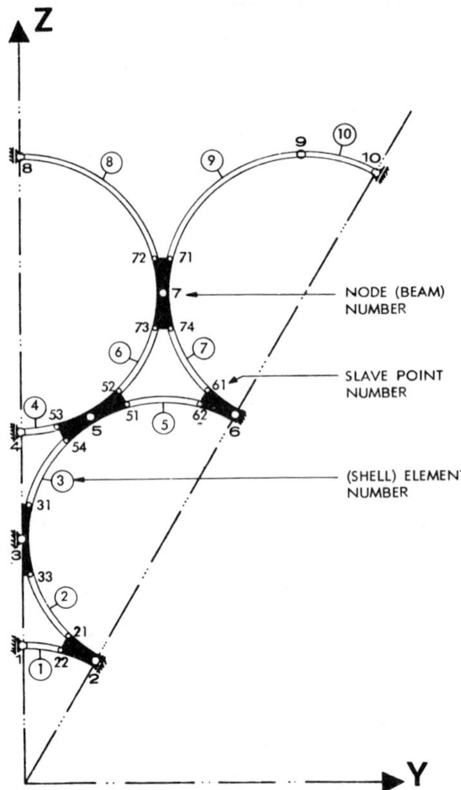

Figure 6 Modeling of the example structure in Fig. 5.

Table 1

CHARACTERISTIC DATA OF RAFT ANALYSIS

Model	Fig. 6	Refined
Shell segment types	1	2
Shell element types	3	4
Shell elements	10	24
Beam element types	3	3
Beam elements (nodes)	10	24
Degrees of freedom	26	88

brane action. The narrow shell elements (numbered 2 to 7 in Fig. 6) behave like clamped arches with large negative and positive bending moments. The wider shell elements, 8 and 9, have a moment distribution similar to that for short shell roofs. The rigid connection between the cells prevents development of the full membrane hoop force, except in the midportion of the wide shells.

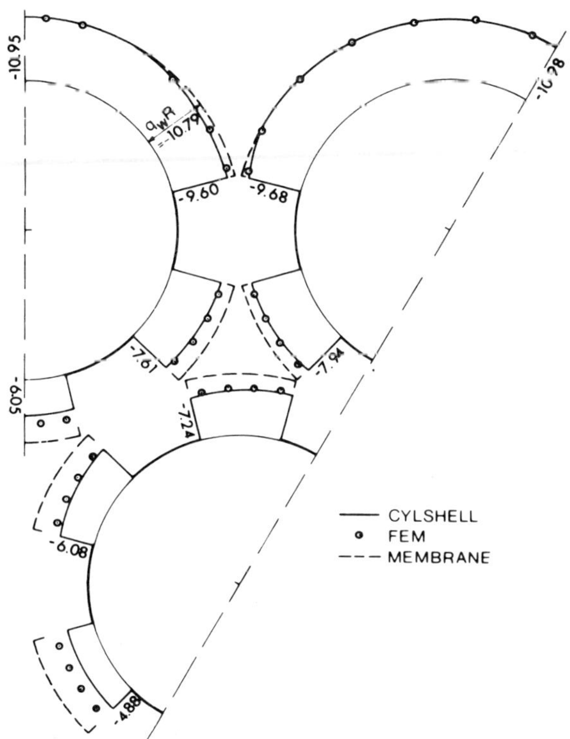

Figure 7 Hoop forces N_s (MN/m) 19 m above the bottom.

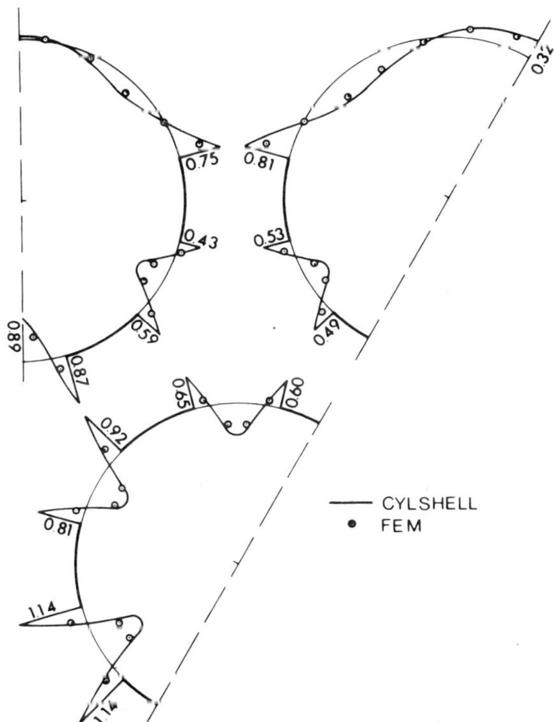

Figure 8 Hoop moments M_s (MN) 19 m above the bottom.

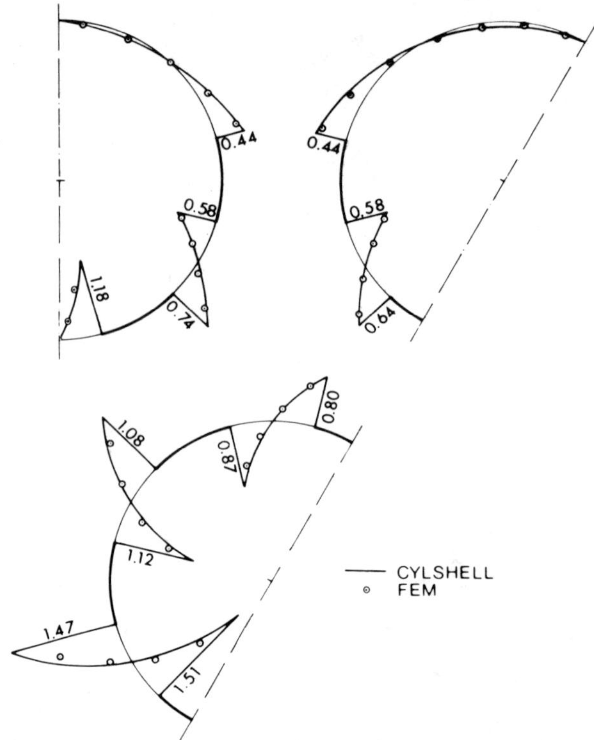

Figure 9   Shear forces Q_s (MN) 19 m above the bottom.

Figures 7, 8, and 9 also include results from a finite-elements analysis accomplished by CDC Data Centers in Stockholm and Oslo, on behalf of A. S. Höyer-Ellefsen, Oslo. The results are included here by courtesy of A. S. Höyer-Ellefsen. For this analysis the MSC/NASTRAN program was used and adapted for quadrilateral shell elements, beam and spring elements, and rigid connections. The model included the complete structure.

As a whole, the results obtained by the two models agree remarkably well. Some discrepancies appear in the narrow shell panels in the interior part of the raft (elements 2, 3, and 4, Fig. 6). Figure 7 shows that these shells develop larger hoop forces in the finite element model than in the shell model. The difference can be explained by the compression of the upper and lower dome systems (see Fig. 3) that occurred in the former. That effect is not accounted for in the shell analysis, wherein end sections were assumed to be rigid in their plane. A correction for the effect is possible, if desired.

An advantage of the shell model is that it describes accurately the gradients of moments and shear forces in the boundary areas.

Thus far, the forces and moments at midheight have been discussed.

Experience from analysis of cylindrical shell roofs might lead to the conclusion that these stress resultants govern the design. Both finite element and CYLSHELL program analyses reveal, however, that hoop moments increase toward the cell ends. To investigate that phenomenon a number of analyses of single panels were conducted using the CYLSHELL program; these analyses are described in [7]. Some of the results quoted here refer to panels with dimensions that are typical for CONDEEP structures. The panels are simply supported at the ends and clamped at two vertical edges. Uniform radial pressure is applied as the external load. To obtain convergence, as many as 40 terms were included in the series. Results were checked by the finite difference program STAGS [1], and were essentially the same, except that the peaks were less accurately described by the finite difference mesh. As indicated in the figures, symmetry was included in the computations.

Figure 10 shows the distribution of hoop moments along a vertical edge. From a fairly constant level in the middle portion the moments rise to a peak more than twice as large close to the end. This end effect is typical when radial deflection is restrained. It should not, however, be mistaken

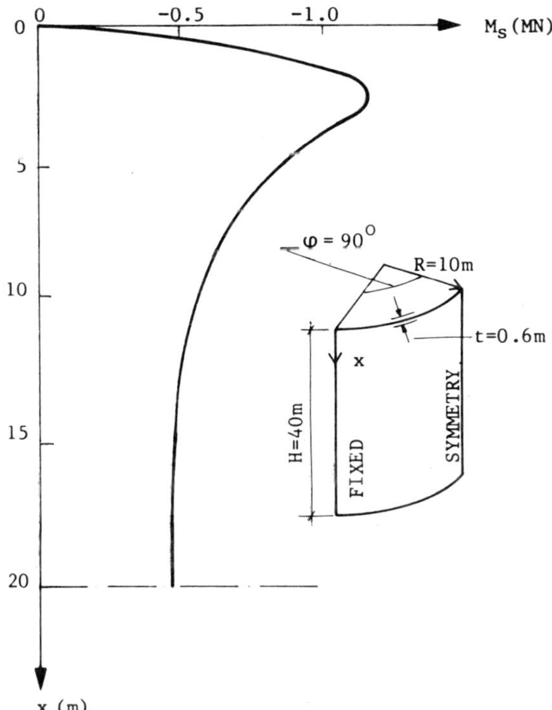

Figure 10 Hoop moments M_s along a fixed vertical edge. External radial pressure 1 MPa.

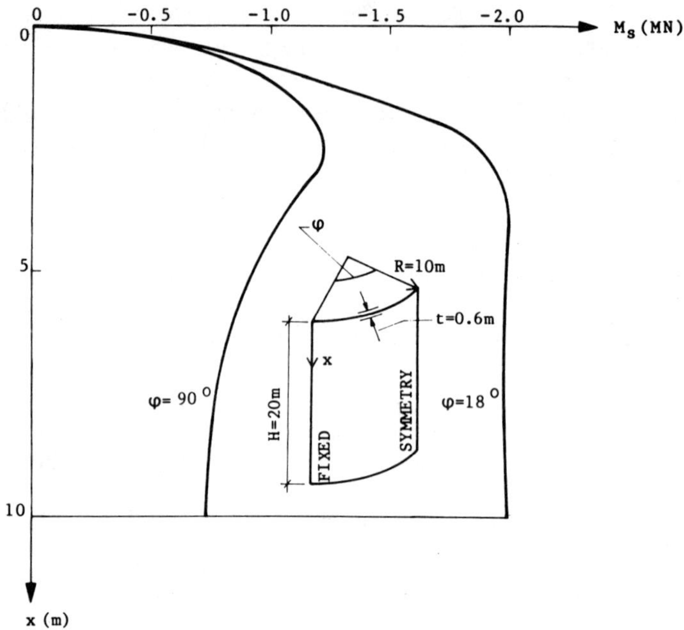

Figure 11 Hoop moments M_s along a fixed vertical edge. External radial pressure 1 MPa.

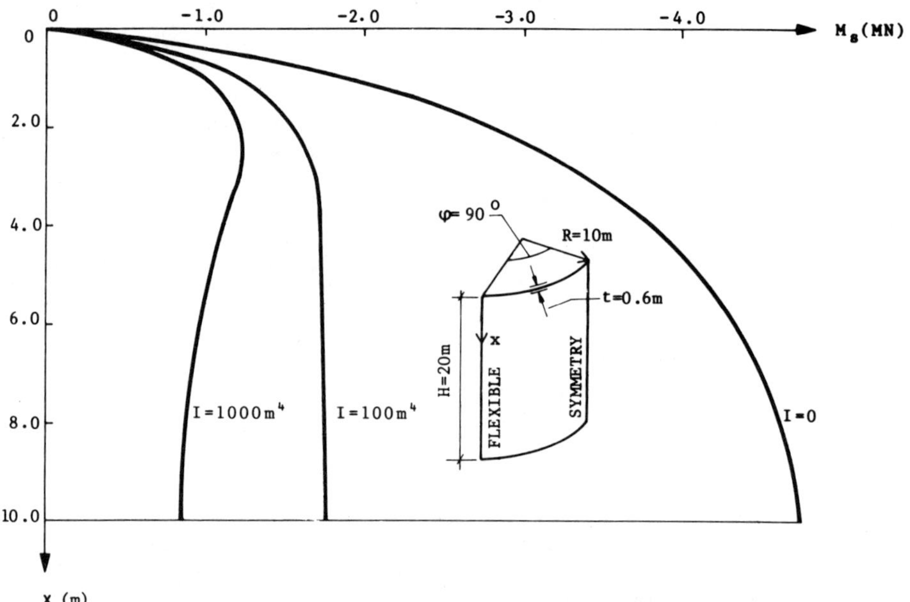

Figure 12 Hoop moments M_s along a flexible vertical edge. External radial pressure 1 MPa.

for the Poisson effect that accompanies the longitudinal end moment that would appear in an axisymmetric case. That effect amounts to about only 10% of the moment shown in Fig. 10.

Figure 11 shows panels with opening angles of different size. In the case of smaller openings ($\phi = 18°$), the whole level of hoop moments increases, and the end peak disappears. Thus, narrow panels act more like arches, and shell effects become secondary.

The reason similar effects are unknown in cylindrical shell analysis is illustrated by Fig. 12. This figure shows the distribution of hoop moments along a vertical edge where all displacement and rotation components except the tangential displacement are zero. This displacement depends on the flexural stiffness of an edge beam. Increased moments toward the ends appear only for a very rigid edge beam ($I = 1000$ m^4). For more flexible edge beams the hoop moments increase and the end peaks disappear. Ordinary roof shells are of this type.

CONCLUSION

By using cellular structures with circular cylindrical walls, one may take advantage of direct arch action to transfer radial water pressure. However, a pure membrane action with hoop forces $p_w R$ can develop only when free motions are allowed. Prevention of displacements along vertical edges causes considerable reduction in hoop forces, accompanied by hoop moments and shear forces. This effect is particularly pronounced in narrow panels. In panels with large opening angles hoop moments are particularly large close to the ends. These effects necessitate that shell analysis of cellular rafts be accurate, particularly if limited cracking is a primary aim.

Considerable insight into the structural behavior of a particular design can be gained by inexpensive analyses based on traditional shell theory. Final designs have so far been based on comprehensive finite element analyses.

NONLINEAR ANALYSES

To assess accurately the ultimate limit capacity, a nonlinear analysis should be carried out. With the dimensions used in CONDEEP structures, nonlinear geometric effects are secondary and can be estimated by rather simple analytical considerations [11]. Nonlinear material effects are, however, important at high load levels. They can be considered accurately only in a complete nonlinear shell analysis that involves numerous difficulties both in describing material properties and in computing response. Results are, however, available for axisymmetric concrete tanks. Some results were reported at the recent conference on Structural Mechanics in Reactor Technology [5].

REFERENCES

[1] ALMROTH, B. O., ET AL., "Collapse Analysis for Shells of General Shape. Vol. II, User's Manual for the STAGS-A Computer Code." *Lockheed Palo Alto Research Laboratory Technical Report AFFDL TR-718*, Palo Alto, California, Mar. 1973.

[2] APELAND, K., "Offshore Concrete Structures: Design Aspects of Offshore Gravity-Base Structures." Chapter of the present book.

[3] ARALDSEN, P. O., "SESAM-69. A General Purpose Finite Element Method Program." *Computers and Structures*, Vol. 4, No. 1, Jan. 1974, pp. 41–68.

[4] BERG, S., "A Model for Capacity Control of a Reinforced Concrete Shell Section." Paper presented at the 4th International Conference on Structural Mechanics in Reactor Technology, San Francisco, 1977.

[5] BERG, S., AND LØSETH, S., "Non-linear Analysis of Prestressed Concrete Reactor Pressure Vessels." Paper presented at the 4th International Conference on Structural Mechanics in Reactor Technology, San Francisco, 1977.

[6] BERGAN, P. G., AND ÅLDSTEDT, E., *SESAM. Et programmeringssystem for elementproblemer*. Division of Structural Mechanics, The Norwegian Institute of Technology, Trondheim, Feb. 1968.

[7] BRYNHILDSEN, T., *Sylinderskall i kontinentalsokkelkonstruksjoner*. Division of Structural Mechanics, The Norwegian Institute of Technology, 1976 (unpublished thesis).

[8] DONNELL, L. H., "Stability of Thin-Walled Tubes Under Torsion." *NACA Report 479*, 1934.

[9] DONNELL, L. H., "A Discussion of Thin Shell Theory." *Proceedings of the Fifth International Congress for Applied Mechanics*, Cambridge, Massachusetts, 1938, p. 66, Van Nostrand, New York, 1939.

[10] FISKVATN, A., AND LØSETH, S., "CYLSHELL—User's Manual." *SINTEF Report STF71 A75034*, Trondheim, 1975.

[11] HOLAND, I., "Ultimate Capacity of Concrete Shells." *BOSS '76 Proceedings*, Vol. 1, The Norwegian Institute of Technology, Trondheim, 1976.

[12] HOLAND, I., FISKVATN, A., AND LØSETH, S., "Analysis of Cylindrical Shell Systems." *IABSE Proceedings P-4/77*.

[13] LARSEN, P. K., "Offshore Concrete Structures: Wave Loads on Gravity Platforms." Chapter of the present book.

[14] LINDVIK, H., "Finite Element Structural Analysis of CONDEEP Offshore Oil Production Platform by SESAM-69." Paper presented at the Finite Element Congress, Baden-Baden, Nov. 18–19, 1974.

OFFSHORE CONCRETE STRUCTURES: DESIGN ASPECTS OF OFFSHORE GRAVITY-BASE STRUCTURES

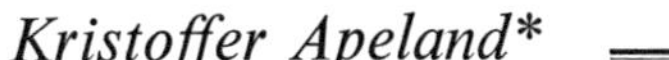

*Kristoffer Apeland**

BRIEF HISTORICAL REVIEW

Although some experience had been gained earlier with offshore concrete lighthouses the major breakthrough in the field of design of offshore gravity base structures came in 1971 when the French C. G. Doris concept for the Ekofisk Concrete Tank for a water depth of 70 m was coupled with the construction technique and construction site of the Norwegian contracting company, Norwegian Contractors.

By the end of 1978 twelve gravity-base production platforms, in addition to the Ekofisk Tank, will have been installed in the North Sea in depths varying from 100 to 150 m. Some characteristics of these platforms are as follows (see [1]).

Ordered by:		Designed by:		Fabricated by:		Installed in:		Year of Installation	
Shell	5	CONDEEP	5	Norway	6	UK	8	1975	3
ELF/Total	4	Doris	3	UK	4	Norway	1	1976	3
Mobil	2	Seatank	3	Sweden	1	UK/Norway,			
						Frigg field	3	1977	4
Chevron	1	Andoc	1	Holland	1			1978	2

*Professor, dr. techn., Partner of Multiconsult, Oslo, and Professor of Building Technology, Oslo School of Architecture.

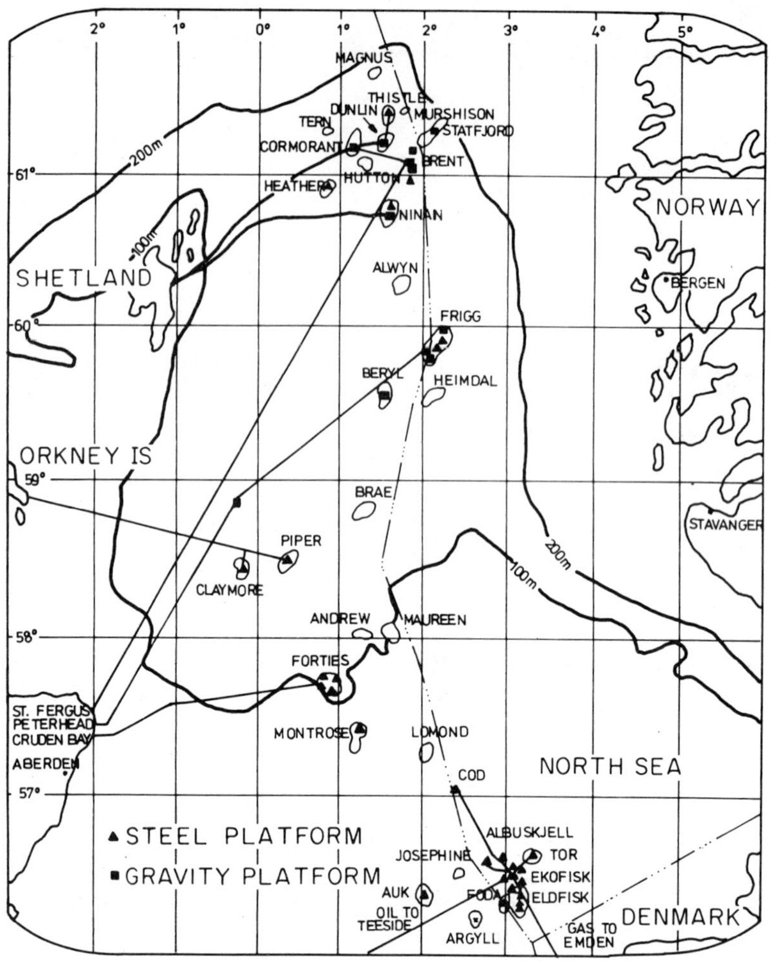

Figure 1   North Sea region.

Although the exploitation rate of the North Sea fields has been slowed by political decisions, there is reason to believe that development will continue, albeit at a more moderate rate.

Because a majority of North Sea oil fields discovered so far are located in deep waters (Fig. 1), it is anticipated that gravity-base platforms for greater water depths will find a market in the reasonably near future.

FUNCTIONAL FEATURES
OF GRAVITY-BASE PLATFORMS

Because temporary oil storage is necessary for continuous operation in offshore regions with rough weather conditions, and because massive concrete structures can easily accommodate oil storage chambers, oil storage capability may be considered the admission ticket to the oil market for concrete structures. When compared with steel structures, concrete structures offer a number of other advantages characterized particularly by safety and economy.

Economics play a decisive part in the operation of an oil field. However, when human life is endangered by fire, explosion, blowouts, or other mischance, operation is stopped irrespective of cost. A properly designed concrete structure offers considerable safety advantages over a steel structure. We see decisions being made along these lines of thought today, for example in connection with the exploitation of the Statfjord field on the Norwegian shelf.

Low maintenance cost is often considered to be a valuable feature of concrete structures. This claim is not obvious, since maintenance must be assessed and related to the lifetime of the structure. Since the problem of corrosion of concrete under offshore conditions has not been fully clarified, the problem of maintenance is still disputed and requires further research and experience.

Owing to the high cost of installation in deep and rough waters, the capability of a concrete platform to carry payload during tow-out to the field has emerged as one of the more important features of the platform. Because the platform during tow-out functions as a barge, it may carry cargo provided that it is designed for such use.

In recent years installations have become increasingly more integrated and correspondingly heavy. This may be illustrated by three examples (see [6]):

1. The CONDEEP platform for Mobil's Beryl field carried a deck load of 15,000 tons when towed out in 1975.
2. The CONDEEP platform for the Statfjord A field carried a deck load of 20,000 tons when towed out in summer 1977.
3. For the Statfjord B field a tender has been offered with a platform carrying 48,000 tons (including the deck), when towed out to the field.

It is obvious that if these large installments can be carried out in a quiet fjord, instead of under rough weather conditions in the field, a considerable savings will result. Moreover, the installation time, which in turn influences the start of production, will be under much better control.

It should be recognized, however, that installation of deep-water platforms in a fjord naturally requires deep water in the fjord itself and along the towing route.

ASPECTS OF CONCEPTUAL DESIGN

General approach

The design of a concrete platform is a challenging optimization problem. One of the main difficulties of the problem is that the structure during its lifetime will go through a number of phases, which, for classification purposes, are normally defined as follows:

1. Construction.
2. Towing.
3. Installation.
4. Operation.

During any of these phases the structure can be subjected to loads and environmental conditions which may differ completely from those that occur during another phase. Moreover, the conditions of any phase may subsequently govern the design of one or more elements of the structure.

In fact, the logical solution in one phase may prove unsuitable in another. As an example, concepts that are ideal for wave-load conditions during the operational phase, have proved to be invalid for towing conditions, and vice versa.

In addition to variations in loads and other conditions during the various phases, a number of major design parameters change from case to case, e.g.:

1. Water depth, soil conditions, and general environmental conditions vary from one location to another.
2. Required payload capacity during towing and the building site, as well·as the towing route, may vary from project to project.
3. Oil storage and deck load capacities during operation will vary from case to case.
4. Operational requirements, functional field layouts, and corresponding safety requirements will vary from case to case.

In view of the vast number of design parameters, and because a number of detailing problems lie in the forefront of research, there is as yet no computer program available by which this general optimization problem may be solved. So far, the conceptual design problem has been solved by an iterative trial-and-error method. In view of the amount of work involved in analyzing various aspects of the problem, however, there is usually not enough time or resources available to complete more than a couple of steps in the iterative process.

Another important consideration should be noted in this connection. During the design process it is of utmost importance to account for the construction problem. Large concrete structures, such as those designed for 150 m of deep water, pose special problems. Consider the construction of inclined cylinders or shafts, which have been proposed in a design that will be discussed later. Experience has shown that joints in concrete structures exposed to sea water are unsuitable, and thus inclined shafts designed by the slip-forming method are now preferred. When the use of inclined shafts was

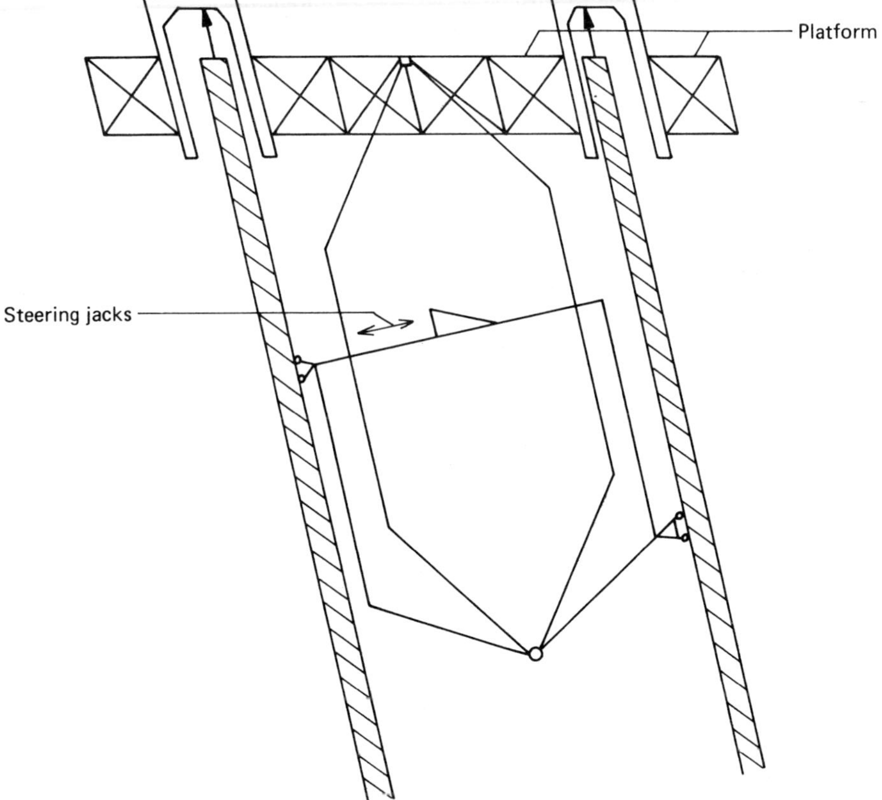

Figure 2 Inclined slip forming.

proposed, the Norwegian contracting company F. Selmer found it necessary to develop a design method and to test it on a large scale (see Fig. 2 and [2]). Not until the method had been proved feasible was it practical to formulate a concept for inclined shafts.

Ideal concepts

Although platforms of the caisson type, such as the CONDEEP platform, have been quite successful, the concepts themselves are far from being optimal. The main drawback is their limited adaptability for various kinds of operation layouts for the production process.

The ideal concept should:

1. Offer adaptable deck layouts with flexible spacing and location of shafts for drilling conductors and oil pipeline risers.
2. Be easily adjustable for variations in payloads during tow out.

3. Be easily adaptable to varying oil storage requirements.
4. Be easily adjustable to variations in soil conditions.
5. Require reliable construction methods.
6. Yield an economical and safe structure.

Contradictions and conflicts clearly exist between these requirements. However, the history of the concrete platform is still young, and much development is ahead.

For illustration of the problems, several projects will be briefly presented.

The Selmer Tripod concept

Consider the Selmer Tripod concept based on the idea of making a more adaptable concept than that of the typical caisson type.

The platform is a monolithic structure with three tapered or bottle-shaped towers as main components. The towers are rigidly connected near the seabed by three heavy box girders, and are also connected through the deck structure, which may be a concrete or steel structure (see Figs. 3 and 4).

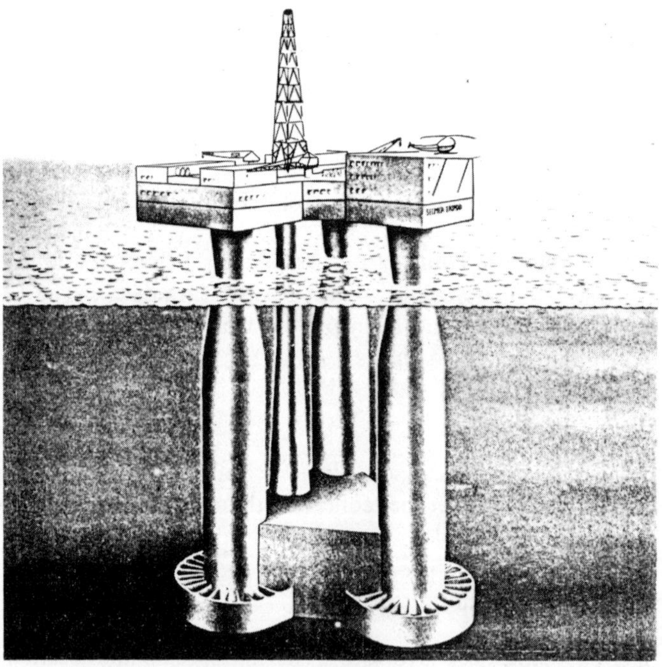

Figure 3   Selmer Tripod.

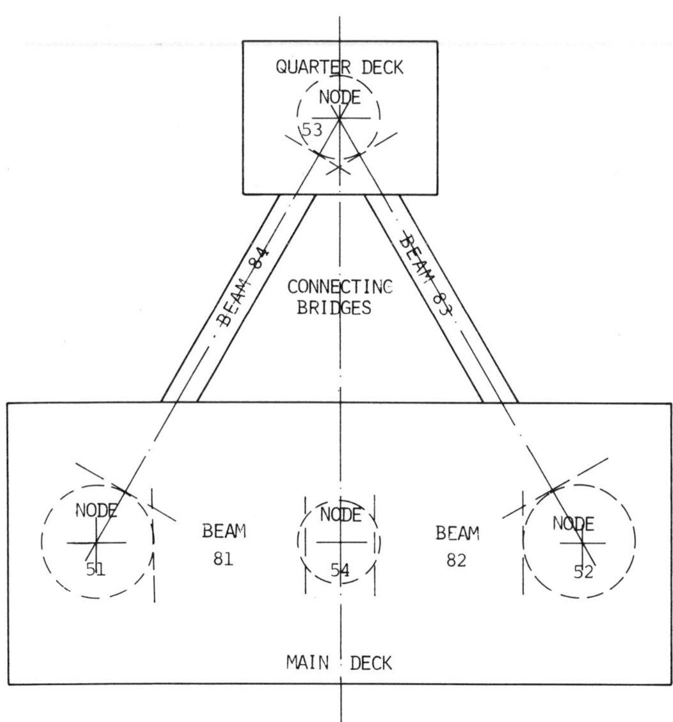

Figure 4 Selmer Tripod, plan of steel deck.

Adaptability, as far as functional requirements are concerned, will be obtained by the following means:

1. Layout flexibility may be obtained by changing the distance between the towers, offering varying deck areas and shaft locations.
2. Variation in payload capability may be obtained by varying either the size and/or the tapering of the towers, or by varying the distance between the towers.
3. Oil storage requirements may be satisfied by altering the diameters of the towers and/or applying the box girders as oil storage chambers.
4. Increase of foundation area for poor soil conditions may be obtained by cantilevering foundations from the towers and/or by utilizing the floor of the box girders as a foundation area.

Conceptual studies for projects, accompanied by tenders, have shown that the concept will be competitive with respect to safety as well as overall economy.

It seems fair to point out that for layout flexibility the triangular plan concept may be a little rigid, and that a concept having a rectangular plan with four towers might have been better. Moreover, for very poor soil conditions it may be difficult to obtain sufficient foundation area without rather large cantilevering, which in turn may result in costly foundations.

The Selmer Tripod 300 concept

Some of the basic ideas of the Selmer Tripod concept for shallow waters (up to 150 m water depth) have been expanded in the Selmer Tripod 300 concept for deeper waters (i.e., up to 300 m water depths; Figs. 5, 6).

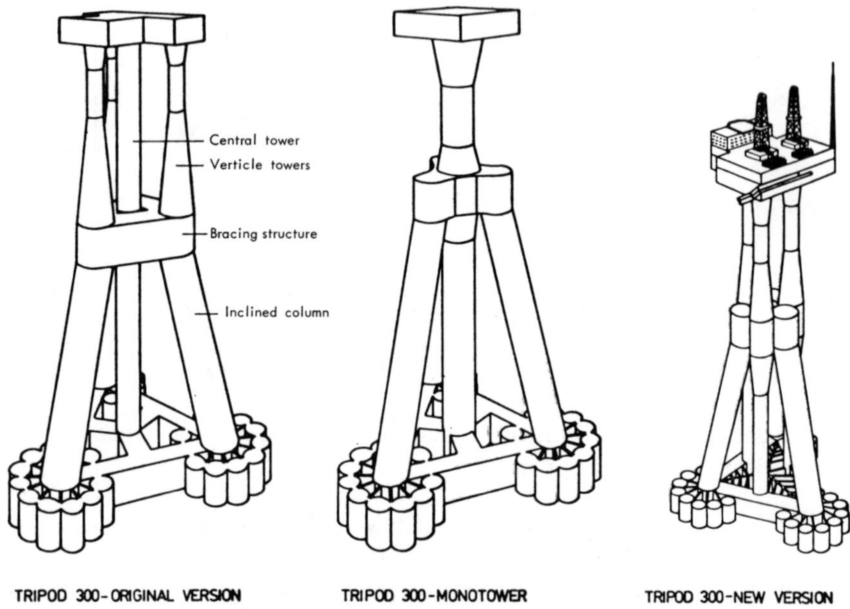

Figure 5   Selmer Tripod 300, three versions.

The Selmer Tripod 300 platform concept is a space frame structure in which the towers of the Tripod have been replaced by partly inclined legs or shafts. As mentioned earlier, the construction technique for slipforming inclined shafts has been developed and tested on a large scale by Selmer. From a static/dynamic point of view the space frame or truss shape has proved favorable.

The Selmer Tripod 300 is more or less as functionally adaptable as the Tripod concept. Thus far the concept has been analyzed for configurations with relatively short distances between the towers at deck level, and several versions have been presented, among them a monotower version for upper levels. Because the shaft diameter dimensions are relatively small, separate cellular caisson foundations may be provided for each shaft.

With the enormous rise in costs for steel platforms as the water depth increases, the Selmer Tripod 300 seems to be a very promising concept for greater water depths, and the Norwegian fjords are deep enough to serve as construction sites for platforms which later are to be installed on the continental shelf. In addition, the concept offers considerable oil storage capability.

Investigations have shown that of the typical caisson-type CONDEEP platform and the Tripod 300 space frame structure, the latter is more economical for water depths of about 200 m.

STRUCTURAL DESIGN ASPECTS

Design criteria and structural codes

Until North Sea development started, offshore activity involving fixed structures had primarily been centered around the U.S. Gulf of Mexico area. The design of the structures was normally based on the API-RP2A code, which was developed with the purpose of covering the field of fixed steel template or jacket structures supported by piles. The API code offers considerable freedom as far as load specification is concerned; see the discussion by Moses in [3].

When concrete gravity-base concepts were presented around 1970, the design criteria were not agreed upon, and it has been necessary to define design criteria from case to case. Over the last few years considerable effort has been put into the development of reliable structural codes. A number of documents have been presented, one of the more recent being "Design Regulations for Fixed Offshore Structures," issued by the Norwegian Petroleum Directorate [4].

The development of a reliable code for gravity-base structures is complicated by the urgency of its need at a time when designers are still in the middle of general discussions on the application of probabilistic analysis and the development of reliability-based code formats. So far, codes for offshore structures have been based on the limit-state method and the application of characteristic values combined with appropriate partial coefficients. This is comforting to the designer and may also turn out to be acceptable in the long run. Lind predicts that most structural designs will be made with the lowest level (level I) methods of safety checking, whereas higher-level safety checking based on exact probabilistic analysis for whole structural systems will be used very rarely (i.e., for checking special structures [5]). This does not mean that a simple deterministic approach will suffice in every case. An example is the determination of dynamic response, for which it will be necessary to perform a stochastic analysis, at least for slender structures.

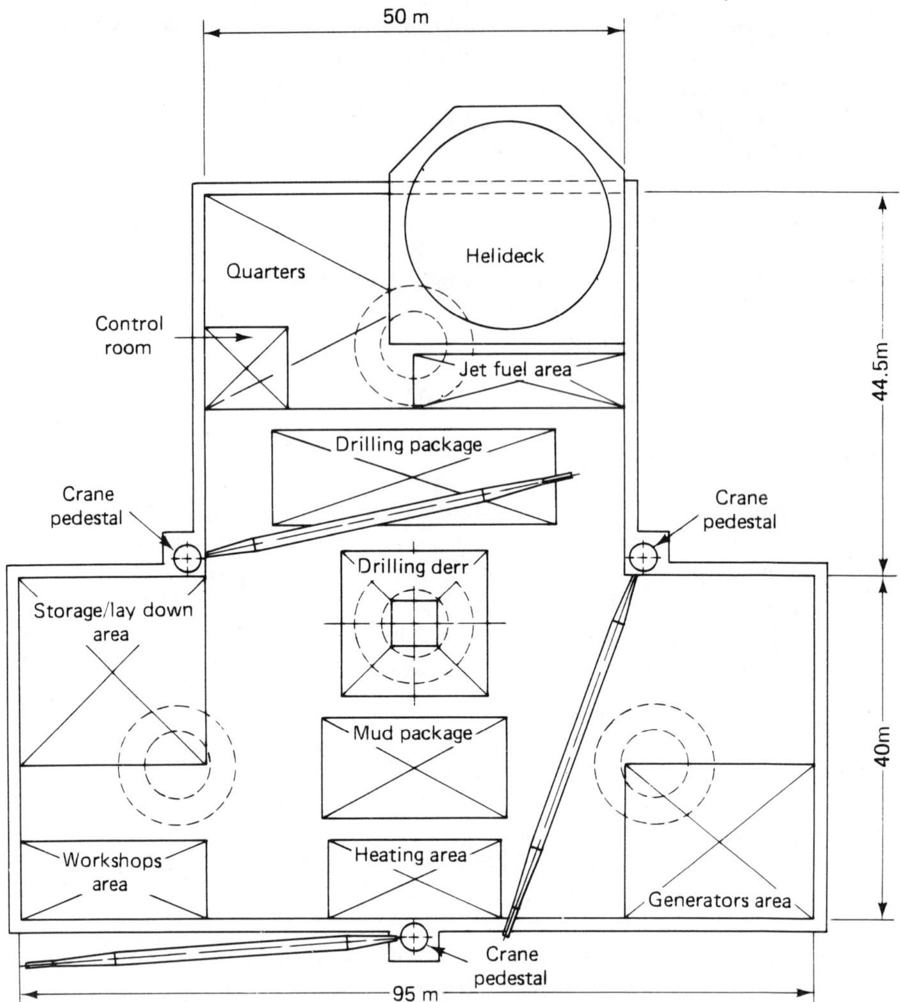

Figure 6　Selmer Tripod 300, deck proposals.

One may conclude that the code format for gravity-base structures may be of a semiprobabilistic type. The problem of code formats is outside the scope of this discussion, however.

Important design aspects for slender structures

The design of a gravity-base concrete platform involves most aspects of advanced analysis and design for concrete structures. For deep waters and correspondingly slender structures the following aspects are of particular

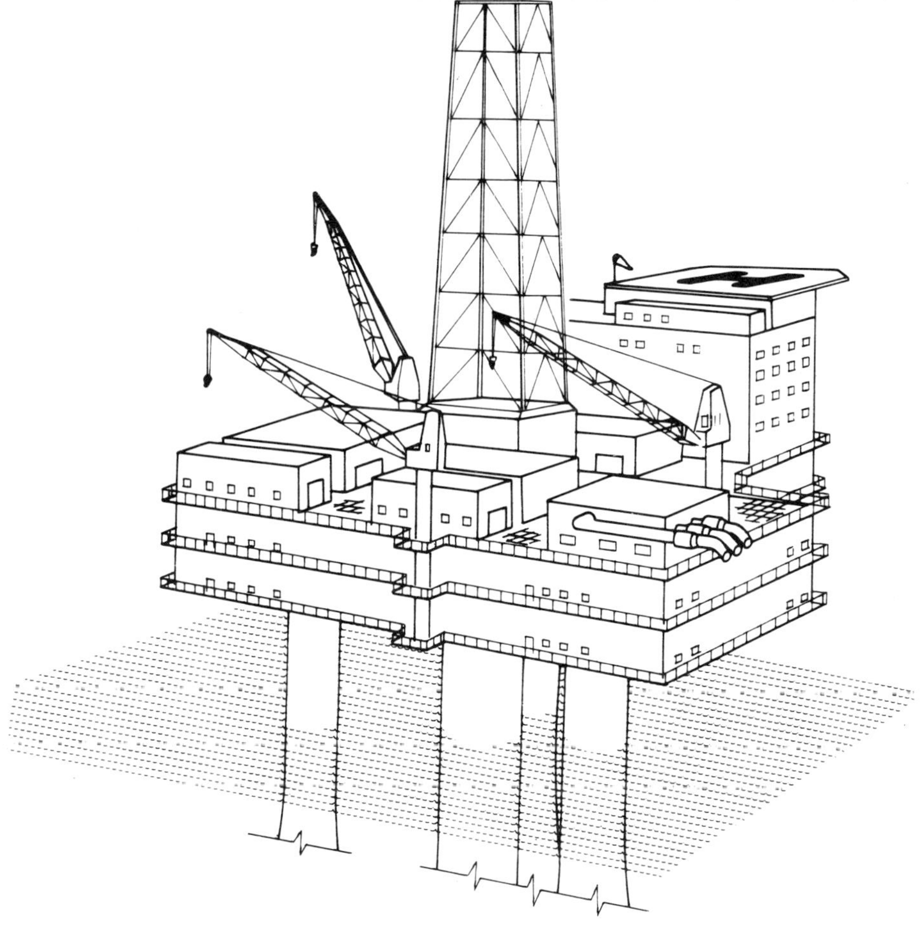

Figure 6 (Continued)

importance:

1. Determination of wave loads and performance of static analysis.
2. Interaction of soil and structure.
3. Wave spectrum and dynamic response analysis.
4. Structural response to seismic activity.
5. Fatigue.
6. Structural stability of thin components (shell elements).

For illustration, some results presented by Bull-Hansen and coworkers in [6] for the Selmer Tripod 300 are referred to below:

For a simplified deterministic approach three characteristic waves are used:

1. An extraordinary load, defined as a storm wave with a 100-year return period, a wave height of 30.0 to 32.0 m, and corresponding wave periods of 14 to 17 s.
2. An ordinary load, defined as a wave with an approximately 1 month return period, a wave height of 20.5 m and a wave period of 12.8 s.
3. A resonance wave of height 2.15 m and wave period of 6.4 s.

For the spectral approach a wave spectrum according to ISSC is used, with a significant wave height $H_{1/3} = 15.0$ m and a mean period of 12 to 14 s, corresponding to a 6-hour storm with a maximum wave height of about 29 m.

A number of mathematical models and methods are applied to analyze structural behavior. The state of equilibrium is studied by means of a global space frame analysis, by which major load effects are determined. The load effects thus found are applied in the detailed analysis of the various members of the structure. In view of the fact that the structures are prestressed, the stiffnesses used in the space frame analysis are determined from the net concrete sections assuming linearly elastic behavior.

For the detailed analysis of local elements of the structure (e.g., stability), more advanced methods that account for nonlinearities, shape imperfections, and cracking of the concrete are applied. As an example, nonconservative loads that result from changes in load direction has been found to affect unfavorably the capacity of components (see [7]).

The interaction between the structure and the seabed soil is recognized by introducing a set of springs representing the flexibility of the soil, which is considered to be an elastic half-space continuum. The dynamic behavior of the structure is studied by analyzing the free vibration of the structure and the dynamic response to environmental loads.

The mathematical model used for one conceptual alternative is shown in Fig. 7. A typical response diagram is shown in Fig. 8. The dynamic amplification has proved to be on the order of 20 to 80% percent of the static response for the most severely stressed sections of the structure.

For poor or moderate soil conditions, the flexibility of the soil is probably the most significant parameter for dynamic behavior. Figures 9 and 10 show the natural period and first-mode shape of a structure versus the undrained shear strength of the soil.

The increase in main-load effects resulting from dynamic effects and structure/soil flexibilities is on the order of 40% for $s_u = 75$ kN/m² and 25% for $s_u = 150$ kN/m².

Studies of earthquake response have been performed on the basis of assumed base rock acceleration spectra, which have been modified by the

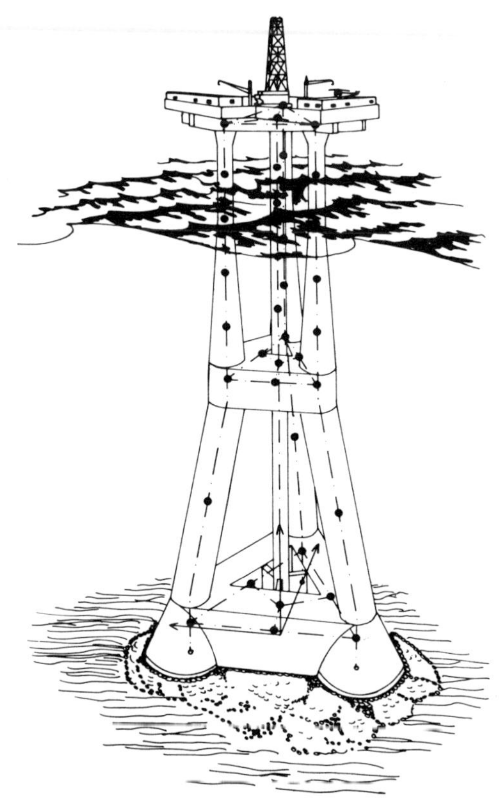

Figure 7   Mathematical model.

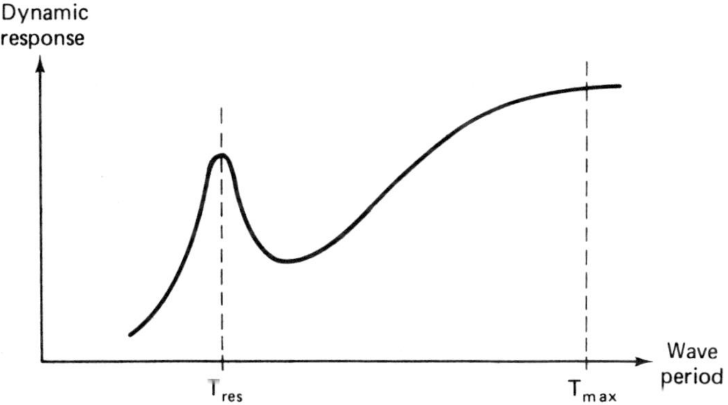

Figure 8   Typical dynamic response diagram.

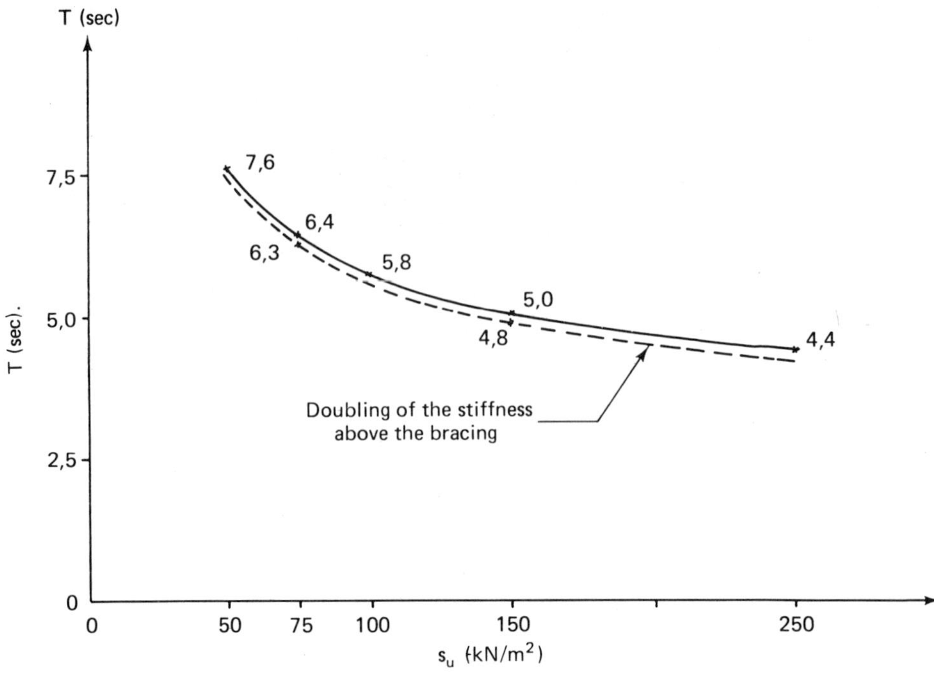

Figure 9   Natural period versus shear strength.

Norwegian Geotechnical Institute by the application of finite element analysis of the soil, in which properties were accounted for (see Fig. 11).

Control of the fatigue resistance of the structure is performed on the basis of stress ranges obtained from the dynamic response analysis, and with the application of the hypothesis of cumulative damage (Miner's rule). It should be emphasized that knowledge of the properties of structural concrete with regard to fatigue is meagre and the problem requires further research.

For more detailed information about these analyses, see [6].

CONCLUSIONS ON DESIGN ASPECTS

The importance of the functional features for gravity-base structures has been stressed as an important design aspect. Also, the importance of the application of practical construction methods has been pointed out as a concern of the designer.

A number of structural design aspects for gravity-base offshore concrete structures have been presented and important experiences have been mentioned.

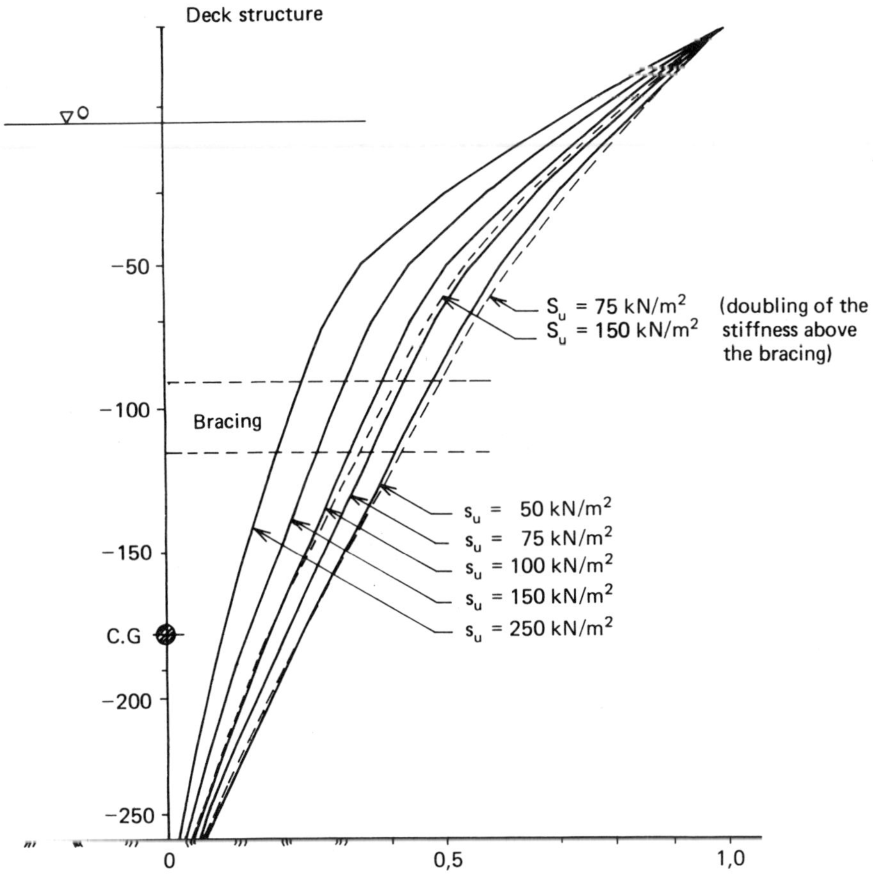

Figure 10   Mode shape of first mode for varying shear strength of soil.

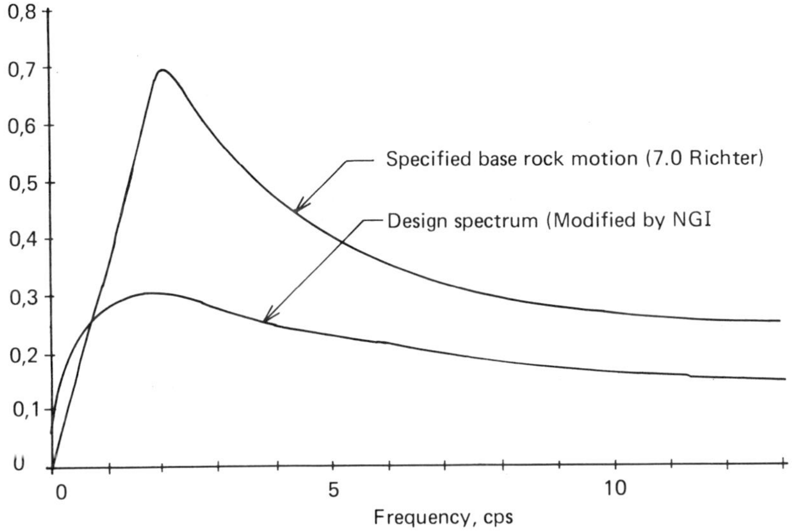

Figure 11   Proposed design acceleration spectrum.

REFERENCES

[1] SJOERDSMA, G. W., "Present and Future Development of Offshore Structures in over 100 m Water Depth," BOSS 76, Norwegian Institute of Technology, 1976.

[2] F. SELMER, "Inclined Slipforming," Oslo, 1976.

[3] MOSES, F., "Safety and Reliability of Offshore Structures," International Research Seminar on Safety of Structures Under Dynamic Loading, Norwegian Institute of Technology, 1977.

[4] Norwegian Petroleum Directorate, "Regulations for the Structural Design of Fixed Structures on the Norwegian Continental Shelf," 1977.

[5] LIND, N. C., "Reliability-Based Structural Codes," International Research Seminar on Safety of Structures under Dynamic Loading, Norwegian Institute of Technology, 1977.

[6] BULL-HANSEN, B., HAGBERG, T., AND GRANHEIM, G. O., "Tripod 300, An Offshore Concrete Platform for 300 m Water Depth," Conference on the Behaviour of Slender Structures, University of London, September 1977.

[7] ÅLDSTEDT, E., "Nonlinear Analysis of Reinforced Concrete Frames," Division of Structural Mechanics, Norwegian Institute of Technology, 1975.

RESISTANCE OF HYPERBOLIC COOLING TOWERS TO WIND AND EARTHQUAKE LOADING

W. B. Krätzig and W. Zerna**

INTRODUCTION

In a thermal power station (Fig. 1) heated steam drives the turbogenerator, which produces electric energy. To create an effective heat sink at the end of this process, the steam is condensed and recycled into the boiler. This requires a large amount of cooling water, which is heated and may be recooled artificially in a cooling tower.

Even in the most efficient fossil fuel thermal (nuclear) power plants, only about 40 (30)% of the generated heat is turned into electric energy. The remaining 60 (70)% is discharged into the environment through the smokestack and the cooling water circuit.

In order to avoid thermal pollution of rivers, lakes, and seashores, natural draft cooling towers are effective and popular corrective measures. These engineering constructions are able to balance environmental factors and investment and operating costs of the power station with the demands of a reliable electric energy supply.

*Dr.-Ing., o. Prof. of Structural Engineering; Institut für Konstruktiven Ingenieurbau, Ruhr-Universität Bochum, Germany.

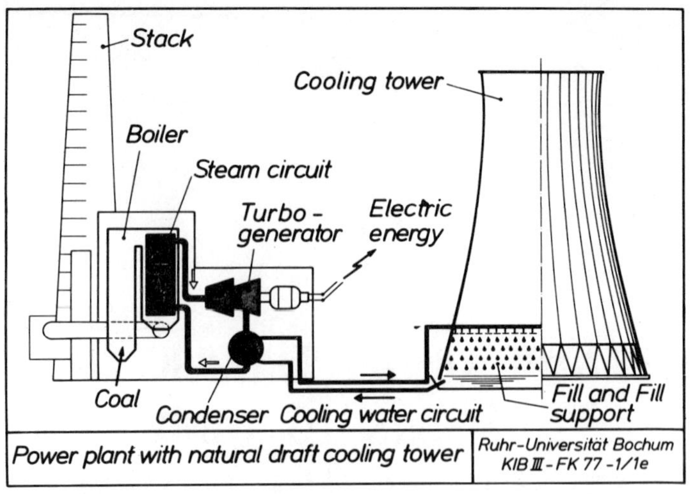

Figure 1 Power plant with natural draft cooling tower.

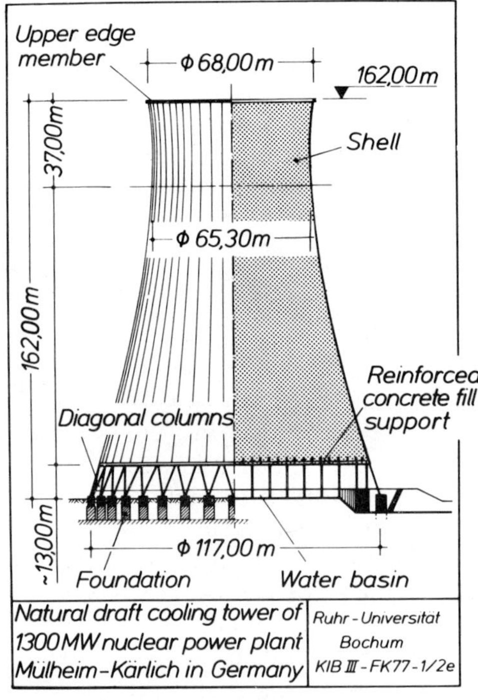

Figure 2 Natural draft cooling tower of 1300-MW nuclear power plant at Mülheim-Kärlich in Germany.

In a wet, natural draft cooling tower, as shown in Fig. 2, hot water is distributed evenly through channels and pipes above the fill (packing). As the hot water flows and drops through the fill sheets, it is met by cooler rising air. Evaporative cooling occurs and the cooled water is collected in the water basin to be recycled into the condenser.

The difference in density of the warm air inside and the cold air outside the tower creates the natural draft. This upward flow of warm air, which leads to a continous stream of fresh air through the air inlets into the tower, is protected against atmospheric turbulence by a huge reinforced concrete shell. This cooling tower shell is supported by diagonal columns bridging the air inlet and by the foundation.

Dry cooling towers that operate on the basis of convective cooling replace the water distribution, the fill, and the basin by a closed pipe system around the air inlet.

Although cooling tower shells form the largest existing shell structures, their design and construction over the years have mainly followed approved rules of craftsmanship (Fig. 3). This changed rapidly, however, after the

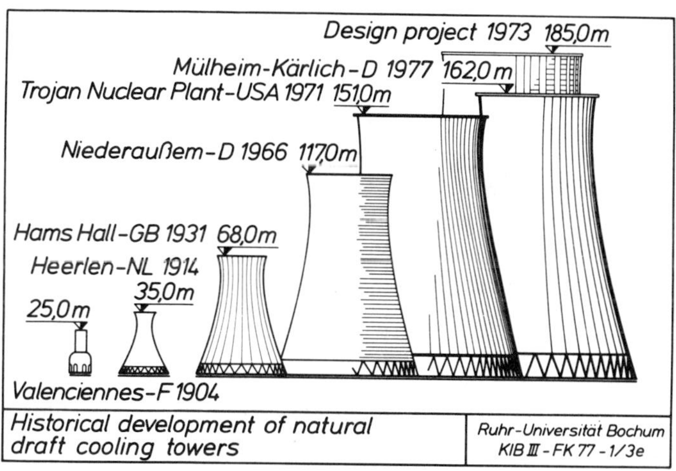

Figure 3 Historical development of natural draft cooling towers.

Ferrybridge collapses in 1965 [4,26]. Within a few years the response phenomena of these structures were studied in detail and safety concepts with improved design rules were developed [2,14,27]. These international research activities were intensified after a new collapse in 1973 in Ardeer [15] and concentrated on dynamic and stability effects.

In the past 10 years many cooling tower research projects, public and private, have been carried out at the Institut für Konstruktiven Ingenieurbau-

KIB at the Ruhr-University Bochum.[1] This paper summarizes some of the results concerning wind and earthquake hazards. Although both effects are time dependent, their different dynamic characteristics permit a quasi-static treatment of wind effects but require a dynamic analysis of earthquake loading.

The Institute KIB had the opportunity to influence the design of more than 50 large cooling towers constructed since 1965. All these towers are today free of cracks and damage, proving the advantages of close connections between engineering research and practice.

WIND LOADING OF COOLING TOWERS

The cooling tower shell, the supporting columns, and the foundation are stressed by gravity loads, and subject to wind and thermal effects, earthquake forces, differential settlements, forces from the scaffolding, and forces from the central crane.

For wind effects, pressure components normal to the shell surface are given by

$$q(z, \theta^1) = (1 + \phi)K(z)c(\theta^1)q_{10}$$

where z = coordinate of height above ground level
$\quad \theta^1$ = circumferential coordinate
$\quad K(z)$ = vertical profile of wind pressure
$\quad c(\theta^1)$ = circumferential pressure distribution
$(1 + \phi)$ = gust response factor incorporating the dynamic response of the tower to atmospheric turbulence
$\quad q_{10}$ = basic wind pressure at a height of 10 m above ground level

The formula mentioned above does not describe the physical effects of wind flow in nature; it is an engineering estimate. The Hellmann exponential law of $K(z) = q/z)/q_{10}$, proved for mean hourly wind speeds, is specified in [2] with reference to [6]. In the new German code DIN 1055.4, $K(z)$ is given in the simplified form of Fig. 4 [8]. The basic wind pressure q_{10} is computed from the largest 5-s mean velocity with a return period of 50 years. For this specification the gust response factor $(1 + \phi)$ related to the membrane forces takes values between 1.00 and 1.10, depending on the lowest natural frequency and the damping ratio (see, e.g., [18] and the contribution by Niemann in [16], p. 38). Gust response factors for circumferential bending moments are remarkably higher (Fig. 5).

[1]The authors wish to thank their coworkers and colleagues named in the references for their efforts. Special thanks are due to the Deutsche Forschungsgemeinschaft and the Landesamt für Forschung NRW for financial support.

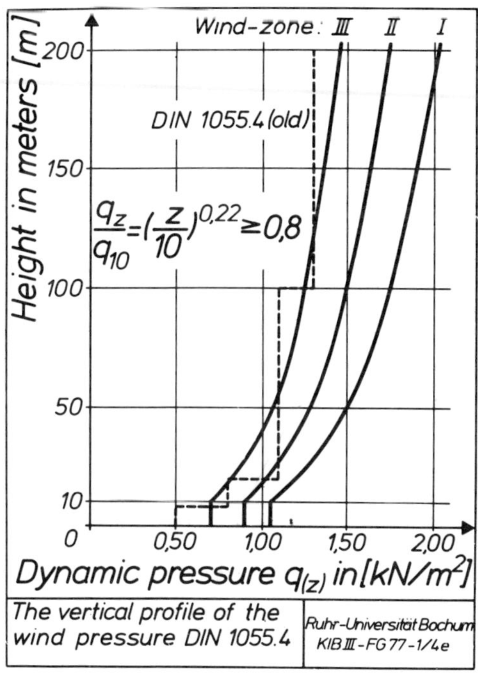

Figure 4   The vertical profile of the wind pressure [8].

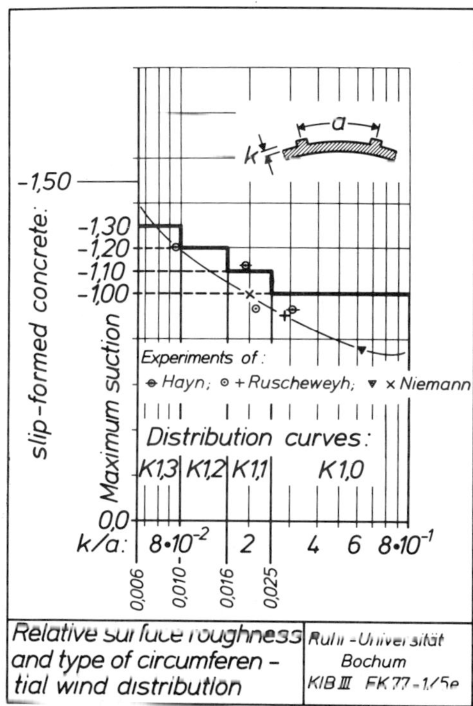

Figure 5   Relative surface roughness and type of circumferential wind distribution.

The largest mile concept with a 100-year wind return as used in the United States should lead to higher factors $(1 + \phi)$. If the design is based on a wind speed of 100 mph, the amplification is about 15 to 25% [25].

The circumferential pressure distribution $c(\theta^1)$ is of considerable importance for the state of stress in the cooling tower shell, and it depends on the Reynolds number and the surface roughness. The roughness is caused artificially by vertical wind ribs varying in number, distance a (taken at $\frac{1}{3}$ of the tower height) and height k in Fig. 6.

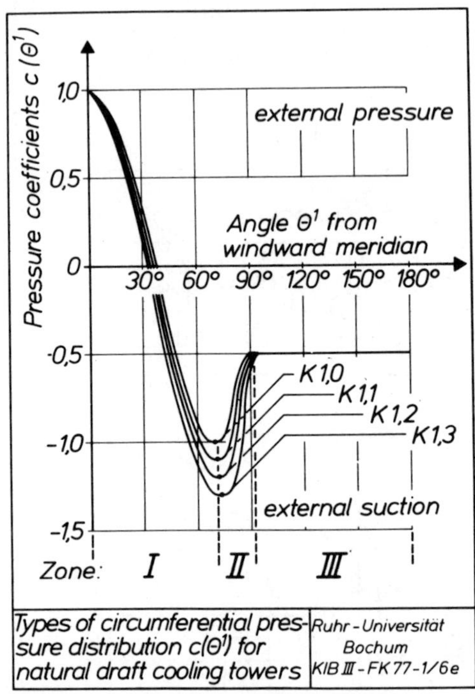

Figure 6 Types of circumferential pressure distribution $c(\theta^1)$ for natural draft cooling towers.

Reliable data from wind tunnel tests and full-scale experiments led to a correlation of relative surface roughness and type of pressure distribution given in Figs. 6 and 7. The different distribution of Fig. 7 may end the era of uncertainty and ignorance regarding this basic question (see, e.g., [4,12,26]). If these recommended curves and their analytic and numerical expressions in Tables 1 and 2, evaluated by Niemann and based on the best available experiments, are used, spatial effects caused by wind flow around the tower can be corrected for, giving upper bounds for the maximum and the wake suction; they also lead to actual values for the drag coefficient c_w [16,23].

The distribution of external pressure $q(z, \theta^1)$ should be supplemented by an internal suction of 0.5 of the dynamic pressure on the top of the tower.

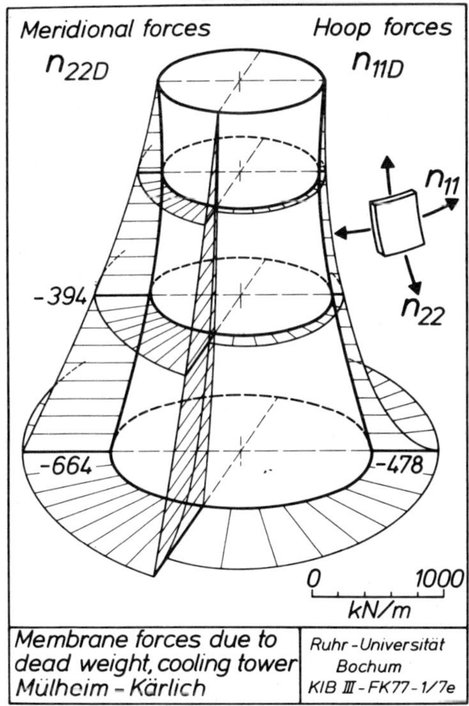

Figure 7 Membrane forces due to dead weight, cooling tower at Mülheim-Kärlich.

Table 1

ANALYTICAL FORMULAS FOR THE PRESSURE DISTRIBUTION OF FIG. 6

Curve	Drag coeffi- cient c_w	Minimum suction	Zone I	Zone II	Zone III
K 1,0	0,66	-1,0	$1-2{,}0(\sin\frac{90}{70}\varphi)^{2,267}$	$-1{,}0+0{,}5\{\sin[\frac{90}{21}(\varphi-70)]\}^{2,395}$	-0,5
K 1,1	0,64	-1,1	$1-2{,}1(\sin\frac{90}{71}\varphi)^{2,239}$	$-1{,}1+0{,}6\{\sin[\frac{90}{22}(\varphi-71)]\}^{2,395}$	-0,5
K 1,2	0,60	-1,2	$1-2{,}2(\sin\frac{90}{72}\varphi)^{2,205}$	$-1{,}2+0{,}7\{\sin[\frac{90}{23}(\varphi-72)]\}^{2,395}$	-0,5
K 1,3	0,56	-1,3	$1-2{,}3(\sin\frac{90}{73}\varphi)^{2,166}$	$-1{,}3+0{,}8\{\sin[\frac{90}{24}(\varphi-73)]\}^{2,395}$	-0,5

The suction may be assumed rotationally symmetric and constant over the height of the tower.

INTERNAL FORCES AND COUPLES

The middle surface of the shell will be described by coordinates θ^1 in the circumferential and θ^2 in the meridional direction. In the scope of bending theory, the shell responds to external forces and boundary displacements

Table 2

NUMERICAL VALUES OF THE PRESSURE DISTRIBUTIONS OF FIG. 6

Angle Θ^1	Pressure coefficients for curve			
	K 1,0	K 1,1	K 1,2	K 1,3
0°	1,00	1,00	1,00	1,00
5°	0,99	0,98	0,98	0,98
10°	0,93	0,93	0,92	0,92
15°	0,84	0,83	0,82	0,81
20°	0,70	0,69	0,67	0,65
25°	0,52	0,50	0,48	0,46
30°	0,31	0,29	0,26	0,23
35°	0,09	0,06	0,02	-0,01
40°	-0,14	-0,18	-0,22	-0,26
45°	-0,37	-0,42	-0,46	-0,51
50°	-0,58	-0,63	-0,69	-0,74
55°	-0,75	-0,82	-0,88	-0,95
60°	-0,89	-0,96	-1,04	-1,11
65°	-0,97	-1,06	-1,14	-1,23
70°	-1,00			-1,29
71°		-1,10		
72°			-1,20	
73°				-1,30
75°	-0,96	-1,07	-1,19	-1,29
80°	-0,80	-0,92	-1,05	-1,19
85°	-0,61	-0,70	-0,82	-0,95
90°	-0,50	-0,53	-0,59	-0,68
95°	-0,50	-0,50	-0,50	-0,52
100°	-0,50	-0,50	-0,50	-0,50
⋮	⋮	⋮	⋮	⋮
180°	-0,50	-0,50	-0,50	-0,50

with a state of internal force variables as follows:

Normal and in-plane shear forces n_{11}, n_{22}, n_{12}.
Bending and twisting moments m_{11}, m_{22}, m_{12}.
Transverse shear resultants q_1, q_2.

In the supporting columns we find mainly axial forces, tensile as well as compressive, and very small couples and transverse shear resultants. For this reason such columns are often idealized as rods linked to the shell and the foundation.

Experience has shown that membrane theory (n_{11}, n_{22}, n_{12}) is a satisfactory approximation for cooling tower shells, provided that it is completed by an edge-effect theory in regions of local bending (zone of upper and lower edge ring). This enables us to avoid a time-consuming finite element analysis and to use an appropriate, but simpler method of computation. The method of characteristics works incomparably quickly for all membrane shells of negative Gaussian curvature and is coded easily. It was primarily developed in a computerized version for cooling tower shells in [21], using a finite difference technique. Meanwhile there exists a complete computer program [13] for the automatic design of hyperbolic, conical, cylindrical, or mixed cooling tower shells based on this method [20].

In Figure 7 an example is illustrated of the computed membrane forces due to dead weight (specific weight of the reinforced concrete: 2.50 t/m³) for the cooling tower shown in Fig. 2. This rotationally symmetric state of stresses is superposed by the internal wind forces given in Figs. 8 to 10. The wind specification herein uses the pressure distribution $K1.0$, the wind pressure of the old code DIN 1055.4 given in Fig. 4, and $(1 + \phi) = 1.00$. Tension is characterized by a positive, compression by a negative sign.

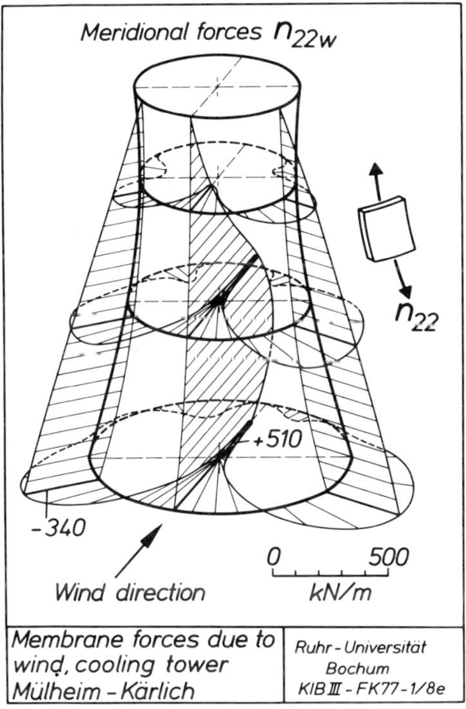

Figure 8 Meridional membrane forces due to wind, cooling tower at Mülheim-Kärlich.

DESIGN CRITERIA FOR COOLING TOWER SHELLS

A properly shaped cooling tower shell remains in a state of complete compression for most of its lifetime (Fig. 7). In the case of severe storms this compression in the flanks of the tower will increase dramatically and

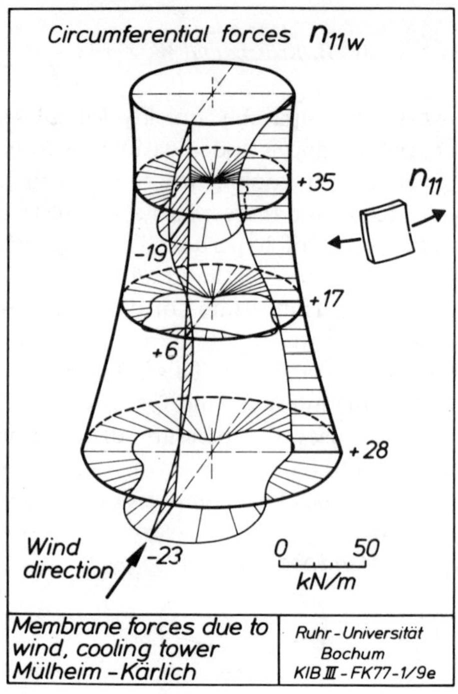

Figure 9 Circumferential membrane forces due to wind, cooling tower at Mülheim-Kärlich.

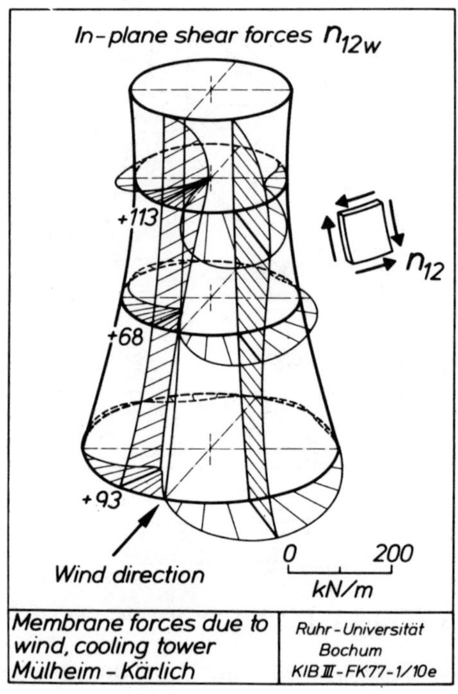

Figure 10 Membrane shear forces due to wind, cooling tower at Mülheim-Kärlich.

decrease or change into tension near the windward meridian, caused mainly by the meridional membrane wind forces of Fig. 8. Thus the shell must resist tensile forces, as well as instability effects due to compression.

Evaluation of tensile reinforcement should follow the combination rules of the ACI-Standard 318–71:

$$D + W + T$$

$$0.9D + 1.30(W + T)$$

with extreme internal forces due to dead weight (D), wind (W), and temperature (T). In particular, the latter combination is a severe but adequate restriction necessary to achieve sufficient structural safety.

In Germany the following combinations are employed:

$D + W + T$ using a safety factor of $v_s = 1.75$ against yield stress

$D + \epsilon W$ using $1.50 \leq \epsilon \leq 1.75$, $v_s = 1.00$

For very large cooling towers such as that shown in Fig. 2 the load factor ϵ may be increased up to 2.00. It has been shown recently (see the contributions of Krätzig and Niemann in [16], pp. 81 and 85) that a factor of about 2.00 leads to a probability of failure of less than 10^{-4}, which is considered to be sufficient for the tensile side of the shell and the supports.

Many types of shells buckle at very low loads as compared with theoretical bifurcation results. Before buckling experiments were initiated, the imperfection sensitivity of cooling tower shells had to be studied. From detailed theoretical and numerical investigations it was determined that cooling tower shells with perceptible meridional curvature have a negligible sensitivity to initial imperfections smaller than the thickness of the shell.

Experiments with large plastic models of hyperbolic shells, carefully conducted over the years by Mungan [22], and simultaneous numerical computations showed an interesting first result: when a structure is stiffened by upper and lower ring beams, buckling is chiefly local.

The proposed stability criterion is shown in Figs. 11 to 13. It postulates that any critical combination given in Fig. 12 of the effective stresses σ_{11}, σ_{22} due to dead load, wind, internal suction, probable seismic effects, and the buckling stresses $\sigma_{11cr}, \sigma_{22cr}$ at any point of the middle surface is responsible for the load-bearing capacity. Buckling stresses σ_{11cr} and σ_{22cr} are experimentally evaluated and numerically corrected by a bifurcation theory. The parameters k_{G11} and k_{G22} in Fig. 13, which contains the influence of tower shapes other than those tested, stem also from a bifurcation context.

It should be mentioned that the factor of buckling safety required by the German Standard DIN 1045 is $v_B > 5$. Based on computations with the

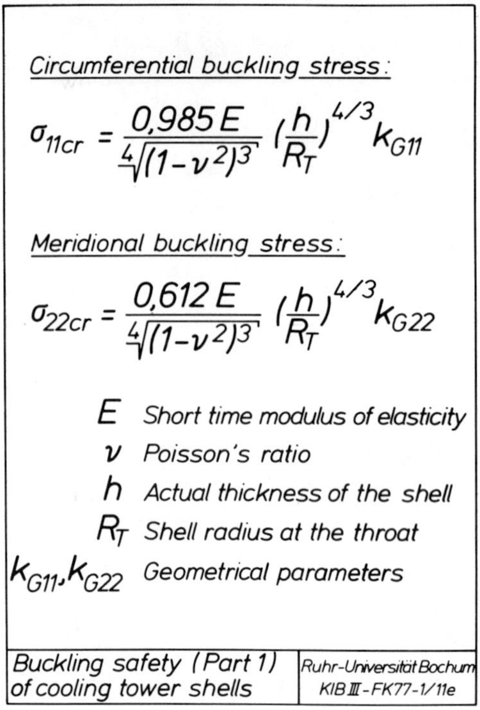

$$\sigma_{11cr} = \frac{0{,}985\,E}{\sqrt[4]{(1-\nu^2)^3}} \left(\frac{h}{R_T}\right)^{4/3} k_{G11}$$

$$\sigma_{22cr} = \frac{0{,}612\,E}{\sqrt[4]{(1-\nu^2)^3}} \left(\frac{h}{R_T}\right)^{4/3} k_{G22}$$

Figure 11   Buckling stresses for hyperbolic shells.

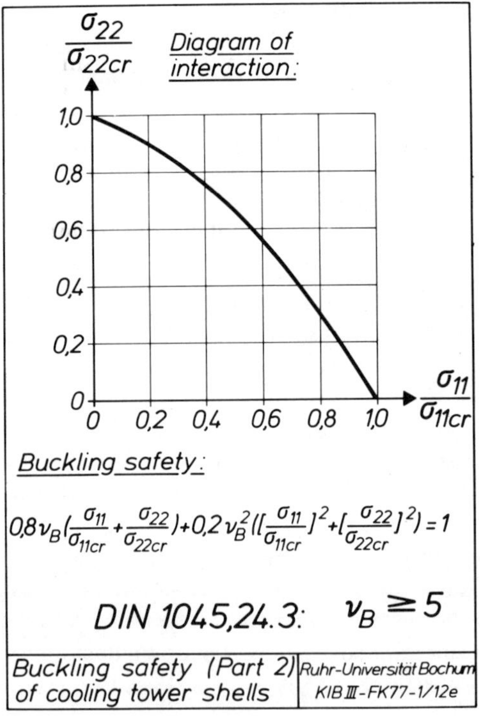

$$0{,}8\,\nu_B\left(\frac{\sigma_{11}}{\sigma_{11cr}} + \frac{\sigma_{22}}{\sigma_{22cr}}\right) + 0{,}2\,\nu_B^2\left(\left[\frac{\sigma_{11}}{\sigma_{11cr}}\right]^2 + \left[\frac{\sigma_{22}}{\sigma_{22cr}}\right]^2\right) = 1$$

Figure 12   Proposed diagram of interaction for hyperbolic shells.

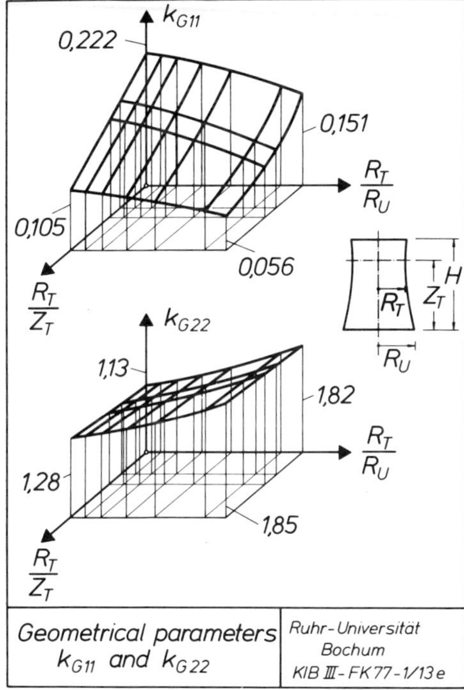

Figure 13 Geometrical parameters used in Fig. 11.

short-time modulus of elasticity E of the concrete, this high value ensures a probability of failure of less than 10^{-5}, considering hairline cracks and shrinkage.

During construction the shell must survive without an upper ring beam. Inextensile deformations as studied by Der and Fidler must be avoided [2,5,7]. Finally, it should be confirmed that a proper design concept includes many more details [2,14,27]: two layers of reinforcement in both directions at each face, an adequate limitation of hairline cracks, a sufficient minimum reinforcement, and so on.

WIND RESPONSE AND TOWER SHAPE

Apart from considerations of thermodynamics, the most economic shape for a cooling tower is often questioned, varying the basic sizes within certain limits. Should preference be given to slim or compact structures? Which is the best form of the meridional curve?

To answer these or similar questions, a parametric study was carried in which three types of cooling towers were compared: B0 160, B0 180, B0 200 [29]. As shown in Fig. 14, all differ in height but are equal in volume,

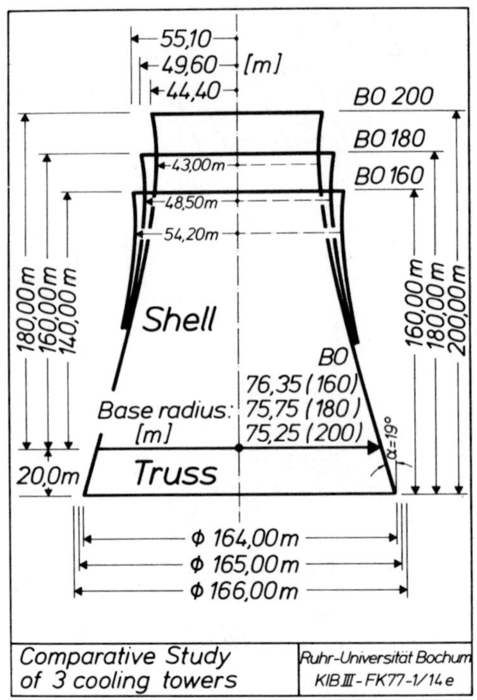

Figure 14   Shape and size of cooling tower types BO 160, BO 180, and BO 200.

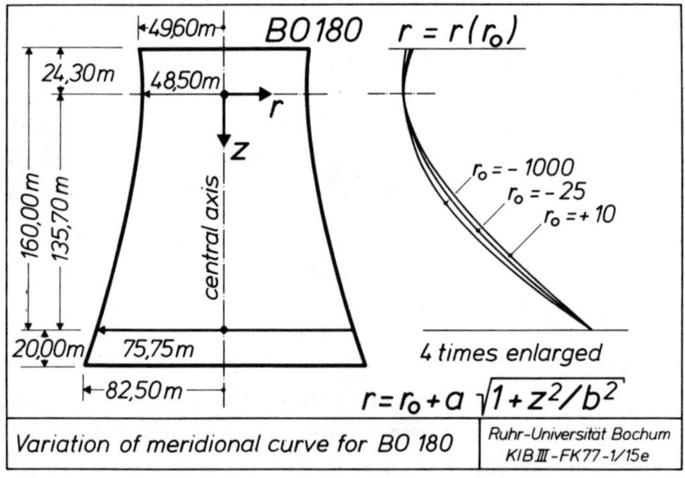

Figure 15   Variations of the meridional curve for the BO 180 type.

their base diameters are nearly equal. Thus the thermal efficiency of the towers is nearly identical.

For each tower a variation of the analytic form of the meridional curve using the parameter r_0 was carried out. As shown in Fig. 15, $r_0 = 0$ describes a pure hyperbola rotating about its axis. For values of r_0 greater (smaller) than zero, the axis of hyperbola lies between (outside) the meridional curve and the central axis.

This parameter r_0 significantly influences the response of the tower as seen in the graphs of the internal wind forces in Figs. 16 to 19. These

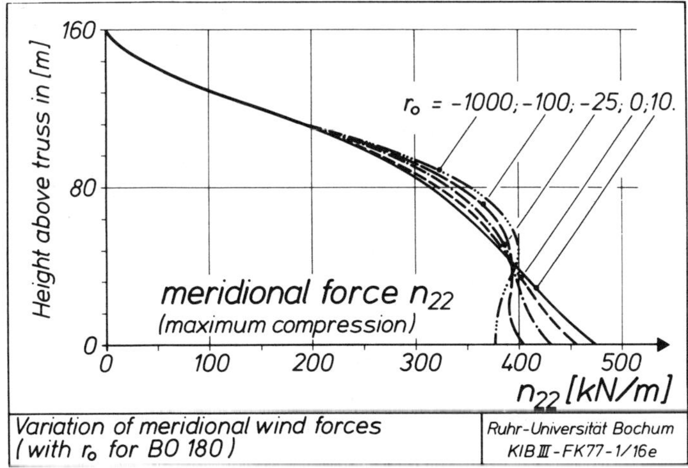

Figure 16 Variation of meridional wind forces (maximum compression, BO 180).

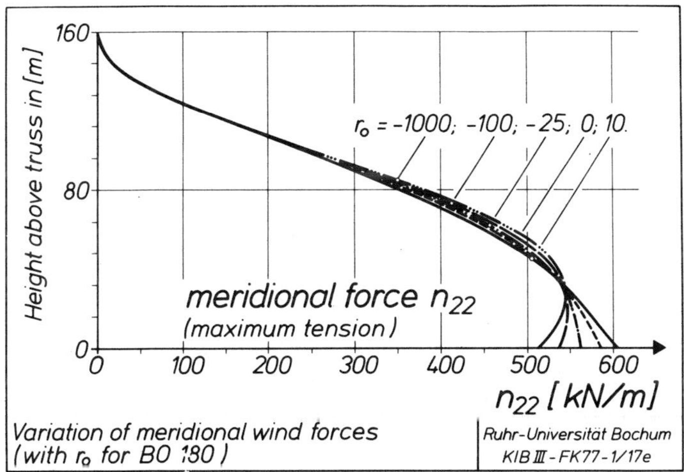

Figure 17 Variation of meridional wind forces (maximum tension, BO 180).

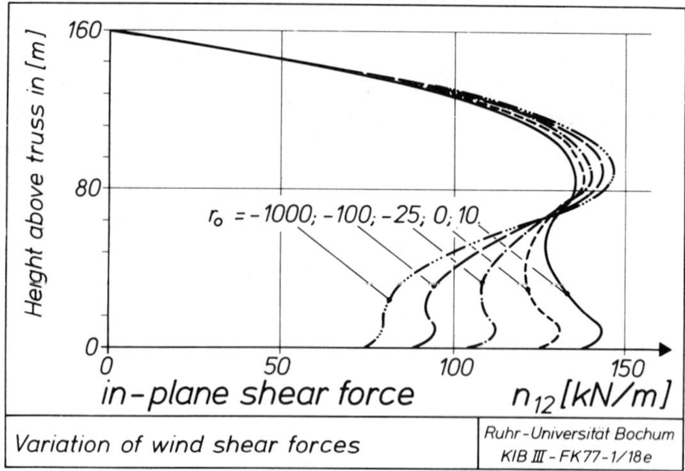

Figure 18   Variation of in-plane wind shear forces (BO 180).

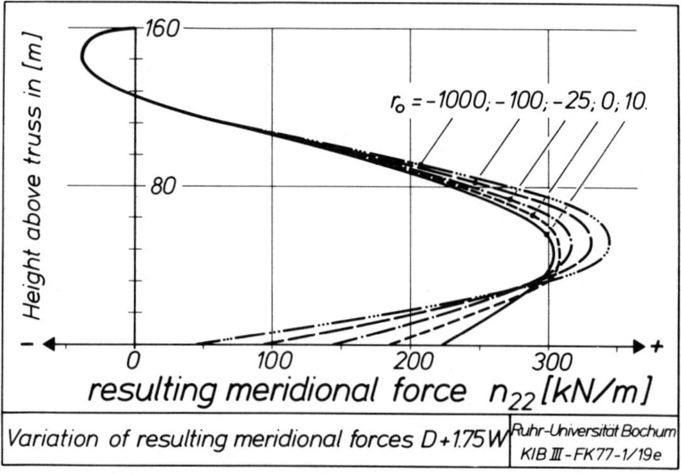

Figure 19   Variation of resulting meridional forces, $D + 1.75W$ (BO 180).

results verify the original finding (see contribution by Krätzig in [17], p. 84) that the amount of reinforcement (Fig. 19) in the shell, as well as the magnitude of axial forces in the columns and the foundation, can be influenced strongly by the parameter r_0.

To determine the dependence of the buckling safety v_B and the lowest natural frequency $f_{\min}$ (the reader may be reminded that $f_{\min}$ determines the dynamic gust response factor $1 + \phi$), the variation of thickness over the height of the shell was assumed equal for all towers: starting with 16 cm at the top of the tower, the thickness was increased linearly to 35 cm at the base.

The computation was based on a linear dynamic and bifurcation shell theory using a conical finite element described in the literature [3]. For simplification the maximum wind pressure was approximated to be rotationally symmetric and added to the internal suction. The factor of buckling safety ν_B in Fig. 20, based on a modulus of elasticity of 3.10^7 kN/m^2 for the concrete shell, was defined as a multiple of this "wind load," supplemented by the dead weight of the shell. In the analysis a free, beam-stiffened edge at the top and linked boundary conditions without displacements on the bottom were assumed.

The results of the investigation presented in Fig. 20 demonstrate clearly that there is no advantage to designing slim or compact towers. Slim cooling towers are expected to have a slightly higher buckling safety, but their lowest natural frequencies are smaller. Thus, compact towers are affected less intensely by atmospheric turbulence.

In any case, cooling towers with the large negative values of the parameter r_0 behave best: they show the most uniformly curved meridians leading to the highest factors of buckling safety and natural frequencies. The same trend can be seen in the internal wind forces in Fig. 17 and from their combinations with dead load (Fig. 19). The integers n in Fig. 20 indicate the

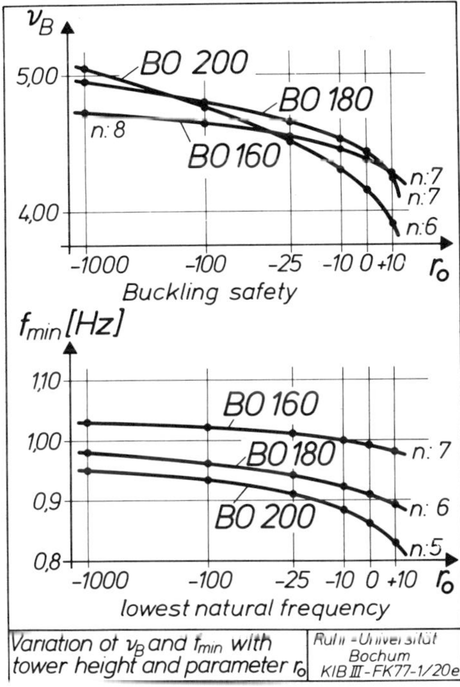

Figure 20 Buckling safety ν_B and lowest natural frequency f_{min} for cooling tower types BO 160, BO 180, and BO 200.

numbers of circumferential waves of the buckling and vibration pattern. A column simulation, instead of the idealized boundary conditions on the bottom of the shell, would reduce the natural frequencies to the range of 10 to 15%.

MODEL OF COMPUTATION
FOR SEISMIC RESPONSE

While the lowest natural frequencies of a cooling tower lie very close to the peak of the energy spectrum for earthquake effects but far for wind loading, the responses due to both loadings and their methods of computation differ considerably. Whereas a quasi-static analysis is required for the response to wind loads, seismic effects require a complete dynamic analysis.

The computation of the dynamic response of cooling towers due to seismic loadings is generally based on finite element analysis using conical shell elements or elements of general shells of revolution [1,11]. If the velocity of the seismic shear and compression waves in the ground below the foundation is assumed to be of a higher order of magnitude than that of the material elements of the structure, all displacements and internal forces of the rotationally symmetric tower will follow sine or cosine functions in a circumferential direction θ^1. If an integration procedure according to θ^1 is applied to a membrane or bending theory finite element the number of degrees of freedom can be reduced and a one-dimensional element similar to a Timoshenko beam (including shear deformations) element is created.

In the same way, the calculation procedure (see contributions by Meskouris, Krätzig, and Harnach in [16] simplifies the ring of columns as horizontal, vertical, and rotational springs, all with linearly elastic characteristics. The cooling tower shell is finally replaced by 10 to 20 improved beam elements, as shown in Fig. 21 [19]. The model neglects all bending effects in the structure, thereby describing a membrane state of stress in an exact manner. Additionally, all forces and displacements deviating from the sine/cosine distribution are neglected.

The earthquake analysis is based on the response spectrum method. After calculating the natural frequencies and modes and their participation factors, the internal forces n_{11} and n_{12} as well as the displacements of the tower axis, can be computed for each mode shape based on a normalized ground acceleration. For any damping ratio (e.g., 2% of critical damping) the respective magnification factor of each mode can then be found from the given design spectrum (e.g., Fig. 22). Finally, the root mean square of the response can be evaluated, which is usually considered to be an adequate measure of observed seismic stresses.

The method described is simple and efficient. The results obtained lie

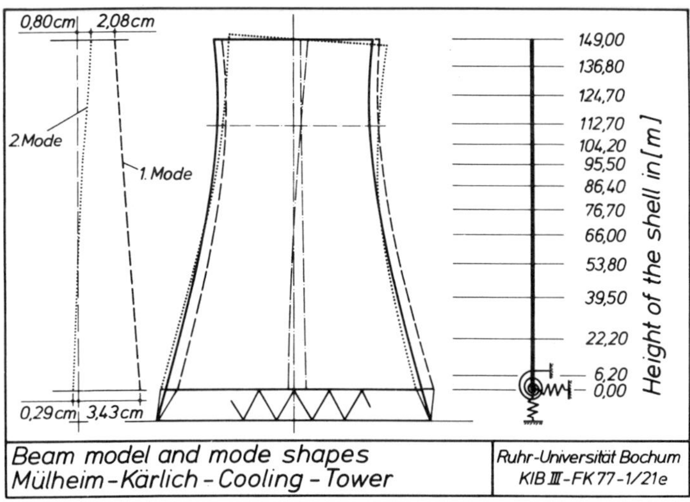

Figure 21 Beam model and mode shapes for Mülheim-Kärlich cooling tower.

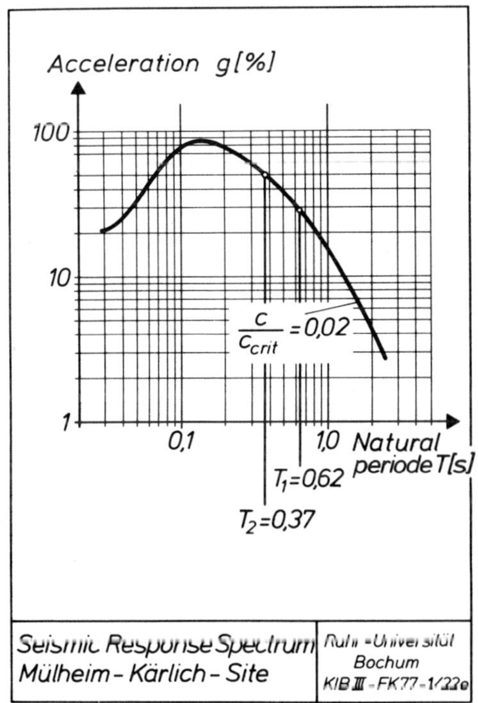

Figure 22 Seismic response spectrum for Mülheim-Kärlich site.

within a range of 3 % as compared with those of [11,24] obtained using a refined shell element.

SEISMIC RESPONSE
OF THE MÜLHEIM-KÄRLICH COOLING TOWER

As an example, the seismic response is computed by the method described for the 1300-MW cooling tower at Mülheim-Kärlich, whose basic data were given in Fig. 2. The shell is supported by 36 pairs of columns of 11.80 m vertical height, connected with the foundation by a pedestal 1.20 m high. The cross section of each column is circular, with a diameter of 0.85 m. The pairs of columns form a V and meet the lower edge beam at 72 equidistant points.

The thickness of the cooling tower shell varies from 85 cm to 16 cm near the upper edge beam, reaching a constant buckling safety of $v_B = 5.0$ according to Fig. 12. The modulus of elasticity for all reinforced concrete members is 3×10^7 kN/m².

The horizontal design response spectrum for the Mülheim-Kärlich site, based on a damping ratio of 2 %, is given in Fig. 22. This is a 0.2-g spectrum similar to the standard 87 % design spectrum of the USAEC. Figure 22 also shows the natural periods of the first and second seismic mode; their axial displacement w in the direction of the wave velocity and their internal forces n_{22} and n_{12} are given in Figs. 23 and 24 (both in different scales). The RMS superposition of displacements and internal forces may be found in Fig. 25.

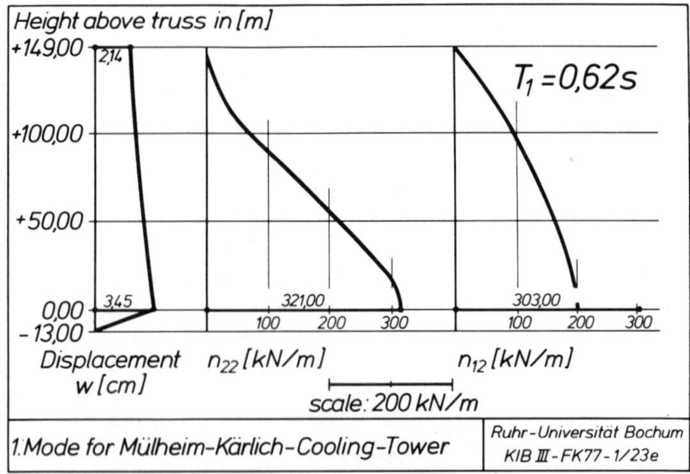

Figure 23 First mode seismic response for the Mülheim-Kärlich cooling tower.

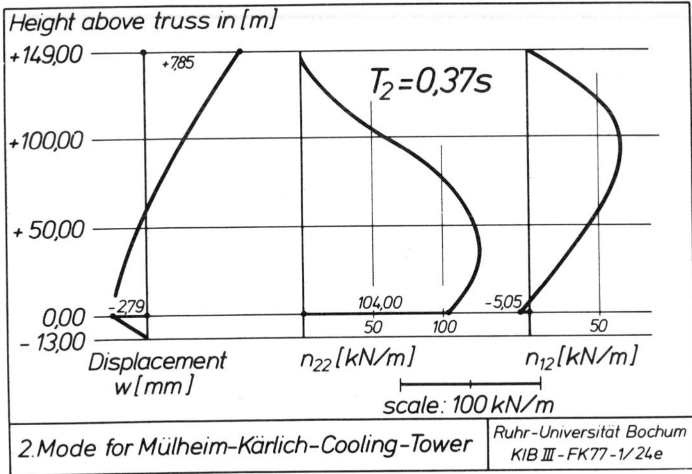

Figure 24 Second mode seismic response for the Mülheim-Kärlich cooling tower.

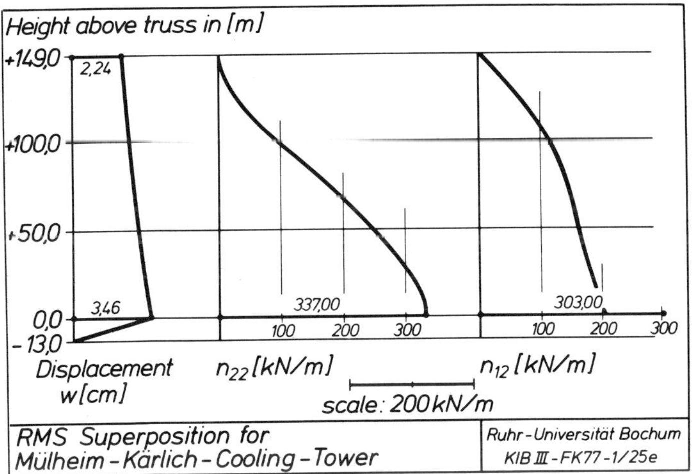

Figure 25 RMS superposition of the first and second modes (Figs. 23 and 24).

INFLUENCE OF FILL CONSTRUCTION

Figures 22 and 25 confirm that natural draft cooling towers may be stressed severely by earthquake effects [1,10]. In general, estimation of seismic response is the last step in a design process; internal forces are added to a geometrically fixed structure. Is fill construction able to reduce seismic stresses and displacements in the cooling tower shell and its supports if difficulties appear there in the final state?

The fill construction of a typical 1300-MW wet cooling tower consists of 400 to 600 reinforced concrete columns, connected by prefabricated beams to a spatial framework. This framework, loaded by the fill sheets and the pipes of the water distribution, is generally separated from the shell and is therefore stiffened by additional frames, trusses, or diaphragms.

For the Mülheim-Kärlich cooling tower, the fill construction consists of 436 reinforced concrete columns, stiffened by 74 truss frames in each of two orthogonal directions in the ground plan, as shown in Fig. 26. The mass distribution of the complete fill, including the water, has an average value of 1.30 t/m². If the fill were connected to the lower edge member of the shell, an additional spring and concentrated mass would have to be added to the beam model of Fig. 21. Because of the higher ductility of the framework, this mass is assumed to affect the shell within the two bounds given in Table 3 (cases 2 and 3). Finally, 68 supplementary frames were designed as shown in Fig. 26 and considered to be additional stiffening elements (cases 4 and 5 in Table 3).

The longest natural periods for the original cooling tower (case 1) with

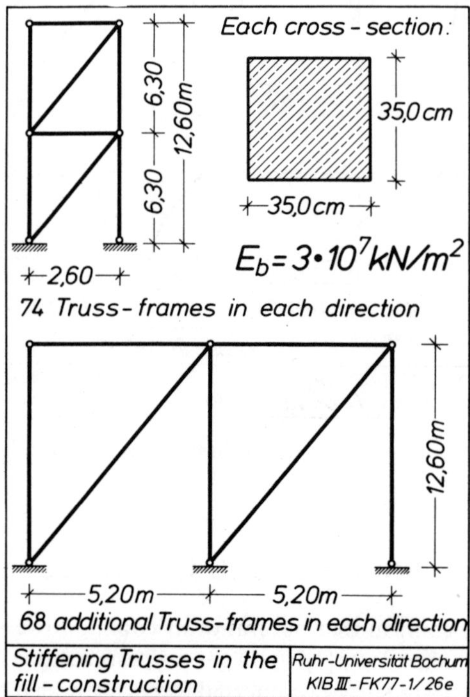

Figure 26   Stiffening trusses of the fill construction.

Table 3

NATURAL PERIODS OF THE SHELL-COLUMN MODEL
FOR THE VARIANTS OF THE FILL CONSTRUCTION

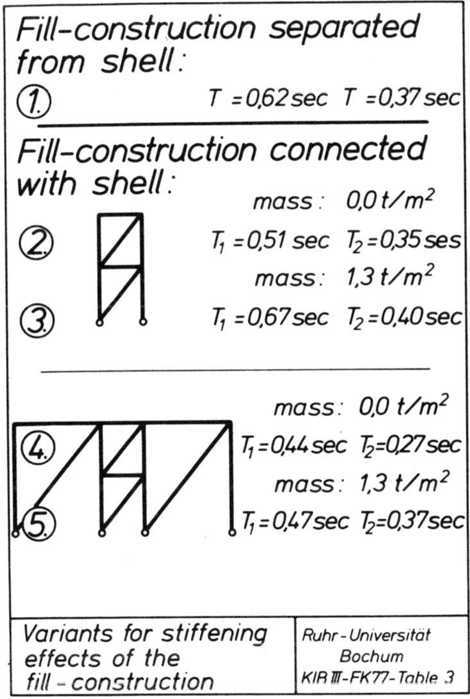

a separated framework and all further construction variants mentioned earlier are given in Table 3. The resulting displacements w of the lower edge of the shell, presented in a RMS superposition in Fig. 27, can be reduced significantly as the stiffness of the horizontal spring is increased. The displacements of the upper edge of the shell grow larger simultaneously, which considerably increases the internal forces n_{22} (in Fig. 28) and n_{12}.

As a preliminary result it is certain that the fill construction would affect the response of the cooling tower shell if connected to it. The minimum stresses in the tower appear to occur for quasi-rigid body translations of the shell.

CONCLUSIONS

In the present paper we have attempted, to summarize knowledge of the response of natural draft cooling towers to wind and earthquake effects. The response due to dead weight and wind load is fairly well known, and appro-

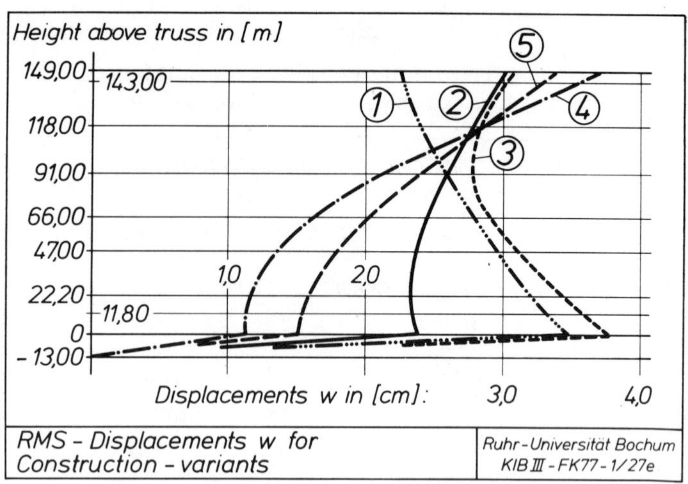

Figure 27   Graph of the displacements w for different variants of the fill construction of the Mülheim-Kärlich cooling tower.

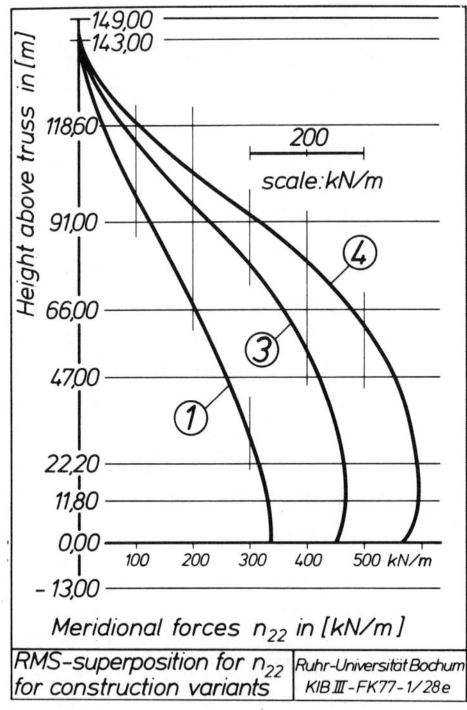

Figure 28   RMS superposition for the meridional forces n_{22}.

priate design rules have been recommended and employed in practice. On the other hand, seismic response is treated in a secondary manner, although cooling towers can be heavily damaged by earthquakes. Further research on the influence of tower shape and column configuration on earthquake response should be carried out in order to rank wind and seismic effects equally in the design process of a cooling tower.

REFERENCES

[1] ABU-SITTA, S. H., AND DAVENPORT, A. G., Earthquake Design of Cooling Towers. *Journal of the Structural Division, ASCE*, Vol. 96, 1970, pp. 1889-1902.

[2] ACI-ASCE COMMITTEE 334, Recommended Practice for the Design and Construction of Reinforced Concrete Cooling Tower Shells. Subcommittee on Hyperbolic Cooling Towers of the ASCE-ACI Task Committee on Concrete Shell Design and Construction, ASCE, New York, 1975.

[3] BENZ, H.-J., Linearisierte Stabilitäts- und Schwingungsberechnung bei versteiften Rotationsschalen. Institut f. Konstruktiven Ingenieurbau, Ruhr-Universität Bochum, *Technical Report 76-10*.

[4] CEGB, Report of the Committee of Inquiry into Collapse of Cooling Towers at Ferrybridge. Central Electricity Generating Board, London, 1966.

[5] COLE, P. P., ABEL, J. F., AND BILLINGTON, D. P., Buckling of Cooling-Tower Shells, *Journal of the Structural Division, ASCE*, Vol. 101, 1975, Paper 11364 and 11365, pp. 1185-1222.

[6] DAVENPORT, A. G., The Relationship of Wind Structure to Wind Loading. *Proc. Congr. Wind Effects*, Teddington, 1963, Her Majesty's Stationery Office, London, 1965.

[7] DER, T. J., AND FIDLER, R., A Model Study of the Buckling Behaviour of Hyperbolic Shells. *Proc. Institution of Civil Eng.*, Vol. 41, London, Jan. 1968.

[8] FACHNORMENAUSSCHUSS BAUWESEN (FNBAU), Entwurf DIN 1055, Blatt 4, Lastannahmen für Bauten: Windlast, July 1975.

[9] GIONCU, V., Stability of Hyperboloidal Shells. Discussion of Paper 11446. *Journal of the Structural Division, ASCE*, Vol. 102, 1976, pp. 1512-1514.

[10] GOULD, P. L., SURYOUTOMO, H., AND SEN, S. K., Stresses in Column-Supported Hyperboloidal Shells Subject to Seismic Loading. *Earthquake Engineering and Structural Dynamics*, Vol. 5, 1977, pp. 3-14.

[11] GOULD, P. L., SEN, S. K., AND SURYOUTOMO, H., Dynamic Analysis of Column-Supported Hyperboloidal Shells. *Earthquake Engineering and Structural Dynamics*, Vol. 2, 1974, pp. 269-279.

[12] GURFINKEL, G., AND WALSER, A., Analysis and Design of Hyperbolic Cooling

Towers. *Journal of the Power Division, ASCE*, Paper 8997, June 1972, pp. 133–152.

[13] HARNACH, R., ANKE—Ein Programmsystem für den automatischen Naturzugkühlturm-Entwurf. Institut f. Konstruktiven Ingenieurbau, Ruhr-Universität Bochum, *Technical Report 77-4*.

[14] I.A.S.S., Recommendations for the Design of Hyperbolic or Other Similarly Shaped Cooling Towers. International Association for Shell and Space Structures, Working Group No. 3, Brussels, 1977.

[15] ICI, Report of the Committee of Inquiry into Collapse of Cooling Tower at Ardeer Nylon Works. Imperial Chemical Industries Limited, London, 1974.

[16] INSTITUT KIB, Kühlturm-Symposium 1977, Theorie, Konstruktion, Bauausführung. *konstruktiver ingenieurbau berichte*, Heft 29/30, Vulkan-Verlag, Essen, 1977.

[17] INSTITUT KIB, Naturzug-Kühltürme, ihre Festigkeitsberechnung und Konstruktion. *konstruktiver ingenieurbau berichte*, Heft 1, Vulkan-Verlag, Essen, 1968.

[18] KÖNIG, G., AND ZILCH, K., Zur Windwirkung auf Gebäude. *Beton- und Stahlbetonbau*, Vol. 67, 1972, pp. 32–42.

[19] KRATZIG, W. B., AND MESKOURIS, K., Approximate Earthquake Response of Cooling Towers by a Beam-like model. Contribution to ASCE, Power Division Specialty Conference, Phoenix/Arizona, Feb. 1978.

[20] KRATZIG, W. B., ET AL., Kühltürme. Kapitel VIII in: *Jahrbuch Bautechnik im Kraftwerksbau*, 1, Ausgabe 1876/77. Vulkan-Verlag, Essen, 1976.

[21] KRATZIG, W. B., Schnittgrößen und Verformungen windbeanspruchter Naturzugkühltürme. *Beton- und Stahlbetonbau*, Vol. 61, 1966, pp. 247–255.

[22] MUNGAN, I., Buckling Stress States of Hyperboloidal Shells. *Journal of the Structural Division, ASCE*, Vol. 102, 1976, Paper 12465, pp. 2005–2020.

[23] NIEMANN, H. J., Zur stationären Windbelastung rotationssymmetrischer Bauwerke im Bereich transkritischer Reynoldszahlen. Institut f. Konstruktiven Ingenieurbau, Ruhr-University Bochum, *Technical Report 71-2*.

[24] SEN, S. K., GOULD, P. L., AND SURYOUTOMO, H., High Precision Static and Dynamic Finite Element Analysis of Shells of Revolution. *Research Report 23*, Structural Division, Washington University, St. Louis, 1973.

[25] SINGH, M. P., AND GUPTA, A. K., Gust Factors for Hyperbolic Cooling Towers. *Journal of the Structural Division, ASCE*, Vol. 102, 1976, pp. 371–386.

[26] THE INSTITUTION OF CIVIL ENGINEERS, Natural Draught Cooling Towers—Ferrybridge and After. *Proceedings*, Conference held at The Institution of Civil Engineers, June 12, 1967, London, 1967.

[27] VIK, Technische Anleitung für die Berechnung und Ausführung von großen Kühltürmen. Vereinigung Industrielle Kraftwirtschaft, Essen, 1970.

[28] WIEGEL, R. L., *Earthquake Engineering*. Prentice-Hall, Englewood Cliffs, N.J., 1970.